Lecture Notes in Physics

Edited by J. Ehlers, München, K. Hepp, Zürich
R. Kippenhahn, München, H. A. Weidenmüller, Heidelberg
and J. Zittartz, Köln
Managing Editor: W. Beiglböck, Heidelberg

88

Kolumban Hutter
Alphons A. F. van de Ven

Field Matter Interactions in Thermoelastic Solids

A Unification of Existing Theories of Electro-Magneto-Mechanical Interactions

Springer-Verlag Berlin Heidelberg GmbH 1978

Authors
K. Hutter
Federal Institute of Technology
CH-8092 Zürich

A. A. F. van de Ven
Technological University
Eindhoven
The Netherlands

ISBN 978-3-540-09105-9 ISBN 978-3-540-35546-5 (eBook)
DOI 10.1007/978-3-540-35546-5

Originally published by Springer-Verlag Berlin Heidelberg New York in 1978

2153/3140-543210

To our wives
Barbara and Maria

PREFACE

The last two decades have witnessed a giant impetus in the formulation of electrodynamics of moving media, commencing with the development of the most simple static theory of dielectrics at large elastic deformations, proceeding further to more and more complex interaction models of polarizable and magnetizable bodies of such complexity as to include magnetic dissipation, spin-spin interaction and so on and, finally, reaching such magistral synthesis as to embrace a great variety of physical effects in a relativistically correct formulation. Unfortunately, the literature being so immense and the methods of approach being so diverse, the newcomer to the subject, who may initially be fascinated by the beauty, breadth and elegance of the formulation may soon be discouraged by his inability to identify two theories as the same, because they look entirely different in their formulation, but are suggested to be the same through the description of the physical situations they apply to.
With this tractate we aim to provide the reader with the basic concepts of such a comparison. Our intention is a limited one, as we do not treat the most general theory possible, but restrict ourselves to non-relativistic formulations and to theories, which may be termed deformable, polarizable and magnetizable thermoelastic solids. Our question throughout this monograph is basically: What are the existing theories of field-matter interactions; are these theories equivalent, and if so; what are the conditions for this equivalence?
We are not the first ones to be concerned with such fundamental ideas. Indeed, it was W.F. Brown, who raised the question of non-uniqueness of the formulation of quasistatic theories of magnetoelastic interactions, and within the complexity of his theory, he could also resolve it. Penfield and Haus, on the other hand, were fundamentally concerned with the question how electromagnetic body force had to be properly selected. This led them to collect their findings and to compare the various theories in an excellent monograph, in which they rightly state that equivalence of different formulations of electrodynamics of deformable continua cannot be established without resort to the constitutive theory, but at last, they dismissed the proper answer, as their treatment is incomplete in this regard. For this reason the entire matter was re-investigated in the doctoral dissertation of one of us (K. Hutter), but this work was soon found unsatisfactory and incomplete in certain points, although the basic structure of the equivalence proof as given in Chapter 3 of this tractate, was essentially already outlined there. Moreover, Hutter was still not able to compare certain magnetoelastic interaction theories so that what he attempted remained a torso anyhow.

The difficulties were overcome by van de Ven in a series of letters, commencing in fall of 1975, in which we discussed various subtleties of magnetoelastic interactions that had evolved from each of our own work. The correspondence was so fruitful that we soon decided to summarize our efforts in a joint publication. It became this monograph, although this was not our initial intention. Yet, after we realized that a proper treatment required a presentation at considerable length, we decided to be a little broader than is possible in a research report and to write a monograph, which would be suitable at least as a basis for an advanced course in continuum mechanics and electrodynamics (graduate level in the US). We believe that with this text this goal has been achieved. We must at the same time, however, warn the reader not to take this tractate as a basis to learn continuum mechanics and/or electrodynamics from the start. The fundamentals of these subjects are assumed to be known.

Our acknowledgements must start with mentioning Profs.J.B. Alblas (Technological University Eindhoven) and Y.H. Pao (Cornell University). They were the ones who initiated our interest in the subject of magnetoelastic interactions. While performing the research for this booklet and during our preparation of the various draughts we were supported by our institutions, the Federal Institute of Technology, Zürich and the Technological University, Eindhoven, and were, furthermore, encouraged by Prof. J.B. Alblas, Eindhoven, Prof. D. Vischer, Zürich, Prof. I. Müller, Paderborn, Prof. H. Parkus, Vienna, Dr. Ph. Boulanger, Brussels and Dr. A. Prechtl, Vienna. The support and criticism provided by them, directly or indirectly, were extremely helpful. We are grateful to these people not only for their keen insight and willingness to discuss the issues with us, but also for their encouragement in general.

During the initial stage and again towards the end of the write-up of the final draught of this monograph K. Hutter was financially supported in parts by the Technological University, Eindhoven, to spend a total of a two months period (September 1976 and April 1978) at its Mathematics Department. Without the hospitality and the keen friendship of the faculty and staff members of this department and especially of Prof. J.B. Alblas and his group, the work compiled in these notes would have barely have been finished so timely. The burden of typing the manuscript was taken by Mrs. Wolfs-Van den Hurk. It was her duety to transform our hand-written draughts into miraculously looking typed sheets of over 200 pages. Her effort, of course, is gratefully acknowledged.

Eindhoven and Zürich
in the summer of 1978

K. Hutter
A.A.F. van de Ven

TABLE OF CONTENTS

1. BASIC CONCEPTS

1.1 PREVIEW

With the boundaries of physical science expanding rapidly, present day engineers must necessarily assimilate information and knowledge of subjects that continue to become more and more complex. In the study of the nature and mechanical behavior of engineering materials, however, their knowledge seldom progresses beyond the level of elementary theory of elasticity and simple ideas of the theory of plasticity. More esoteric theories of material behavior and the interaction with various fields are generally left out of consideration or soon abolished as being mathematically intractable or economically unjustifyable.

This book is an account on one of the above mentioned more esoteric theories. What we have in mind is the response of deformable bodies to electromagnetic fields. Indeed, the interaction of electromagnetism with thermoelastic fields is not only a challenging scientific problem, but it is increasingly attracting also engineers in the nuclear power and electronic industry from a purely applied point of view.

The subject of electrodynamics of moving media has always been a controversal one. This book will not end or resolve all controversies, because we can answer some, but not all the relevant questions in connection with a complete thermodynamic theory of electromagnetism.

The basic difficulties in the description of electromechanical interaction models are manifold. A first difficulty is concerned with the invariance properties of the electromagnetic field equations. As is well-known, Maxwell's equations are invariant under Lorentz transformations, while the balance laws of classical mechanics are invariant under Euclidean transformations and frame indifferent under Galilei transformations. Clearly, a proper derivation should also treat the mechanical equations relativistically. This is true, but for most problems of technical relevance, relativistic effects are negligible.

It is therefore customary, in general, to treat the mechanical equations classically, while the equations of electrodynamics are handled relativistically. In so doing it might in these theories become uncertain what transformation properties some variables are based upon. Yet the knowledge of such transformation properties is important, because they give us indications as to what variables are comparable among different theories.

A second and even more serious difficulty can be found in the definitions of electromagnetic body force, body couple and energy supply. The roots of this difficulty lie in the separation of the electromagnetic field quantities in near and far field effects. This separation has been and still is the root of

controversies, because almost every author separates the total fields differently. In other words, near and far fields are not unique.
A third difficulty is connected with the Maxwell equations, which for deformable moving matter were first derived by Minkowski. Apart from the Maxwell equations in Minkowski's form there exists a variety of other forms of the Maxwell equations in deformable media, all of which are motivated from particular models. The "action" of the electromagnetic fields upon the material is described hereby by quantities referred to as polarization and magnetization. However, dependent upon the model of derivation, polarization and magnetization of one theory may be and in general are different from polarization and magnetization of another theory. Hence, while all formulations of electromagnetism of deformable continua are equally valid - we know of five different descriptions - special care must be observed that variables of one theory are not confused with those of another.
As a final difficulty, we mention that the equations of a theory of electromechanical interaction are highly nonlinear. Generally, they defy any exact analysis even for the most simple problems that are of physical relevance. As a result, linearization procedures are needed.
To render the above statements more precise, consider the equations of motion which may be derived by formulating the balance law of momentum to an arbitrary part of the body. The local form of this balance law states that "mass times acceleration equals divergence of stress plus body force". In ordinary classical mechanics the body force is either set equal to zero, or else given by the gravitational force. A body couple hardly occurs in applications, in which case the balance law of moment of momentum implies the symmetry of the (Cauchy) stress tensor. When the body under consideration is interacting with electromagnetic fields, however, body force and body couple are given by electromagnetic quantities.
The total force and the total moment excerted on a body by electromagnetic fields may be separated into a long range and a short range effect. The long range effect is expressed as a body force and body couple. The short range effect, on the other hand, manifests itself as surface tractions which can be combined with the mechanical tractions giving rise thereby to the definition of the stress tensor. This decomposition is not unique, thus leading to non-unique body force expressions and non-unique stress tensors. As a consequence, the electromagnetic body couple cannot be unique either.
Although this non-uniqueness might be quite striking to the novel reader it is nontheless not disturbing at all if looked upon from the right point of view.

Indeed, it is not important that the above separation into force and stress is unique, because differences in the body forces can always be absorbed in the stress tensors, provided that they are expressible as a divergence of a stress. A variety of mutually incompatible formulas for the force expressible in terms of stresses are therefore equivalent with respect to the total force. Only this force is physically observable. Thus the incompatibility is not physical, but metaphysical or semantic.

The incompatibility expressed above also occurs in the energy equation (first law of thermodynamics). This equation states that the time rate of change of the internal energy is balanced by the power of working of the stresses, the divergence of the heat flux and the energy supply due to electromagnetic effects and due to heat. Since stress was already said to be non-unique, it follows that internal energy, heat flux and electromagnetic energy supply cannot be determined uniquely either. Likewise, the electromagnetic energy supply might contain a term that is the divergence of a vector which could be absorbed in the heat flux vector. As an immediate consequence, it cannot be assured that heat flux is energy flux of thermal nature. We shall therefore prefer the term energy flux instead.

As was the case for the momentum equation, seemingly incompatible expressions for internal energy, energy flux and energy supply of electromagnetic origin do not prevent two theories from being equivalent. However, it is easily understandable that a proof of equivalence must be difficult in general for, stress, internal energy and entropy (and also some electromagnetic field vectors) are interrelated by thermodynamic conditions. More explicitly, thermodynamic requirements make stress and entropy (and other quantities) derivable from a so called free energy. If two theories are different in the body force, body couple and energy supply, therefore, the condition that the momentum equation and energy equation remains the same must amount to an interrelation between the free energies of two theories. Hence, equivalence of two theories of electromechanical interactions is a thermodynamic statement in general.

The question of equivalence of two theories lies at the center of the different formulations of electromechanical interaction theories. Although there is a valid point behind the statement that equivalence of different theories need not be proved, because these theories describe different physical situations, we nevertheless take the position that different formulations of electromechanical interaction theories should yield the same results for physically measurable quantities, if the theories are claimed to be applicable to a certain class of material response. For instance, if we call a material a thermoelastic polarizable and magnetizable solid and if there is more than one formulation for such a solid, one should expect that, irrespective of all differences in the details, these theories will in any initial boundary value problem deliver the same results for physically measurable quantities. Measurable or observable quantities are all those which can be measured uniquely by two different observers. All kinematical quantities that are derivable from the motion are measurable in principle and so is the (empirical) temperature. Regarding electromagnetic field quantities, we take the position that they are not measurable except in vacuo where they can be observed by measuring the force on a test charge. There exist variables not observable by any means. These are all those which are not defined except by the mathematical properties laid down for them.

To demonstrate the equivalence of the different formulations of electromechanical interaction theories it is necessary to prove that physically measurable quantities in two formulations assume the same values in every point of the body for any initial boundary value problem. This does not only mean that the field equations of one theory must be transformable into those of the other, but this condition also includes the boundary and initial conditions. One of the major goals of this monograph is to give an exposition of the existing theories of polarizable and magnetizable electrically and thermally conducting materials and to show in what sense they can be called equivalent.

The reasons behind this non-uniqueness of electrodynamics in moving media are twofold. For one, the action of the body on the electromagnetic fields is generally described by adding to the field variables occurring in vacuo two other electromagnetic field vectors. This addition is not unique and results in different forms of the Maxwell equations. Second, even when we restrict ourselves to a particular form of the Maxwell equations, the electromagnetic forces, couples and energy supply terms need not be unique. More precisely, we mention that the two electromagnetic field vectors describing the interaction of a ponderable body can be introduced for instance by postulating that every material point is equipped with a number of non-interacting electric and magnetic dipoles. These dipole moments then form the two additional electromagnetic field vectors which are called polarization and magnetization. When the calculations with these postulated dipoles are carried through consistently, a certain set of Maxwell equations (now called the Chu-formulation) emerges. These equations are different from those which follow from the postulation that magnetization is modeled as an electric circuit which follows the motion of the material particle in question (statistical and Lorentz formulations).

As far as electromagnetic body forces are concerned these are not even unique when one is restricting oneself to a particular interaction model. Indeed, in the Chu-formulation we shall present two versions of body force expressions and we shall prove that both are not distinguishable by any measurements. This proof will also be given for all other formulations. However, we shall not present the models as such, because they are amply treated in the pertinent literature.

Although the proof of the equivalence of various theories of electrodynamics in deformable continua is a very important achievement, we want to state here clearly that we have performed this proof only on the level of non-relativistic theories. The exact definitions of the term "non-relativistic" will be made precise in the respective Chapters. It may suffice to mention that it essentially means that in MKSA-units terms containing a c^{-2}-factor are neglected. Here, c is the speed of light in vacuo. There exists a number of other theories of electromechanical interactions in which it is claimed that terms of order V^2/c^2 are neglected (V = velocity of the particle in the body) while those containing a c^{-2}-factor are kept. We term such approximations "semi-relativistic". Quasistatic theories (terms, containing a c^{-1}-factor are neglected) will not be treated here.

It is a well-known fact that fluids are best handled in the spatial description. It is also known that electrodynamics is usually only formulated in the spatial description. Yet for a theory of solids it would be advantageous when all equations could be referred to the reference configuration. This is indeed possible and it essentially amounts to the introduction of new electromagnetic field variables. It turns out that these so called Lagrangian field variables are much more convenient to describe the theory of solids, because many thermodynamic formulas appear in a more condensed form this way. Another reason for the introduction of the material description is its advantage in the linearization of the governing equations. This linearization procedure is substantially easier when performed in the material rather than in the spatial description.

This brings us naturally to the linearization procedure of the various theories. In principle, there are two alternatives open to extract some useful information from these complicated equations. One is to find numerical solutions for the nonlinear equations and the other is to linearize the equations on the basis of a sequence of consistent approximations. We shall follow the latter, because it provides a better access to the real physics of the problem. The linearization procedure is analogous to situations referred to as "small fields superimposed upon large fields". The difference between these general treatments and ours is that the deformations are assumed to be small. This assumption is not necessary, and indeed the formal expansion procedures we shall apply also hold true for the general case. When the restriction to small deformations is used, however, it means physically that large external fields primarily induce strong electromagnetic fields within the body, but only small deformations. Therefore in the first step of evaluating the induced electromagnetism, the deformations may be neglected alltogether. A set of zeroth order equations which formally agrees with rigid body electrodynamics, is thus obtained. In the second step small strains are considered which add small but important corrections to the zeroth order electromagnetic fields. Thus, the second set consists of linear field equations, the coefficients of which generally depend upon the zeroth order electromagnetic fields. These field equations may then be applied to solve problems like magnetoelastic buckling, wave propagation in a material subject to electromagnetic fields, etc.

Clearly, because we shall prove that all electromechanical interaction theories of polarizable and magnetizable solids are equivalent, the linearization procedure mentioned above need only be performed for one particular theory which can be selected according to our needs. Moreover, it should be clear that this equivalence must amount in the statement how the free energy as a function of its independent variables in one theory is related to the free energy of another theory. The set of independent variables in this second free energy may very well be different from the first one. In other words, the correspondence relations for equivalence of various theories are dependent on which set of independent variables is chosen in the constitutive relations, but it is quite clear that the equivalence as such should not depend on the choice of the independent fields. From a theoretical point of view the problem just raised is not a serious one, because, in principle, equivalence of different constitutive formulations in one single theory can be established quite easily. It then suffices to prove equivalence of two different electromagnetic descriptions of deformable bodies with the aid of just one constitutive formulation in each of them.

The problem to find the free energy of a particular formulation from that of another one is a very difficult problem in practice, however. It amounts to solving a functional differential equation the solutions to which are not known to date. Nevertheless, special cases are straightforward to handle. They serve as explicit examples which should demonstrate that equivalence is possible. Mathematically this is important, because it serves as an explicit demonstration that the functional differential equations mentioned above do admit exact solutions. That these correspond to a reasonable physical situation is a nice additional property. The general question of existence and nonexistence of solutions will not be attacked here. Instead we look at the approximations in the way described below.

It is customary in applications to write for the free energies polynomial expressions, and it is generally assumed that these polynomials can be truncated at a certain level. When polarization and magnetization are amongst the independent constitutive variables the free energy will be a polynomial expression in the deformation tensor, the temperature, polarization and magnetization, and the coefficients in this polynomial expression give rise to effects like magnetic and electric anisotropy, magnetostriction, electrostriction etc. The coefficients bear the names electric and magnetic susceptibilities etc. The same theory could be derived also with the electric field strength and the magnetic induction as independent fields instead of magnetization and polarization. The free energy of this formulation would again be expressed as a polynomial of its variables and it would again be truncated at a certain level. This polynomial would again give rise to effects like electric and magnetic anisotropies, magnetostriction and electrostriction etc, but it is evident that the coefficients of this polynomial must be different from those of the other, if the two formulations aim at describing the same phenomena. The literature is full of confusion in this regard, mainly because different coefficients bear the same name. From the above it is, however, quite clear that there must be relations between the above mentioned coefficients. We shall show what these relations look like and in what sense the emerging approximate theories can be regarded to be equivalent. The findings can be summarized as follows: Two formulations, in which the free energies are represented by polynomials of a certain order in their variables can only be equivalent to within terms that were omitted in the expansion process. Only on the basis that these terms are negligibly small can we claim two theories to be equivalent. An analogous statement also holds for one single formulation in which certain constitutive quantities are interchanged as dependent and independent variables.

We end with an outlook on problems this theory may be applied to.

1.2 KINEMATICS

As is common in continuum mechanics, we regard a body as a three-dimensional manifold embedded in Euclidean 3-space. Its elements are called particles. Let $\mathcal{R}_R$ be its reference configuration and $\mathcal{R}_t$ its configuration at time t. Instead of $\mathcal{R}_t$ we shall subsequently write $\mathcal{R}$, and we shall refer to $\mathcal{R}$ as the present configuration of the body. Parts of the body will be denoted by V_R and $\mathcal{V}$, dependent on whether they are referred to the reference configuration and the present configuration, respectively. The boundary of V_R and of $\mathcal{V}$ will be denoted by ∂V_R and $\partial \mathcal{V}$, respectively. The position of a particle in $\mathcal{R}_R$ will be designated by $\underline{X}$ (X_α, $\alpha = 1,2,3$), whereas the one in the present configuration $\mathcal{R}$ is $\underline{x}$ (x_i, $i = 1,2,3$). A motion of the body is then described by the mapping

$$(1.1) \qquad x_i = \chi_i(X_\alpha, t), \qquad (i,\alpha = 1,2,3) .$$

We assume $\underline{\chi}$ to be invertible and this is tantamount to assuming that the functional determinant

$$(1.2) \qquad J := \det(\partial\chi_i/\partial X_\alpha) ,$$

never vanishes and may without loss of generality be assumed to be positive. In the above and throughout this monograph symbolic and Cartesian tensor notation is used. Greek indices refer to the material coordinates $\underline{X}$, and Latin indices to the spatial coordinates $\underline{x}$. Summation convention will be used over doubly repeated indices and commas preceding indices indicate differentiations with respect to space variables. The symbol d/dt or the superimposed dot will designate differentiation with respect to time t, holding the particle $\underline{X}$ fixed, i.e.

$$(1.3) \qquad \frac{d\Phi}{dt} \equiv \dot{\Phi} := \frac{\partial\Phi(\underline{X},t)}{\partial t} .$$

$\dot{\Phi}$ is called the material time derivative. Likewise, $\partial/\partial t$ will denote differentiation with respect to time t, holding the spatial position $\underline{x}$ fixed. Hence,

$$(1.4) \qquad \frac{\partial\Phi}{\partial t} := \frac{\partial\Phi(\underline{x},t)}{\partial t} .$$

$\partial\Phi/\partial t$ is called the local, or partial, time derivative, and it is easy to see that (1.3) and (1.4) may be combined to yield

$$(1.5) \qquad \frac{d\Phi}{dt} = \dot{\Phi} = \frac{\partial\Phi}{\partial t} + \dot{x}_i \frac{\partial\Phi}{\partial x_i} .$$

The local deformation of a body in the neighborhood of a particle $\underline{X}$ may be characterized to first order by the deformation gradient $F_{i\alpha}$, defined by

$$(1.6) \qquad F_{i\alpha} := \frac{\partial \chi_i(\underline{X},t)}{\partial X_\alpha} = x_{i,\alpha} \ .$$

Its inverse exists and is written as

$$(1.7) \qquad F^{-1}_{\alpha i} = X_{\alpha,i} \ .$$

Of particular interest are objective combinations of $F_{i\alpha}$. The right Cauchy-Green deformation tensor $C_{\alpha\beta}$ is defined by

$$(1.8) \qquad C_{\alpha\beta} := F_{i\alpha}F_{i\beta} \ .$$

In applications it is more convenient to use the Lagrangian strain tensor, or the deformation tensor,

$$(1.9) \qquad E_{\alpha\beta} := \frac{1}{2}(C_{\alpha\beta} - \delta_{\alpha\beta}) \ ,$$

instead. Here, $\delta_{\alpha\beta}$ is the Kronecker delta.

1.3 EQUATIONS OF BALANCE

1. The Balance Laws of Mechanics

It is the ultimate goal of any thermodynamic theory, be it a theory of a simple fluid or a fairly complicated description of electromechanical interaction phenomena, to calculate within a body the independent fields of this theory as functions of space and time. For the determination of these fields, we need field equations, which are obtained when the balance laws of mechanics and electrodynamics are combined with constitutive equations. Basic ingredients of any theory are thus the balance equations which are discussed below.

We start with the balance laws of mechanics. The integral expressions of the laws of conservation of mass, balance of momentum, moment of momentum and energy in the spatial and material description are well-known. For the former they are

$$(1.10) \qquad \frac{d}{dt}\int_V \rho \, dv = 0 \ ,$$

$$(1.11) \qquad \frac{d}{dt}\int_V \rho \dot{x}_i dv = \int_{\partial V} t_{ij} da_j + \int_V \rho F_i dv \ ,$$

$$(1.12)\qquad \frac{d}{dt}\int_V \rho x_{[i}\dot{x}_{j]}dv = \int_{\partial V} x_{[i}t_{j]k}da_k + \int_V \rho(L_{ij} + x_{[i}F_{j]})dv ,$$

and

$$(1.13)\qquad \frac{d}{dt}\int_V (\tfrac{1}{2}\rho\dot{x}_i\dot{x}_i + \rho U)dv = \int_{\partial V}(\dot{x}_i t_{ij} - q_j)da_j + \int_V (\rho\dot{x}_i F_i + \rho r)dv .$$

Here, ρ is the mass density per unit volume, t_{ij} is the Cauchy stress tensor, F_i the total body force due to electromagnetic fields and external actions, L_{ij} the dual of the body couple L_k, i.e.

$$(1.14)\qquad L_{ij} = \frac{1}{2} e_{ijk}L_k .$$

U is the internal energy per unit mass, q_i the energy flux, consisting of heat flux and non-thermal energy flux, and r the total energy supply, due to the electromagnetic fields and due to heat. Bracketed indices indicate anti-symmetric tensors. Note that the stress tensor t_{ij} and the stress vector $t_i^{(n)}$ on a surface element with exterior unit normal $\underline{n}$ are related by

$$(1.15)\qquad t_i^{(n)} = t_{ij}n_j .$$

Often, the stress tensor is defined through

$$(1.16)\qquad t_i^{(n)} = \bar{t}_{ji}n_j .$$

This definition implies $\bar{t}_{ij} = t_{ji}$. Moreover, t_{ij} is not symmetric in general. It is assumed that body force, body couple and energy supply can be decomposed into two parts; one is due to the electromagnetic fields, the other is supposed to be externally applied and known from the outset. Hence,

$$\rho F_i = \rho F_i^e + \rho F_i^{ext} ,$$

$$(1.17)\qquad \rho L_{ij} = \rho L_{ij}^e ,$$

$$\rho r = \rho r^e + \rho r^{ext} .$$

Here, ρF_i^e, ρL_{ij}^e and ρr^e are thought to be expressed in terms of electromagnetic field quantities, while ρF_i^{ext} and ρr^{ext} are known. We have assumed that there are no externally applied body couples.

For sufficiently smooth fields, the balance laws (1.10)-(1.13) assume the form

$$\dot{\rho} + \rho\dot{x}_{i,i} = 0 \ ,$$

$$\rho\ddot{x}_i = t_{ij,j} + \rho F^e_i + \rho F^{ext}_i \ ,$$

(1.18)

$$t_{[ij]} = \rho L^e_{[ij]} \ ,$$

$$\rho\dot{U} = t_{ij}\dot{x}_{i,j} - q_{i,i} + \rho r^e + \rho r^{ext} \ ,$$

where use has been made of (1.17). Here, and throughout this work by $\dot{x}_{i,j}$ is meant

(1.19) $$\dot{x}_{i,j} = (\dot{x}_i)_{,j} = F^{-1}_{\alpha j}\dot{F}_{i\alpha} \ .$$

As follows from $(1.18)^1$, the present density ρ is related to ρ_o, the density in the reference configuration, by

(1.20) $$\rho = \frac{\rho_o}{|J|} \ .$$

2. The Maxwell Equations

As is well-known, there are several formulations of electrodynamics, all of which may be derived from different postulations. For reference we refer the reader to Penfield and Haus ([1], Ch. 7). It is not our intention to derive the electromagnetic balance laws from various charge and electric circuit models, because we aim at describing the equivalence of these formulations rather than emphasizing their differences.

We start with the Conservation of Magnetic Flux.

Let B_i be the magnetic flux density (or magnetic induction) and let $\mathcal{E}_i$ be the effective electric field strength, sometimes also called the electromotive intensity. The conservation law of magnetic flux (Faraday law) may then be expressed as

(1.21) $$\frac{d}{dt}\int_S B_i da_i + \int_{\partial S} \mathcal{E}_i d\ell_i = 0 \ .$$

This relation must hold for any material surface S with boundary ∂S in 3-space.

By applying (1.21) to a closed surface $S = \partial V$, one can derive the Gauss-Faraday law, which states that

$$(1.22) \qquad \int_{\partial V} B_i da_i = 0 .$$

For sufficiently smooth fields, the balance laws (1.21) and (1.22) can be written in local form. We assume the underlying technique to be well-known and only give the results:

$$(1.23) \qquad e_{ijk}E_{k,j} + \overset{\star}{B}_i = 0, \qquad B_{i,i} = 0 .$$

Here, the convective derivative $\overset{\star}{\psi}_i$ of a vector ψ_i is defined by

$$(1.24) \qquad \overset{\star}{\psi}_i := \frac{\partial \psi_i}{\partial t} + \psi_{i,j}\dot{x}_j + \psi_i \dot{x}_{j,j} - \psi_j \dot{x}_{i,j} .$$

The second basic law of electromagnetism is the law of Conservation of Charge. Let Q be the electric charge density and J_i the conductive current. The conservation law of electric charges may then be expressed by the global balance law

$$(1.25) \qquad \frac{d}{dt} \int_V Q \, dv + \int_{\partial V} J_i da_i = 0 .$$

This equation holds for any material volume V with boundary ∂V.
It is a purely formal matter to introduce a field D_i such that

$$(1.26) \qquad \int_V Q \, dv = \int_{\partial V} D_i da_i ,$$

holds for any material part V of the body. This introduction of D_i suggests to call it charge potential; however, we shall refer to D_i as the dielectric displacement.
It follows from (1.25) and (1.26) that the conservation of charge is satisfied if a field H_i exists such that

$$(1.27) \qquad \int_S J_i da_i + \frac{d}{dt} \int_S D_i da_i = \int_{\partial S} H_i d\ell_i$$

holds for any material surface S with boundary ∂S. Indeed, if we choose $S = \partial V$ in (1.27), then $\partial S = \emptyset$ and (1.27) may be reduced to (1.25). We shall call H_i the effective magnetic field strength or the magnetomotive intensity. Note that H_i and D_i as introduced by (1.26) and (1.27) are not unique.

Again, assuming sufficient smoothness of the fields, the balance laws (1.25), (1.26) and (1.27) may be brought into local form. From (1.26) and (1.27) we then obtain

$$(1.28) \qquad D_{i,i} = \mathcal{Q}, \qquad e_{ijk}\mathcal{H}_{k,j} - \overset{*}{D}_i = \mathcal{J}_i \ ,$$

and from (1.25)

$$(1.29) \qquad (\mathcal{J}_i + \mathcal{Q}\dot{x}_i)_{,i} + \frac{\partial \mathcal{Q}}{\partial t} = J_{i,i} + \frac{\partial \mathcal{Q}}{\partial t} = 0,$$

which is the local conservation law of electric charges. For convenience we have also introduced here the non-conductive current J_i defined by

$$(1.30) \qquad J_i := \mathcal{J}_i + \mathcal{Q}\dot{x}_i \ .$$

The equations (1.28) are called the inhomogeneous Maxwell equations, in contrast to the homogeneous ones,(1.23), because on the right-hand sides they contain electric charge and current densities.

Note that not all of the above equations are independent. Indeed, the conservation law of charge may be derived from the inhomogeneous Maxwell equations. In other words, when integrating the Maxwell equations, the conservation law of electric charges must be fulfilled along with the Maxwell equations. As a result, at most seven of the above electromagnetic field variables can be considered basic, while for the remaining ones constitutive equations must be established. Of course, this does not mean that one cannot establish constitutive equations for more electromagnetic field variables. Indeed, there is a valid point if both $\mathcal{J}_i$ and $\mathcal{Q}$ are considered as dependent constitutive variables. If this is done,however, further integrability conditions must be satisfied.

In the above presentation we have not made any appeal to a specific model, how the material moving body may contribute to the electromagnetic field vectors B_i, $\mathcal{E}_i$, D_i and $\mathcal{H}_i$. We shall not do this here either and refer to the pertinent literature (Penfield and Haus [1], Fano, Chu and Adler [2], Truesdell and Toupin [3], etc.). The reader may be puzzled, however, by the occurrence of the convective time derivatives in (1.23) and (1.28), which contain implicitly the motion. If he so desires, he may eliminate the latter by simply defining two new fields through

$$(1.31) \qquad \mathcal{E}_i = E_i + e_{ijk}\dot{x}_j B_k, \qquad \mathcal{H}_i = H_i - e_{ijk}\dot{x}_j D_k \ .$$

E_i and H_i are the well-known Minkowskian electric and magnetic field strengths. We note that the relations (1.31) are not the only ones through which the motion can be eliminated.

We might further mention that of the four fields B_i, $\mathcal{E}_i$, D_i and $\mathcal{H}_i$ only two are considered basic, while the remaining two will have to be described by constitutive assumptions. In practice, some of the above field quantities are replaced by others, which may allow a better physical insight, but this does not change the fundamental fact that there are two basic vectorial field quantities, while the remaining ones must be expressed in terms of the former. We shall come back to this at a later stage.

3. Material Description

The balance laws listed in the previous two Sections are written in the spatial or Eulerian formulation. Corresponding to these equations there is a material or Lagrangian description.

Let $d\upsilon$, da_k and $d\ell_k$ be the volume-, area- and length-increments in the spatial description, while dV, dA_α and dL_α are the corresponding increments in the material description. Then the following well-known identities hold between these increments:

(1.32) $$d\upsilon = |J|dV, \qquad da_k = JF^{-1}_{\alpha k}dA_\alpha, \qquad d\ell_k = F_{k\alpha}dL_\alpha .$$

Here a remark concerning the second equation is in order. Since the normal vector on a closed surface is defined as the outward normal and outward normals given by $(1.32)^2$ are transformed into inward normals when sgn $J < 0$, we must replace in $(1.32)^2$ J by $|J|$ whenever the surface is closed. Hence, if dA_α and da_k are area-elements of a closed surface, we have

(1.33) $$da_k = |J|F^{-1}_{\alpha k}dA_\alpha .$$

Although we may restrict J to positive values, use of $|J|$ instead of J in the following equations is preferred in order to make the respective quantities objective under the full group of (Euclidean) transformations, (see section 1.6). With the relations (1.32) the mechanical balance laws (1.10)-(1.13) may be transformed into the following forms

(1.34) $$\frac{d}{dt}\int_{V_R} \rho_o dV = 0 ,$$

(1.35) $$\frac{d}{dt}\int_{V_R} \rho_o \dot{x}_i dV = \int_{\partial V_R} T_{i\alpha}dA_\alpha + \int_{V_R} \rho_o F_i dV ,$$

$$(1.36)\qquad \frac{d}{dt}\int_{V_R} \rho_o x_{[i}\dot{x}_{j]}dV = \int_{\partial V_R} x_{[i}T_{j]\alpha}dA_\alpha + \int_{V_R} \rho_o(L_{ij} + x_{[i}F_{j]})dV ,$$

$$(1.37)\qquad \frac{d}{dt}\int_{V_R} (\tfrac{1}{2}\rho\dot{x}_i\dot{x}_i + \rho U)dV = \int_{\partial V_R} (\dot{x}_i T_{i\alpha} - Q_\alpha)dA_\alpha + \int_{V_R} (\rho_o\dot{x}_i F_i + \rho_o r)dV .$$

Here, integration is over reference volume and surface, respectively. $T_{i\alpha}$ is the first Piola-Kirchhoff stress tensor and Q_α is the material energy flux vector, which are related to t_{ij} and q_i according to

$$(1.38)\qquad \begin{aligned} T_{i\alpha} &= |J|F^{-1}_{\alpha j}t_{ij}, & t_{ij} &= |J^{-1}|F_{j\alpha}T_{i\alpha} ,\\ Q_\alpha &= |J|F^{-1}_{\alpha i}q_i, & q_i &= |J^{-1}|F_{i\alpha}Q_\alpha . \end{aligned}$$

Assuming for the external source terms decompositions similar to those listed in (1.17) and supposing sufficient smoothness of the fields involved (1.35)-(1.37) imply

$$(1.39)\qquad \begin{aligned} &\rho_o\ddot{x}_i = T_{i\alpha,\alpha} + \rho_o F^e_i + \rho_o F^{ext}_i ,\\ &T_{[i\alpha}F_{j]\alpha} = \rho_o L_{ij} ,\\ &\rho_o\dot{U} = T_{i\alpha}\dot{F}_{i\alpha} - Q_{\alpha,\alpha} + \rho_o r^e + \rho_o r^{ext} , \end{aligned}$$

whereas (1.34) integrates to yield $\rho_o = \rho_o(\underline{X})$.
Apart from the stress tensors t_{ij} and $T_{i\alpha}$ we shall occasionally also use the so called second Piola-Kirchhoff stress tensor defined by

$$(1.40)\qquad T^P_{\alpha\beta} := T_{i\alpha}F^{-1}_{\beta i} = |J|t_{ij}F^{-1}_{\alpha j}F^{-1}_{\beta i} .$$

There exists also a material formulation of the electromagnetic balance laws (see [4]). To derive the material counterpart of (1.21), (1.22) and (1.25)-(1.27), recall that the integrals occurring in these equations must only be transformed back to the reference configuration. With the aid of (1.32) we obtain straightforwardly

$$(1.41)\qquad \frac{d}{dt}\int_{S_R} \mathbb{B}_\alpha dA_\alpha + \int_{\partial S_R} \mathbb{E}_\alpha dL_\alpha = 0 ,$$

$$(1.42)\qquad \int_{\partial V_R} \mathbb{B}_\alpha dA_\alpha = 0 ,$$

$$(1.43)\qquad \int_{\partial V_R} \mathbb{D}_\alpha dA_\alpha = \int_{V_R} \mathbb{Q}\, dV ,$$

$$(1.44)\qquad -\frac{d}{dt}\int_{S_R} \mathbb{D}_\alpha dA_\alpha + \int_{\partial S_R} \mathbb{H}_\alpha dL_\alpha = \int_{S_R} \mathbb{J}_\alpha dA_\alpha ,$$

$$(1.45)\qquad \frac{d}{dt}\int_{V_R} \mathbb{Q}\, dV + \int_{\partial V_R} \mathbb{J}_\alpha dA_\alpha = 0 .$$

Here, hollow quantities are the material counterparts of B_i, E_i, D_i, H_i, J_i and Q and they are related to these according to the transformation rules

$$(1.46)\qquad \begin{array}{ll} \mathbb{Q} = Q|J| , & Q = |J^{-1}|\mathbb{Q} , \\ \mathbb{J}_\alpha = J_i|J|F^{-1}_{\alpha i} , & J_i = |J^{-1}|F_{i\alpha}\mathbb{J}_\alpha , \\ \mathbb{D}_\alpha = D_i|J|F^{-1}_{\alpha i} , & D_i = |J^{-1}|F_{i\alpha}\mathbb{D}_\alpha , \\ \mathbb{H}_\alpha = H_i F_{i\alpha}\operatorname{sgn} J , & H_i = F^{-1}_{\alpha i}\mathbb{H}_\alpha \operatorname{sgn} J , \\ \mathbb{B}_\alpha = B_i J F^{-1}_{\alpha i} , & B_i = J^{-1}F_{i\alpha}\mathbb{B}_\alpha , \\ \mathbb{E}_\alpha = E_i F_{i\alpha} , & E_i = F^{-1}_{\alpha i}\mathbb{E}_\alpha . \end{array}$$

The quantities $\mathbb{E}_\alpha$,$\mathbb{B}_\alpha$ etc. will be called the material or Lagrangian electromagnetic fields.

Again, although J may be assumed positive we have written these formulas such that the hollow quantities transform under the Euclidean transformation group as scalars (see Section 1.6).

For sufficiently smooth fields the global laws (1.41)-(1.45) can be written in local form. Then they are

$$(1.47)\qquad \begin{array}{ll} e_{\alpha\beta\gamma}\mathbb{E}_{\gamma,\beta} + \dot{\mathbb{B}}_\alpha = 0, & \mathbb{B}_{\alpha,\alpha} = 0 , \\ e_{\alpha\beta\gamma}\mathbb{H}_{\gamma,\beta} - \dot{\mathbb{D}}_\alpha = \mathbb{J}_\alpha, & \mathbb{D}_{\alpha,\alpha} = \mathbb{Q} , \\ \mathbb{Q} + \mathbb{J}_{\alpha,\alpha} = 0 , & \end{array}$$

where all differentiations are with respect to the material coordinates. As was mentioned already before, this set of equations is a dependent one since the conservation of charge is already implicitly contained in the Maxwell equations. As a consequence, at most seven variables are independent, while

the remaining ones must be given by constitutive equations.
At the moment, the Lagrangian fields are introduced purely formally. That they are of advantage will be demonstrated in the Chapters 4 and 5.

1.4 THE ENTROPY PRODUCTION INEQUALITY

Statistical mechanics suggests that there exists an additive quantity, called the entropy, which satisfies the balance law

$$(1.48)\qquad \frac{d}{dt}\int_{V} \rho\eta \, dv + \int_{\partial V} \phi_i da_i - \int_{V} \rho s \, dv = \int_{V} \rho\gamma \, dv .$$

Here, ϕ_i is the entropy flux, s the entropy supply and γ the entropy production. For sufficiently smooth fields (1.48) implies

$$(1.49)\qquad \rho\dot{\eta} + \phi_{i,i} - \rho s = \rho\gamma ,$$

and it is the expression of the second law of thermodynamics that

$$(1.50)\qquad \gamma \geq 0 ;$$

hence

$$(1.51)\qquad \rho\dot{\eta} + \phi_{i,i} - \rho s \geq 0 .$$

We shall set the entropy supply s equal to the external energy supply divided by the absolute temperature θ, i.e.

$$(1.52)\qquad s = \frac{r^{ext}}{\theta} ,$$

but we shall, in general, not assume that entropy flux equals heat flux divided by absolute temperature. Thus, (1.51) becomes

$$(1.53)\qquad \rho\dot{\eta} + \phi_{i,i} \geq \frac{\rho r^{ext}}{\theta} .$$

The material counterpart of (1.48) can easily be derived. In its local form, it reads

$$(1.54)\qquad \rho_o\dot{\eta} + \Phi_{\alpha,\alpha} \geq \frac{\rho_o r^{ext}}{\theta} ,$$

where

$$(1.55)\qquad \Phi_\alpha = |J| F^{-1}_{\alpha i}\phi_i ,$$

is the Lagrangian entropy flux.

Before we proceed it seems to be worthwhile to justify the approach we take regarding the entropy inequality (1.51) and the interpretation we give to the entropy supply and to the entropy flux. Our aim is not the justification of one particular theory against any other one. On the contrary, we shall adopt certain results obtained by using a particular entropy principle in each theory and aim at a comparison of such theories. This comparison should be made on the level of fully developed theories.

1.5 JUMP CONDITIONS

Let

$$\Sigma = \hat{\Sigma}(\underline{X},t) = \hat{\sigma}(\underline{x},t) = 0, \tag{1.56}$$

be a smooth surface, not necessarily material, and let W_N and w_n be its speed of propagation and speed of displacement, respectively, i.e.

$$W_N := -\frac{\partial\hat{\Sigma}/\partial t}{(\hat{\Sigma}_{,\alpha}\hat{\Sigma}_{,\alpha})^{\frac{1}{2}}}, \qquad w_n := -\frac{\partial\hat{\sigma}/\partial t}{(\hat{\sigma}_{,i}\hat{\sigma}_{,i})^{\frac{1}{2}}}. \tag{1.57}$$

Assume further that the electromagnetic field quantities B_i, E_i etc., or $\mathbb{B}_\alpha$, $\mathbb{E}_\alpha$ etc., as well as all mechanical quantities listed in Sections 1.3.1 or 1.3.3 may suffer finite jumps across the surface Σ. In particular we assume $\chi_i(\underline{X},t)$ to be continuous on Σ, but its first derivatives may suffer finite jumps. Hence, it follows that, although

$$\hat{\Sigma}(\underline{X},t) = \hat{\sigma}(\underline{\chi}(\underline{X},t),t)$$

is one and only one surface in V_R, the values for W_N are the same on both sides of the surface only when $\dot{\chi}_i$ and $F_{i\alpha}$ are continuous. Waves in which $\dot{\chi}_i$ and $F_{i\alpha}$ may jump are called shock waves.

By applying the global balance laws to a part of the body (V or V_R) containing the surface of discontinuity, we can derive jump conditions for the fields occurring in these laws. We suppose the methods of derivation to be known and therefore list the results only. The jump conditions obtained thereby will depend on whether one is dealing with the material or the spatial description. In the spatial description, the balance laws (1.21), (1.22) and (1.25)-(1.27) reveal that

$$\begin{aligned} &[\![B_i]\!]n_i = 0, && [\![e_{ijk}E_j n_k + \dot{x}_i B_k n_k - B_i(\dot{x}_k n_k - w_n)]\!] = 0, \\ &[\![D_i]\!]n_i = 0, && [\![e_{ijk}H_j n_k - \dot{x}_i D_k n_k + D_i(\dot{x}_k n_k - w_n)]\!] = 0, \\ &[\![J_i n_i - Q w_n]\!] = 0 \end{aligned} \tag{1.58}$$

Here,

(1.59) $$[\![\Phi]\!] := \Phi^+ - \Phi^- ,$$

denotes the jump of Φ across the surface $\mathcal{S}(\underline{x},t)$ and $\underline{n}$ is the unit normal vector at a point on the singular surface pointing into the positive side of $\mathcal{S}(\underline{x},t)$. We remark that the conditions (1.58) tacitly assume that J_i and Q are finite on the surface of discontinuity. This is a restriction; e.g. it excludes surface distributions of charge and current.

In the material description the counterparts of (1.58) read as follows

(1.60) $$\begin{aligned} &[\![\mathbb{B}_\alpha]\!]N_\alpha = 0, && e_{\alpha\beta\gamma}[\![\mathbb{E}_\beta]\!]N_\gamma + [\![\mathbb{B}_\alpha W_N]\!] = 0 , \\ &[\![\mathbb{D}_\alpha]\!]N_\alpha = 0, && e_{\alpha\beta\gamma}[\![\mathbb{H}_\beta]\!]N_\gamma - [\![\mathbb{D}_\alpha W_N]\!] = 0 , \\ &[\![\mathcal{J}_\alpha]\!]N_\alpha - [\![\mathcal{Q}W_N]\!] = 0 . \end{aligned}$$

If we set $W_N = 0$, or $w_n = \dot{x}_i^+ n_i = \dot{x}_i^- n_i$, then the surface of discontinuity is material. In that case, the jump conditions serve as boundary conditions. We like to note here that, in the sense of the definition given above, the boundary of a body in a vacuum is not a material surface of discontinuity because on this boundary:

(1.61) $$\dot{x}_i^+ = 0, \qquad \text{and} \qquad \dot{x}_i^- n_i = w_n .$$

For the derivation of the jump conditions of momentum and energy, we start with the expressions for the electromagnetic body force and energy supply. We consider it to be known that the electromagnetic momentum and energy supply terms appearing in (1.11), (1.13), (1.35) and (1.37) can always be written in the form

(1.62) $$\begin{aligned} &\rho F_i^e = t_{ij,j}^M + \frac{\partial g_i}{\partial t} , && \rho_o F_i^e = T_{i\alpha,\alpha}^M + G_i , \\ &\rho r^e + \rho F_i^e \dot{x}_i = \pi_{i,i} + \frac{\partial \omega}{\partial t} , && \rho_o r^e + \rho_o F_i^e \dot{x}_i = \Pi_{\alpha,\alpha} + \dot{\Omega} . \end{aligned}$$

In the above relations, $T_{i\alpha}^M$ and t_{ij}^M are electromagnetic stress tensors, which sometimes are called Maxwell stress tensors, G_i and g_i are the electromagnetic momenta in the material and spatial description, respectively, Π_α and π_i are the material and spatial representation of the electromagnetic energy flux and Ω and ω are electromagnetic energy densities. All these quantities are expressible in terms of the electromagnetic fields, the motion and their derivatives. Hence, although we assume the electromagnetic fields to suffer at most a finite jump across the singular surface Σ, such an assumption does not hold in general for ρF_i^e and ρr^e (or $\rho_o F_i^e$ and $\rho_o r^e$), which may become unbounded because of the occurrence of the above mentioned gradients. Apparently, one must write

$$\text{(1.63)}\quad \begin{aligned} \int_{V} \rho F_i^e dv &= \frac{d}{dt}\int_{V} g_i dv + \int_{\partial V} (t_{ij}^M - g_i \dot{x}_j) da_j \,, \\ \int_{V} (\rho r^e + \rho F_i^e \dot{x}_i) dv &= \frac{d}{dt}\int_{V} \omega \, dv + \int_{\partial V} (\pi_i - \omega \dot{x}_i) da_i \,, \end{aligned}$$

in the spatial formulation, and

$$\text{(1.64)}\quad \begin{aligned} \int_{V_R} \rho_o F_i^e dV &= \frac{d}{dt}\int_{V_R} G_i dV + \int_{\partial V_R} T_{i\alpha}^M dA_\alpha \,, \\ \int_{V_R} (\rho_o r^e + \rho_o F_i^e \dot{x}_i) dV &= \frac{d}{dt}\int_{V_R} \Omega \, dV + \int_{\partial V_R} \Pi_\alpha dA_\alpha \,, \end{aligned}$$

in the material description, respectively.

The fields G_i, g_i, $T_{i\alpha}^M$, t_{ij}^M, Π_α, π_i, Ω and ω are related to each other by

$$\text{(1.65)}\quad \begin{aligned} G_i &= |J| g_i \,, \\ T_{i\alpha}^M &= |J| F_{\alpha j}^{-1} (t_{ij}^M - g_i \dot{x}_j) \,, \\ \Pi_\alpha &= |J| F_{\alpha i}^{-1} (\pi_i - \omega \dot{x}_i) \,, \\ \Omega &= |J| \omega \,. \end{aligned}$$

As long as all fields are sufficiently smooth, the left-hand and right-hand sides of (1.63) and (1.64) are equivalent. This is no longer so if ρF_i^e and ρr^e may be unbounded. Then the left-hand sides of (1.63) and (1.64) may not be meaningful at all, whereas the expressions on the right-hand sides still make sense. We therefore postulate the expressions on the right-hand sides to be the appropriate global statements.

Substituting $(1.17)^{1,3}$ into (1.11) and (1.13) and using (1.63) yields the global form of the balance laws of linear momentum and energy appropriate for the derivation of the jump conditions. Under the assumption that the external sources ρF_i^{ext} and ρr^{ext} remain finite at Σ, these global laws, together with the balance of mass (1.10), lead to the following set of jump conditions

$$[\![\rho(\dot{x}_i n_i - w_n)]\!] = 0 ,$$

(1.66) $$[\![(\rho\dot{x}_i - g_i)(\dot{x}_j n_j - w_n)]\!] - [\![t_{ij} + t^M_{ij} - g_i\dot{x}_j]\!]n_j = 0 ,$$

$$[\![(\frac{1}{2}\rho\dot{x}_i\dot{x}_i + \rho U - \omega)(\dot{x}_j n_j - w_n)]\!] - [\![\dot{x}_i t_{ij} - q_j + \pi_j - \omega\dot{x}_j]\!]n_j = 0 .$$

Similarly, in the material description,

$$[\![\rho_o W_N]\!] = 0 ,$$

(1.67) $$[\![(\rho_o\dot{x}_i - G_i)W_N]\!] + [\![T_{i\alpha} + T^M_{i\alpha}]\!]N_\alpha = 0 ,$$

$$[\![(\frac{1}{2}\rho_o\dot{x}_i\dot{x}_i + \rho_o U - \Omega)W_N]\!] + [\![\dot{x}_i T_{i\alpha} - Q_\alpha + \Pi_\alpha]\!]N_\alpha = 0 .$$

Note that the balance of mass simply implies that ρ_o may jump only along with a jump of W_N, or on material surfaces, where $W_N = 0$. For $W_N = 0$, $(1.66)^{2,3}$ constitute the boundary conditions for the tractions and the energy flux of matter and fields, respectively. On the other hand, if ρ_o is continuous on Σ, $(1.67)^1$ implies continuity of W_N.

1.6 MATERIAL OBJECTIVITY

Although we consider the principle of material frame indifference to be known, in general, we would like to call upon it here, mainly because of the complications resulting from the combination of the mechanical balance laws with those of electrodynamics.

Let x_i be the Cartesian coordinates of a particle as measured by a stationary observer and let x_i^* be the Cartesian coordinates of the same particle as measured by another observer in his frame of reference. We call transformations that relate x_i with x_i^* by a rigid-body motion Euclidean transformations. They have the form

(1.68) $$x_i^* = O_{ij}(t)x_j + b_i(t), \qquad t^* = t ,$$

where O_{ij} is an orthogonal (time-dependent) matrix and where b_i is an arbitrary (time-dependent) vector.

A Euclidean transformation for which O_{ij} is not time-dependent and for which $b_i = -V_i t$ is called a Galilean transformation, and it is well-known that the balance laws of classical non-relativistic mechanics are invariant and frame independent under such transformations. They are, however, frame dependent with respect to Euclidean transformations.

In contrast to classical mechanics, the Maxwell equations are neither invariant under Euclidean nor under Galilean transformations. The transformation group here is the extended Lorentz group. This group may be explained as follows. Let (x_i, t) denote the position x_i and time t of a particle as measured by one observer and let (x_i^*, t^*) be those as measured by another observer, who translates relative to the first observer with constant velocity V_i. A transformation of the form

$$x_i^* = x_i + V_i\{\frac{x_k V_k}{V^2}[\frac{1}{\sqrt{1 - V^2/c^2}} - 1] - \frac{t}{\sqrt{1 - V^2/c^2}}\} =$$

(1.69) $$= (x_i - V_i t)(1 + 0(V^2/c^2)) ,$$

$$t^* = \frac{t - \frac{x_k V_k}{c^2}}{\sqrt{1 - V^2/c^2}} = (t - \frac{1}{c^2} x_k V_k)(1 + 0(V^2/c^2)) ,$$

is then called a special Lorentz transformation. Special in this transformation is that the frames x_i and x_i^* are parallel. If they are also rotated with respect to each other then x_i^* on the left-hand side of $(1.69)^1$ must be replaced by $0_{ji} x_j^*$, where 0_{ij} is a constant orthogonal matrix. The group of these transformations is called the extended Lorentz group. In four-dimensional notation

(1.70) $$x_A^* = \Lambda_{AB} x_B, \qquad (A,B = 1,2,3,4) ,$$

where, to within terms of the order of (V^2/c^2), Λ_{AB} is given by

(1.71) $$\Lambda_{AB} = \begin{pmatrix} 0_{11} & 0_{12} & 0_{13} & -0_{1k}V_k \\ 0_{21} & 0_{22} & 0_{23} & -0_{2k}V_k \\ 0_{31} & 0_{32} & 0_{33} & -0_{3k}V_k \\ -V_1/c^2 & -V_2/c^2 & -V_3/c^2 & 1 \end{pmatrix}$$

The Lorentz group is the analogon to the Galilean group and there is no immediate analogon to the Euclidean group.

A quantity that transforms under a particular transformation group as a scalar, vector, axial vector or tensor is called an objective scalar, vector, axial vector or tensor with respect to that transformation group.

The principle of material frame indifference (material objectivity) states that the material response ought not be frame-dependent. But with respect to which transformation group? Obviously, since classical mechanics is invariant under the Galilean group and the special theory of relativity under the Lorentz group, the material response must be invariant with respect to one of these groups, dependent on whether one deals with classical mechanics or with special relativity. We consider it to be known that the principle of material objectivity of classical mechanics as stated by Noll requests the material response to be invariant under Euclidean transformations. There have been objections raised against the general truth of this (see Müller [5]), but this will be of no relevance here, because the results of this monograph will also hold true if we restrict ourselves in those cases to Galilean transformations.

In order to gain some insight into the transformation properties of the electromagnetic field variables, recall that the fields E_i, B_i, H_i and D_i introduced in (1.31) can be written in the form of two skew-symmetric covariant and contravariant four-tensors, namely as

$$(1.72) \qquad \varphi_{AB} = \begin{pmatrix} 0 & B_3 & -B_2 & E_1 \\ -B_3 & 0 & B_1 & E_2 \\ B_2 & -B_1 & 0 & E_3 \\ -E_1 & -E_2 & -E_3 & 0 \end{pmatrix} \quad \text{and} \quad \eta^{AB} = \begin{pmatrix} 0 & H_3 & -H_2 & -D_1 \\ -H_3 & 0 & H_1 & -D_2 \\ H_2 & -H_1 & 0 & -D_3 \\ D_1 & D_2 & D_3 & 0 \end{pmatrix},$$

and that these tensors transform under general transformations of the form $x^{*A}(x^B)$ according to

$$(1.73) \qquad \varphi^*_{AB} = \frac{\partial x^C}{\partial x^{*A}} \frac{\partial x^D}{\partial x^{*B}} \varphi_{CD}, \qquad \eta^{*AB} = \frac{\partial x^{*A}}{\partial x^C} \frac{\partial x^{*B}}{\partial x^D} \eta^{CD} .$$

With

$$(1.74) \qquad \sigma^A := (\mathcal{J}_i + \mathcal{Q}\dot{x}_i, \mathcal{Q})$$

it is then a routine matter to show that the Maxwell equations (1.23) and (1.28) are given by

$$(1.75) \qquad e^{ABCD} \frac{\partial \varphi_{CD}}{\partial x^B} = 0, \qquad \frac{\partial \eta^{AB}}{\partial x^B} = \sigma^A ,$$

where e^{ABCD} is the four-dimensional permutation tensor. The equations (1.73) hold for any transformation $x^{*A} = x^{*A}(x^B)$. If one chooses Euclidean transformations, one can show that

	$\mathcal{Q}$	transforms as an objective scalar ,
(1.76)	$\mathcal{E}_i, D_i, \mathcal{J}_i$	transform as objective vectors ,
	$\mathcal{H}_i, B_i$	transform as objective axial vectors .

For the proof of these statements, the reader may consult Appendix A. In particular, these quantities are objective under Galilean transformations. In the sequel we shall also introduce other electromagnetic variables, such as polarization and magnetization vectors, some of which are also objective vectors in the above sense.

The Maxwell equations (1.75) would formally be invariant under general transformations of the form $x^{*A} = x^{*A}(x^B)$, as can easily be checked by substituting (1.73) into (1.75), if there would not be a relation of the form

$$\eta^{AB} = \eta^{AB}(\varphi_{CD}) \quad , \tag{1.77}$$

that is not invariant under the most general transformations. Indeed, a relation (1.77) even exists in vacuo, in which case it reads

$$D_i = \varepsilon_o E_i , \qquad H_i = \frac{1}{\mu_o} B_i \quad . \tag{1.78}$$

These relations are sometimes referred to as the Maxwell-Lorentz aether relations. Equation (1.78) restricts the class for which the Maxwell equations (1.75) are invariant to the extended Lorentz group. In general, (1.77) gives a model for electromagnetic field interactions with matter.

Based on the properties (1.76) we may then request as is done classically, that the material response be invariant under the Euclidean group. This is an approximation, because the Maxwell equations can never be rendered Lorentz invariant this way. On the other hand, dependent on the choice of electromagnetic body force, body couple and energy supply, the balance laws of mechanics may be Galilean invariant this way. Theories of this nature have been proposed by, amongst others, Toupin [6], Liu and Müller [7], Pao and Hutter [8], Alblas [9], van de Ven [10], Hutter [11] and, implicitly, by De Groot and Suttorp [12], Ch. II. For reasons that will become apparent shortly, we shall call these treatments non-relativistic theories.

Of course, the above selection of the transformation group is not satisfactory, and it should be replaced by one treating the mechanical and electromagnetic equations alike. Such a description must necessarily be relativistic. Hence, the material response must be invariant under the extended Lorentz group. Considering motions only that are small relative to the velocity of light, we may then drop all terms containing a factor c^{-2} (this statement depends on the choice of units and in the way we state it here, MKSA-units are implied). In this way one arrives at Maxwell equations which are Lorentz invariant except for terms with a c^{-2}-factor. The same holds true for the constitutive relations and the mechanical equations. Theories obtained in this fashion may also be called non-relativistic, because we may look at their constitutive treatment in the light of Euclidean transformations, as we shall soon see.

Finally there are formulations in which only the terms of order $\dot{x}^2/c^2$ are dropped, while terms with a c^{-2}-factor (but no $(\dot{x}^2/c^2)$-factor) are kept. Such theories will be called <u>semi-relativistic</u>.

In order to render these ideas a little more precise, consider a stationary frame of Cartesian coordinates and a particle moving with velocity $\dot{x}_i$. An inertial frame, in which the particle is instantaneously at rest, is called the <u>rest frame</u>. Electromagnetic variables as measured by an observer in this frame are called <u>rest frame values</u> and they are related to those in the stationary frame by a Lorentz transformation. Rest frame values will be denoted by a superimposed ring, viz. $\overset{\circ}{E}_i$ etc. Of course, the Maxwell equations also hold in the rest frame (whereby all variables and operations carry the symbol $\circ$). Transforming these equations back to the original frame reveals the transformation rules for the variables B_i, $\overset{\circ}{B}_i$, $\mathcal{E}_i$, $\overset{\circ}{\mathcal{E}}_i$ etc. To within the semi-relativistic approximation the transformation rules are

$$
\begin{aligned}
&\overset{\circ}{\mathcal{B}}_i = \mathcal{B}_i + O(\dot{x}^2/c^2)\,, \qquad \mathcal{B}_i = B_i - \frac{1}{c^2}\, e_{ijk}\dot{x}_j \mathcal{E}_k\,,\\
&\overset{\circ}{\mathcal{E}}_i = \mathcal{E}_i + O(\dot{x}^2/c^2)\,,\\
&\overset{\circ}{D}_i = \mathcal{D}_i + O(\dot{x}^2/c^2)\,, \qquad \mathcal{D}_i = D_i + \frac{1}{c^2}\, e_{ijk}\dot{x}_j \mathcal{H}_k\,,\\
&\overset{\circ}{\mathcal{H}}_i = \mathcal{H}_i + O(\dot{x}^2/c^2)\,,\\
&\overset{\circ}{\mathcal{J}}_i = \mathcal{J}_i + O(\dot{x}^2/c^2)\,,\\
&\overset{\circ}{\mathcal{Q}} = \mathcal{Q} - \frac{1}{c^2}\, \dot{x}_i \mathcal{J}_i + O(\dot{x}^2/c^2)\,.
\end{aligned}
\tag{1.79}
$$

From these definitions as well as (1.31) it follows that $\mathcal{B}_i$, $\mathcal{E}_i$, $\mathcal{D}_i$ etc. are the fields B_i, E_i, D_i etc. as measured by an observer traveling with the particle. Furthermore, under rigid rotations they transform like scalars and vectors, and, as is shown in the theory of relativity (see Møller, [13], page 199), they form within the semi-relativistic approximation the first three components of a set of four-vectors. Hence, $\mathcal{B}_i$, $\mathcal{E}_i$, $\mathcal{D}_i$, $\mathcal{H}_i$, $\mathcal{J}_i$ and $\overset{\circ}{\mathcal{Q}}$ are vectors and scalars, objective under Lorentz-transformations to within the semi-relativistic approximation.

Mere inspection of (1.79) shows, however, that the variables $\mathcal{B}_i$ and $\mathcal{D}_i$ are not objective under Galilean- or Euclidean transformations, because according to (1.76) B_i and D_i are. But, in the non-relativistic limit, where c^{-2}-terms are dropped

$$(1.80)\qquad \mathcal{D}_i \approx D_i \qquad \text{and} \qquad \mathcal{B}_i \approx B_i \ ,$$

and D_i and B_i are under Euclidean transformations an objective vector and an objective axial vector, respectively.

Multiplying both sides of equation (1.79) with μ_o^{-1} shows that

$$(1.81)\qquad \frac{1}{\mu_o}\mathcal{B}_i = \frac{1}{\mu_o} B_i - \varepsilon_o e_{ijk}\dot{x}_j E_k \ ,$$

so that the second term on the right-hand side is no longer of order c^{-2}. Hence, B_i/μ_o cannot be objective non-relativistically (although B_i is). The objective quantity must be $\mathcal{B}_i/\mu_o$ and indeed it can be shown (see Appendix A) that $\mathcal{B}_i/\mu_o$ is objective under Euclidean transformations. A similar argument holds for D_i, because

$$(1.82)\qquad \frac{1}{\varepsilon_o}\mathcal{D}_i = \frac{1}{\varepsilon_o} D_i + \mu_o e_{ijk}\dot{x}_j H_k \ .$$

Here again, $\mathcal{D}_i/\varepsilon_o$ is the quantity that is objective non-relativistically and not D_i/ε_o, although D_i is. Needless to say that Euclidean invariance of $\mathcal{D}_i/\varepsilon_o$ also follows from the general relativistic transformation properties (see Appendix A). This is why we have called theories non-relativistic which use the principle of material objectivity under Euclidean transformations.

Hence, we may conclude by stating that in the non-relativistic approximation (neglecting c^{-2}-terms) the requirement of invariance of the material response under Galilean transformations (a special Euclidean transformation) is equivalent to the requirement of Lorentz invariance. The difference is at most a philosophical one.

1.7 CONSTITUTIVE EQUATIONS

In order to become field equations the balance laws (1.18), (1.23) and (1.28) must be complemented by constitutive equations. Of course, in this regard various degrees of complexity are possible. Here, we shall restrict ourselves to magnetizable and polarizable solids which deform elastically under the action of electromagnetic and thermal fields and which exhibit electrical and thermal conduction. Mechanical dissipation is left out of consideration as is the exchange interaction and magnetic spin.

To obtain field equations it must first be decided which physical variables we suppose to be the independent fields. With regard to thermo-mechanical variables these fields are generally the motion $\underline{\chi}(\underline{X},t)$ and the temperature $\theta(\underline{X},t)$ (and in case of a fluid the density ρ). The basic electromagnetic field variables are generally two electromagnetic field vectors and the free charge. Any constitutive relation must be expressed therefore as a functional of the motion, the temperature, the free charge and two electromagnetic field variables and derivatives thereof. Taking for instance E_i and H_i as the basic fields the conceivably simplest constitutive class exhibiting the above mentioned properties may have the form

$$(1.83) \qquad C(t) = \bar{C}(x_i(t),\dot{x}_i(t),F_{i\alpha}(t),E_i(t),H_i(t),\theta(t),\theta_{,i}(t),Q(t))$$

Here C stands for any scalar, vector or tensor valued quantity a constitutive equation is established for and $\bar{C}(\cdot)$ is a function of the indicated variables. If the value of C at the instant t depends on the motion x_i, the temperature θ and the elctromagnetic fields E_i, H_i and Q at the same instant only, we call the material response to have no memory. Now the function $\bar{C}(\cdot)$ also depends on $\dot{x}_i(t)$. Hence, with regard to x_i the material appears to remember the past history of a process it has undergone for a very short time. Indeed knowing x_i and $\dot{x}_i$ at a time t allows us to approximate the motion arbitrarily close to times $\tau = t$ as close as we please. We shall show, however, that an explicit dependence on $\dot{x}_i$ is not possible.
As is usual in continuum mechanics, we require the constitutive relations to be independent of the observer. In terms of the above quantities C this principle of material objectivity requires that the quantities C do not only transform as objective scalars, vectors and tensors, respectively, but that $\bar{C}(\cdot)$ is frame independent under the transformation group considered. In view of

the fact that we shall restrict ourselves in the following Chapters to a non-relativistic approach we require the constitutive equations to be frame independent scalar, vector and tensor valued functions under the group of Euclidean transformations. (Actually invariance under Galilean transformations is sufficient for constitutive relations of the form (1.83)). Then it can be shown that the constitutive quantities C cannot depend on x_i and $\dot{x}_i$ explicitly. Furthermore, we have seen in the preceding Section that E_i, H_i and $\mathcal{Q}$ are objective, and thus we may set

$$C = \bar{C}(F_{i\alpha}, E_i, H_i, \theta, \theta_{,i}, \mathcal{Q}) \; . \tag{1.84}$$

Of course, since D_i and B_i are also objective, it is always possible to replace in (1.84) E_i by D_i and/or H_i by B_i.
Moreover, one can formally introduce polarization P_i and magnetization M_i by

$$P_i := D_i - \varepsilon_o E_i , \qquad \text{and} \qquad \mu_o M_i := B_i - \mu_o H_i \; , \tag{1.85}$$

and then, assuming for the moment that P_i and $\mu_o M_i$ are an objective vector and an objective axial vector under Euclidean transformations, respectively, we could also use P_i and $\mu_o M_i$ as basic (i.e. independent) field variables. In this way we can choose from nine possibilities for the constitutive relations of the kind (1.84). However, we shall not discuss them all here, but we shall list below only those combinations, which are physically relevant and will appear in the next Chapter. These are

$$(1.86) \qquad \begin{aligned} C &= \hat{C}\left(F_{i\alpha}, \frac{P_i}{\rho}, \frac{\mu_o M_i}{\rho}, \theta, \theta_{,i}, \mathcal{Q}\right) , && \text{(a)} \\ C &= \tilde{C}\left(F_{i\alpha}, E_i, \frac{\mu_o M_i}{\rho}, \theta, \theta_{,i}, \mathcal{Q}\right) , && \text{(b)} \\ C &= \bar{C}(F_{i\alpha}, E_i, H_i, \theta, \theta_{,i}, \mathcal{Q}) , && \text{(c)} \\ C &= \overset{+}{C}\left(F_{i\alpha}, \frac{P_i}{\rho}, B_i, \theta, \theta_{,i}, \mathcal{Q}\right) , && \text{(d)} \\ C &= \check{C}(F_{i\alpha}, E_i, B_i, \theta, \theta_{,i}, \mathcal{Q}) \; . && \text{(e)} \end{aligned}$$

Although all the arguments appearing in the constitutive relations (1.86) are objective quantities, we have not satisfied yet the requirement that $\hat{C}(\cdot)$, $\tilde{C}(\cdot)$, $\bar{C}(\cdot)$, $\overset{+}{C}(\cdot)$ and $\check{C}(\cdot)$ must be objective tensorial, vectorial and scalar functions of their variables with respect to the Euclidean group.

The explicit form of these expressions depends on whether $\hat{C}(\cdot)$ etc. is a scalar, vector or tensor valued function. The representations in all these cases are well-known and are due to Noll. The reader may consult Truesdell and Noll's treatise [14] for an account on the history and for a proof. If C is an objective scalar under the Euclidean transformation group it may be shown that its constitutive function is frame independent if

$$
\begin{aligned}
C &= \hat{C}(C_{\alpha\beta},P_{\alpha},M_{\alpha},\theta,\theta_{,\alpha},\mathcal{Q}) , && \text{(a)}\\
C &= \tilde{C}(C_{\alpha\beta},E_{\alpha},M_{\alpha},\theta,\theta_{,\alpha},\mathcal{Q}) , && \text{(b)}\\
C &= \bar{C}(C_{\alpha\beta},E_{\alpha},H_{\alpha},\theta,\theta_{,\alpha},\mathcal{Q}) , && \text{(c)}\\
C &= \overset{+}{C}(C_{\alpha\beta},P_{\alpha},B_{\alpha},\theta,\theta_{,\alpha},\mathcal{Q}) , && \text{(d)}\\
C &= \check{C}(C_{\alpha\beta},E_{\alpha},B_{\alpha},\theta,\theta_{,\alpha},\mathcal{Q}) . && \text{(e)}
\end{aligned}
\tag{1.87}
$$

Here $C_{\alpha\beta}$ is the right Cauchy-Green deformation tensor, defined by

$$
C_{\alpha\beta} := F_{i\alpha}F_{i\beta} , \tag{1.88}
$$

and

$$
\begin{aligned}
&P_{\alpha} := \frac{1}{\rho} P_i F_{i\alpha}, \qquad M_{\alpha} := \frac{\mu_o}{\rho} M_i F_{i\alpha} \operatorname{sgn} J ,\\
&E_{\alpha} := E_i F_{i\alpha}, \qquad H_{\alpha} := H_i F_{i\alpha} \operatorname{sgn} J, \qquad B_{\alpha} = \frac{1}{\mu_o} B_i F_{i\alpha} \operatorname{sgn} J .
\end{aligned}
\tag{1.89}
$$

If C_i is an objective vector, then

$$
C_i = F_{i\gamma}\hat{C}_{\gamma}(C_{\alpha\beta},P_{\alpha},M_{\alpha},\theta,\theta_{,\alpha},\mathcal{Q}), \text{ etc.} \qquad \text{(a)} \tag{1.90}
$$

If C_i is an objective axial vector, then

$$
C_i = \operatorname{sgn} J \, F_{i\gamma}\hat{C}_{\gamma}(C_{\alpha\beta},P_{\alpha},M_{\alpha},\theta,\theta_{,\alpha},\mathcal{Q}), \text{ etc.} \qquad \text{(a)} \tag{1.91}
$$

Finally, for a second order tensor C_{ij} one obtains

$$
C_{ij} = F_{i\gamma}F_{j\delta}\hat{C}_{\gamma\delta}(C_{\alpha\beta},P_{\alpha},M_{\alpha},\theta,\theta_{,\alpha},\mathcal{Q}), \text{ etc.} \qquad \text{(a)} \tag{1.92}
$$

We now shall outline the procedure that will be used to bring the constitutive equations in their ultimate form. First, we substitute the constitutive relations (1.86) for the dependent variables into the Maxwell equations and into the balance laws of mass, linear and angular momentum and energy. The resulting equations are the field equations for ρ, χ_i, θ, Ω and the two basic electromagnetic fields (e.g. P_i and M_i in case a)). Any solution of these field equations is called a thermodynamic process. Following Coleman and Noll, [15], we request the entropy inequality, or any inequality derived from it, to hold for all smooth thermodynamic processes. This implies that at a particle we may freely choose the independent variables and derivatives thereof as long as the field equations are not violated thereby. In an open system, that is for a body with arbitrary external body force ρF_i^{ext} and energy supply ρr^{ext} no difficulties arise in this regard in the momentum and energy equation, because to any process there exist appropriate force and energy supply terms with the aid of which the momentum and energy equations are satisfied identically. Special care should be observed with regard to the electromagnetic variables, however, since, as can easily be deduced from the Maxwell equations, not all the gradients of the basic electromagnetic fields are arbitrary. Such gradients may occur in the entropy inequality, and if they do, the relations implied by the Maxwell equations must be fulfilled along with the exploitation of the entropy inequality. A detailed explanation of this point is given by Hutter [16]. In any case, once the constitutive relations are substituted into the entropy inequality, the term ρr^{ext} is eliminated (with the aid of the energy equation) and all the above mentioned side conditions are properly taken into account, an inequality results with terms that are explicitly linear in variables that in a thermodynamic process may have any arbitrarily assigned value. Therefore, since otherwise this inequality would be violated, each of the coefficients of these variables must be identically zero. These conditions then imply the constitutive equations in their ultimate form. This procedure, which we assume to be familiar to the reader, will be applied (and described in greater detail) in the next Chapter.

REFERENCES

[1] Penfield, P. and H.A. Haus, *Electrodynamics of Moving Media*, The M.I.T. Press, Cambridge, Massachusettes, 1967.

[2] Fano, R.M., L.C. Chu and R.B. Adler, *Electromagnetic Fields, Energy and Forces*, John Wiley & Sons, Inc., New York, 1960 (Reprinted by the M.I.T. Press).

[3] Truesdell, C. and R.A. Toupin, *The Classical Field Theories*, Encyclopedia of Physics, Vol. III/1, ed. S. Flügge, Springer-Verlag, Berlin, 1960.

[4] Hutter, K., *On Thermodynamics and Thermostatics of Viscous Thermoelastic Solids in the Electromagnetic Fields. A Lagrangian Formulation*, Arch. Rat. Mech. Anal. 58 (1975), 339-368.

[5] Müller, I., *On the Frame Dependence of Stress and Heat Flux*. Arch. Rat. Mech. Anal. 45 (1972), 241-250.

[6] Toupin, R.A., *A Dynamical Theory of Elastic Dielectrics*, Int. J. Eng. Sc. 1 (1963), 101-126.

[7] Liu, I.S. and I. Müller, *On the Thermodynamics and Thermostatics of Fluids in Electromagnetic Fields*, Arch. Rat. Mech. Anal. 46 (1972), 149-176.

[8] Pao, Y.H., and K. Hutter, *Electrodynamics of Moving Elastic Solids and Viscous Fluids*, Proc. I.E.E.E. 63 (1975), 1011-1021.

[9] Alblas, J.B., *Electro-Magneto-Elasticity*, Topics in Applied Continuum Mechanics, eds. J.L. Zeman and F. Ziegler, 71-114, Springer-Verlag, Wien, 1974.

[10] Ven, A.A.F. van de, *Interaction of Electromagnetic and Elastic Fields in Solids*, Dr. of Science Thesis, University of Technology Eindhoven, the Netherlands, 1975.

[11] Hutter, K., *A Thermodynamic Theory of Fluids and Solids in the Electromagnetic Fields*, Arch. Rat. Mech. Anal., 64 (1977), 269-298.

[12] De Groot, S.R. and L.G. Suttorp, *Foundations of Electrodynamics*, North-Holland Publishing Co., Amsterdam, 1972.

[13] Møller, C., *The Theory of Relativity*, Oxford University Press, London, 1972.

[14] Truesdell, C. and W. Noll, *The Nonlinear Field Theories of Mechanics*, Handbuch der Physik Vol. III/3, ed. S. Flügge, Springer-Verlag, Berlin, 1960.

[15] Coleman, B.D. and W. Noll, *The Thermodynamics of Elastic Materials with Heat Conduction and Viscosity*, Arch. Rat. Mech. Anal. 13 (1963), 167-178.

[16] Hutter, K., *Thermodynamic Aspects in Field-Matter Interactions*, in Electromagnetic Interactions in Elastic Solids, ed. by H. Parkus, Springer, Wien, 1978.

2. A SURVEY OF ELECTROMECHANICAL INTERACTION MODELS

2.1 SCOPE OF THE SURVEY

In this Chapter we make an attempt to surveying various electromagnetic interaction models known to date. We do not present all the descriptions of magnetizable and polarizable deformable bodies, but list the ones that have received considerable attention in the recent literature only. A comparison of these models with still other ones will be given in Section 2.6.

Any continuum theory of deformable bodies subject to electromagnetic fields amounts to the presentation of the basic electromagnetic field variables, their relations to the other fields, as well as to the postulation of electromagnetic body force, body couple and energy supply. Once this is done, the Maxwell equations and the balance laws of mechanics and thermodynamics can be expressed in terms of the variables of the model in question. Using thermodynamic arguments, it is then a routine matter to reduce a postulated set of constitutive equations to a form compatible with the second law of thermodynamics.

It is the purpose of this chapter to present the various models of electromechanical interactions, to scrutinize their invariance properties, to derive the constitutive theory in each peculiar case and to present each model such that firstly, a comparison of one model with another can be achieved fairly straightforwardly and secondly, an initial boundary value problem can be solved at least in principle.

2.2 THE TWO-DIPOLE MODELS

As discussed by Fano, Chu and Adler [1], or Penfield and Haus [2], in the <u>Chu-formulation</u> the Maxwell equations for moving matter are expressed in terms of $\mathcal{Q}$, J_i, the velocity field $\dot{x}_i$ and four electromagnetic field quantities E_i^C, H_i^C, P_i^C and M_i^C which are related to the fields D_i, $\mathcal{E}_i$, B_i and $\mathcal{H}_i$ by the following transformation rules

$$
(2.1)\qquad
\begin{aligned}
D_i &= \varepsilon_o E_i^C + P_i^C, \qquad \mathcal{E}_i = E_i^C + \mu_o e_{ijk}\dot{x}_j H_k^C ,\\
B_i &= \mu_o H_i^C + \mu_o M_i^C, \qquad \mathcal{H}_i = H_i^C - \varepsilon_o e_{ijk}\dot{x}_j E_k^C .
\end{aligned}
$$

For reasons that will become apparent shortly we shall occasionally make use of two auxiliary fields defined by

$$B^a_i := \mu_o H^C_i, \quad D^a_i := \varepsilon_o E^C_i . \tag{2.2}$$

As usual, E^C_i and H^C_i are the electric and magnetic field strengths, and P^C_i and M^C_i are the polarization per unit volume and the magnetization per unit volume, respectively. In order to differentiate these fields from those occurring in other formulations, we have used a superscript C to indicate that these fields are those as defined by Chu. Substituting (2.1) and (2.2) into the Maxwell equations (1.23) and (1.28), what results reads as follows

$$\begin{aligned}
&B^a_{i,i} = -\mu_o M^C_{i,i} , \\
&e_{ijk}E^C_{k,j} + \frac{\partial B^a_i}{\partial t} = -\mu_o \frac{\partial M^C_i}{\partial t} + \mu_o e_{ijk}(e_{k\ell m}\dot{x}_\ell M^C_m)_{,j} , \\
&D^a_{i,i} = \mathcal{Q} - P^C_{i,i} , \\
&e_{ijk}H^C_{k,j} - \frac{\partial D^a_i}{\partial t} = J_i + \frac{\partial P^C_i}{\partial t} - e_{ijk}(e_{k\ell m}\dot{x}_\ell P^C_m)_{,j} .
\end{aligned} \tag{2.3}$$

In the above equations μ_o and ε_o are universal constants with $\varepsilon_o\mu_o = c^{-2}$, c being the speed of light in vacuo.

If we regard E^C_i and H^C_i as the basic electromagnetic fields, then, apart from J_i and $\mathcal{Q}$, the terms on the right-hand sides of (2.3) may be interpreted as charge and current densities due to polarization and magnetization. This interpretation is helpful for the derivation of electromagnetic body force, body couple and energy supply.

The above set of electromagnetic variables is based on the postulations that:

i) only two vector quantities are necessary to describe the electromagnetic fields in free space, and

ii) material bodies contribute toward the electromagnetic fields by acting as sources for these fields.

These sources are usually interpreted in terms of electric and magnetic dipoles, each such doublet being composed of a negative and positive electric and magnetic monopole. We would like to deemphasize this interpretation, mainly because of the well-known objections phycisists may rise against it. Nevertheless we accept the Chu-formulation as a proper description, but view it as obtained from the definitions (2.1) by mere variable transformations.

Before we proceed it is advantageous to investigate how the Chu-variables behave under the transformations discussed in Section 1.6. Following an approach analogous to the one outlined by Truesdell and Toupin in [3], Sect. 283, it is not difficult to show that under the Euclidean transformation group

$$\begin{aligned} &\mathcal{Q} \text{ transforms as an objective scalar ,} \\ (2.4) \qquad &E_i,\ P_i^C,\ J_i, \text{ transform as objective vectors ,} \\ &H_i,\ \mu_o M_i^C, \text{ transform as objective axial vectors .} \end{aligned}$$

It can also be shown that

$$\begin{aligned} (2.5) \qquad &D_i^a \text{ transforms as an objective vector, while} \\ &B_i^a \text{ transforms as an objective axial vector} \end{aligned}$$

under the Euclidean transformation group. For the main lines of the proof the reader may consult Appendix A. It is clear that the above transformation rules also apply under the slightly less general Galilean group. The properties laid down in (2.4) and (2.5) are exact in the sense that the Maxwell equations can exactly (that is without the neglect of c^{-2}-terms) be rendered Euclidean invariant. Yet, the relations (2.2) must also be satisfied, and they are not Euclidean invariant, but Lorentz invariant or Galilean invariant only to within terms containing a c^{-2}-factor (see also eqs. (2.6)[7,8]).

Therefore, the Chu-formulation of electromagnetism consisting of the Maxwell equations (2.3) together with the aether relations (2.2) is invariant under the Galilean group only in the sense of the non-relativistic approximation. Hence, in order to remain consistent, we must neglect in the following interaction models all terms preceded by a factor c^{-2}.

In order to investigate semi-relativistic objectivity requirements, let us consider the extended Lorentz group. Under these transformations the Chu variables transform (to within terms of the order $0(V^2/c^2)$) according to

$$(2.6) \qquad \begin{aligned} &\overset{\circ}{\mathcal{Q}} = \mathcal{Q} - \frac{1}{c^2}\dot{x}_i J_i + 0(V^2/c^2), \quad \overset{\circ}{J}_i = J_i + 0(V^2/c^2) , \\ &\overset{\circ}{E}_i = E_i + 0(V^2/c^2), \quad \overset{\circ}{H}_i = H_i + 0(V^2/c^2) , \\ &\overset{\circ}{P}{}_i^C = P_i^C + 0(V^2/c^2), \quad \overset{\circ}{M}{}_i^C = M_i^C + 0(V^2/c^2) , \\ &\overset{\circ}{D}{}_i^a = D_i^a + \frac{1}{c^2} e_{ijk}\dot{x}_j H_k^C + 0(V^2/c^2), \quad \overset{\circ}{B}{}_i^a = B_i^a - \frac{1}{c^2} e_{ijk}\dot{x}_j E_k^C + 0(V^2/c^2) . \end{aligned}$$

Thus it is obvious that the transformation rules (2.4) also hold under the extended Lorentz group, if terms of $0(V^2/c^2)$ are dropped. There is only one exception, namely that $\mathcal{Q}$ has to be replaced by $\dot{\mathcal{Q}}$. In the non-relativistic approximation ($c^{-2} = 0$) this difference is negligible, however, and hence in this approximation we may everywhere replace $\mathcal{Q}$ by $\overset{o}{\mathcal{Q}}$. Moreover, there are reasons to assume that $\| J_i \|$ is of the order of $\| \mathcal{Q}\underline{V} \|$. Then, $\mathcal{Q}$ may even be replaced by $\overset{o}{\mathcal{Q}}$ in the semi-relativistic approximation. As a consequence the Chu-formulation of electromagnetism has the nice feature to obey the non-relativistic as well as the semi-relativistic invariance requirements. Note that, in contrast with (2.5), the auxiliary fields D_i^a and B_i^a do not behave as objective quantities under the Lorentz transformation group, not even in a semi-relativistic sense.

A complete theory of deformable, polarizable and magnetizable bodies consists of a set of electromagnetic and thermomechanical equations, the latter including expressions for the body force, body couple and energy supply due to the electromagnetic fields. On the non-relativistic level there are essentially two distinct formulations both of which use the dipole model not only for the polarization but also for the magnetization.

1. The Two-Dipole Model with a Nonsymmetric Stress Tensor (Model I)

As mentioned above, to complete the description of deformable continua in the electromagnetic fields, expressions for ρF_i^e, ρL_{ij}^e and ρr^e are needed. Penfield and Haus [2] and Pao and Hutter [4] have motivated and derived the following expressions using the two-dipole model

$$\rho F_i^e = \mathcal{Q}E_i^C + \mu_o e_{ijk} J_j H_k^C + P_j^C E_{i,j}^C + \mu_o e_{ijk} \dot{x}_j H_{k,\ell}^C P_\ell^C +$$

$$+ \rho \mu_o e_{ijk} \frac{d}{dt}\left(\frac{P_j^C}{\rho}\right) H_k^C + \mu_o M_j^C H_{i,j}^C +$$

$$(2.7) \qquad - \varepsilon_o e_{ijk} \dot{x}_j E_{k,\ell}^C \mu_o M_\ell^C - \rho \varepsilon_o e_{ijk} \frac{d}{dt}\left(\frac{\mu_o M_j^C}{\rho}\right) E_k^C ,$$

$$\rho L_{ij}^e = P_{[i}^C \mathcal{E}_{j]} + \mu_o M_{[i}^C \mathcal{H}_{j]} ,$$

$$\rho r^e = \mathcal{J}_i \mathcal{E}_i + \rho \mathcal{E}_i \frac{d}{dt}\left(\frac{P_i^C}{\rho}\right) + \rho \mathcal{H}_i \frac{d}{dt}\left(\frac{\mu_o M_i^C}{\rho}\right) .$$

The complete derivation of (2.7) as given by Pao and Hutter, [4], is based on the assumptions that:

i) each material particle is equipped with a number of mutually noninteracting electric and magnetic dipoles,

ii) each monopole suffers an electromagnetic body force as described by the Lorentz force (see below), and

iii) the monopoles of a particular dipole are only a small distance apart so that Taylor series expansions are justified.

The Lorentz force mentioned in assumption ii) is hereby taken in the form

$$\rho F_i^{\text{Lorentz}} = \mathcal{Q}^e E_i^C + \mu_o e_{ijk} \mathcal{Q}^e \dot{x}_j H_k^C + \mathcal{Q}^m H_i^C - \varepsilon_o e_{ijk} \mathcal{Q}^m \dot{x}_j E_k^C , \tag{2.8}$$

where $\mathcal{Q}^e$ and $\mathcal{Q}^m$ are the electric and magnetic charges of the monopoles. Again we do not wish to emphasize the physical aspects of this approach, but we would like to view (2.7) as possible interaction postulates. We refer to the paper of Pao and Hutter, [4], for the details of the derivation. Substituting (2.7) into the balance laws (1.18) yields the local balance laws of mass, linear and angular momentum and energy in the form

$$\begin{aligned}
&\dot{\rho} + \rho \dot{x}_{i,i} = 0 , \\
&\rho \ddot{x}_i = t_{ij,j} + \rho F_i^{ext} + \mathcal{Q} E_i^C + \mu_o e_{ijk} J_j H_k^C + P_j^C E_{i,j}^C + \\
&\qquad + \mu_o e_{ijk} \dot{x}_j H_{k,\ell}^C P_\ell^C + \rho \mu_o e_{ijk} \frac{d}{dt}\left(\frac{P_j^C}{\rho}\right) H_k^C + \\
&\qquad + \mu_o M_j^C H_{i,j}^C - \varepsilon_o e_{ijk} \dot{x}_j E_{k,\ell}^C \mu_o M_\ell^C - \rho \varepsilon_o e_{ijk} \frac{d}{dt}\left(\frac{\mu_o M_j^C}{\rho}\right) E_k^C \\
&t_{[ij]} = P_{[i}^C \mathcal{E}_{j]} + \mu_o M_{[i}^C \mathcal{H}_{j]}, \\
&\rho \dot{U} = t_{ij} \dot{x}_{i,j} - q_{i,i} + \mathcal{J}_i \mathcal{E}_i + \rho \mathcal{E}_i \frac{d}{dt}\left(\frac{P_i^C}{\rho}\right) + \\
&\qquad + \rho \mathcal{H}_i \frac{d}{dt}\left(\frac{\mu_o M_i^C}{\rho}\right) + \rho r^{ext} .
\end{aligned} \tag{2.9}$$

It is a routine though rather elaborate matter to transform $(2.9)^2$ and $(2.9)^4$ to the form listed in (1.62). One obtains

$$
\begin{aligned}
{}^{I}t^{M}_{ij} &= \varepsilon_o E^C_i E^C_j + \mu_o H^C_i H^C_j + \mathcal{E}_i P^C_j + \mu_o \mathcal{H}_i M^C_j + \\
&\quad - \frac{1}{2}\,\delta_{ij}(\varepsilon_o E^C_k E^C_k + \mu_o H^C_k H^C_k) \ ,
\end{aligned}
$$

$$
\begin{aligned}
(2.10)\qquad {}^{I}g_i &= -\frac{1}{c^2}\, e_{ijk} E^C_j H^C_k \ , \\
{}^{I}\pi_i &= -e_{ijk} E^C_j H^C_k + (P^C_i E^C_j + \mu_o M^C_i H^C_j)\dot{x}_j \ , \\
{}^{I}\omega &= -\frac{1}{2}(\varepsilon_o E^C_i E^C_i + \mu_o H^C_i H^C_i) \ .
\end{aligned}
$$

Here we have used a left upper index in order to be able to distinguish these quantities from those of other interaction models.
With the expressions (2.10) the jump conditions of momentum and energy as listed in (1.66) are readily derived. For the sake of easy reference they are given below together with the jump conditions of electromagnetic fields and mass

$$
[\![\mu_o H^C_i + \mu_o M^C_i]\!] n_i = 0, \qquad [\![\varepsilon_o E^C_i + P^C_i]\!] n_i = 0 \ ,
$$

$$
[\![e_{ijk} E^C_j n_k + \mu_o \dot{x}_i M^C_k n_k + \mu_o H^C_i w_n - \mu_o M^C_i(\dot{x}_k n_k - w_n)]\!] = 0 \ ,
$$

$$
[\![e_{ijk} H^C_j n_k - \dot{x}_i P^C_k n_k - \varepsilon_o E^C_i w_n + P^C_i(\dot{x}_k n_k - w_n)]\!] = 0 \ ,
$$

$$
[\![\rho(\dot{x}_i n_i - w_n)]\!] = 0 \ ,
$$

$$
\begin{aligned}
(2.11)\qquad & [\![t_{ij} + \varepsilon_o E^C_i E^C_j + \mu_o H^C_i H^C_j + \mathcal{E}_i P^C_j + \mu_o \mathcal{H}_i M^C_j + \\
&\quad - \frac{1}{2}\,\delta_{ij}(\varepsilon_o E^C_k E^C_k + \mu_o H^C_k H^C_k) + \frac{1}{c^2}\, e_{ik\ell} E^C_k H^C_\ell \dot{x}_j]\!] n_j + \\
&\quad - [\![(\rho\dot{x}_i + \frac{1}{c^2}\, e_{ik\ell} E^C_k H^C_\ell)(\dot{x}_j n_j - w_n)]\!] = 0 \ , \\
& [\![\dot{x}_i t_{ij} - q_j - e_{jk\ell} E^C_k H^C_\ell + (P^C_j E^C_i + \mu_o M^C_j H^C_i)\dot{x}_i + \frac{1}{2}(\varepsilon_o E^C_k E^C_k + \mu_o H^C_k H^C_k)\dot{x}_j]\!] n_j + \\
&\quad - [\![\{\frac{1}{2}\,\rho\dot{x}_i\dot{x}_i + \rho U + \frac{1}{2}(\varepsilon_o E^C_i E^C_i + \mu_o H^C_i H^C_i)\}(\dot{x}_j n_j - w_n)]\!] = 0 \ .
\end{aligned}
$$

Note that the jump conditions (2.11), specialized for material surfaces, have also been obtained by Pao and Hutter, [4]. Their jump conditions for the stress, however, does not contain the term

$$[\![\frac{1}{c^2} e_{ik\ell} \overset{C}{E}_k \overset{C}{H}_\ell \dot{x}_j]\!] n_j \ ,$$

but, as we shall soon see, this term is negligible in a non-relativistic theory as it contains a prefactor c^{-2}.

Before discussing the constitutive equations, it is interesting to look at the invariance properties of the electromagnetic body force, body couple and energy supply. At the beginning of this section the invariance properties of the electromagnetic field variables were listed. From these it is immediately seen that L^e_{ij} transforms under the Euclidean group as an objective skew-symmetric tensor. Under the extended Lorentz group L^e_{ij} still transforms as an objective skew-symmetric tensor, but only to within terms of order $O(V^2/c^2)$. Hence, L^e_{ij} is objective non-relativistically as well as semi-relativistically. If we write the energy supply term $(2.7)^3$ in the form

$$(2.12) \qquad \rho r^e = (J_i + \overset{\star}{P}_i)E_i + \mu_o \overset{\star}{M}_i H_i + (E_i P_j + H_i \mu_o M_j)\dot{x}_{i,j}$$

then we recognize that ρr^e is not an objective scalar under the Euclidean transformation group, this because of the term involving $\dot{x}_{i,j}$. Incidentally, in (2.12) we have written P_i and $\mu_o M_i$ for $\overset{C}{P}_i$ and $\mu_o \overset{C}{M}_i$ in order to stress the invariance properties of $\overset{C}{P}_i$ and $\mu_o \overset{C}{M}_i$. Moreover, it should be noted that the material time derivative of an objective vector is not an objective vector under the Euclidean group; but the convective time derivative is. This is the reason why we have tried to use convective time derivatives in (2.12). The non-objective part of (2.12) is now given by

$$(E_i P_j + \mu_o H_i M_j)\dot{x}_{[i,j]} \ ,$$

or, in view of $(2.9)^3$, by

$$-t_{[ij]}\dot{x}_{[i,j]} \ .$$

Using this result, (2.12) shows that

$$(2.13) \qquad \rho r^e + t_{[ij]}\dot{x}_{[i,j]} = (J_i + \overset{\star}{P}_i)E_i + \mu_o \overset{\star}{M}_i H_i + (E_i P_j + \mu_o H_i M_j)\dot{x}_{(i,j)} \ ,$$

is an objective scalar. Hence, the balance of internal energy can be written as

$$(2.14) \qquad \rho\dot{U} = t_{(ij)}\dot{x}_{(i,j)} - q_{i,i} + (J_i + \overset{*}{P}_i)E_i + \mu_o\overset{*}{M}_iH_i + (E_iP_j + \mu_oH_iM_j)\dot{x}_{(i,j)},$$

a form in which each term is an objective scalar under the Euclidean transformation group. In much the same way it can be shown that each term in the above equation is - to within terms of the order $O(V^2/c^2)$ - also an objective scalar under the extended Lorentz group. Hence, the energy balance law is an invariant equation in the non-relativistic and semi-relativistic sense. Finally, we also investigate the transformation properties of the electromagnetic body force $(2.7)^1$. A straightforward calculation shows that ρF_i^e can be written in the following form

$$(2.15) \qquad \rho F_i^e = (Q - P_{j,j}^C)E_i + e_{ijk}(J_j + \overset{*}{P}{}_j^C)B_k^a +$$

$$- \mu_o M_{j,j}^C H_i - e_{ijk}\mu_o \overset{*}{M}{}_j^C D_k^a + (E_iP_j^C + H_i\mu_oM_j^C)_{,j}$$

Using the transformation properties stated in (2.4) and (2.5) it is readily seen that each term on the right-hand side of this equation is an objective (polar) vector under the Euclidean transformation group. Because of the transformation properties of the auxiliary fields D_i^a and B_i^a, the body force does not enjoy the invariance properties of semi-relativistically correct Lorentz transformations, however.

The above expression of the electromagnetic body force is written in a form which still contains the auxiliary fields. These fields may be eliminated with the aid of the relations (2.2), which, however, are only invariant equations in the non-relativistic approximation. A straightforward calculation, in which the Maxwell equations (2.3) are used and in which terms of order $O(V^2/c^2)$ are neglected, then shows that ρF_i^e can be written in the form

$$(2.16) \qquad \rho F_i^e = QE_i + \mu_o e_{ijk}J_jH_k + P_jE_{i,j} + \mu_o e_{ijk}\overset{*}{P}_jH_k + \mu_oM_jH_{i,j} +$$

$$+ \frac{1}{c^2}[(J_j + \overset{*}{P}_j)(\dot{x}_iE_j - \dot{x}_jE_i) + e_{ik\ell}M_jE_\ell\dot{x}_{k,j} + \rho e_{ijk}\frac{d}{dt}(\frac{M_j}{\rho})E_k^C] \; .$$

In this form the body force is no longer an objective vector under Euclidean transformations, whereby this objectivity is destroyed by the use of the relations (2.2) for the elimination of D_i^a and B_i^a. However, in a non-relativistic approximation the term involving a c^{-2}-term must be dropped. Hence, within a consistent non-relativistic approximation the momentum equation $(2.9)^2$ must be written as

$$(2.17)\qquad \rho\ddot{x}_i = t_{ij,j} + \rho F_i^{ext} + QE_i + \mu_o e_{ijk} J_j H_k + P_j E_{i,j} +$$

$$+ \mu_o e_{ijk} \overset{*}{P}_j H_k + \mu_o M_j H_{i,j} \ ,$$

or since

$$(2.18)\qquad QE_i + \mu_o e_{ijk} J_j H_k + P_j E_{i,j} + \mu_o e_{ijk} \overset{*}{P}_j H_k + \mu_o M_j H_{i,j} =$$

$$= \{\varepsilon_o E_i E_j + \mu_o H_i H_j + E_i P_j + \mu_o H_i M_j - \frac{1}{2}\delta_{ij}(\varepsilon_o E_k E_k + \mu_o H_k H_k)\}_{,j} +$$

$$+ 0(v^2/c^2)$$

as

$$(2.19)\qquad \rho\ddot{x}_i = t_{ij,j} + \rho F_i^{ext} + \{\varepsilon_o E_i E_j + \mu_o H_i H_j + E_i P_j + \mu_o H_i M_j +$$

$$- \frac{1}{2}\delta_{ij}(\varepsilon_o E_k E_k + \mu_o H_k H_k)\}_{,j} \ .$$

Moreover, the c^{-2}-terms in (2.10) and (2.11) must also be neglected so that $^I g_i$ vanishes, i.e.

$$(2.20)\qquad {}^I g_i = 0 \ .$$

Thus the jump conditions for momentum and energy reduce to

$$(2.21)\qquad [\![t_{ij} + \varepsilon_o E_i E_j + \mu_o H_i H_j + E_i P_j + \mu_o H_i M_j - \frac{1}{2}\delta_{ij}(\varepsilon_o E_k E_k + \mu_o H_k H_k)]\!] n_j +$$

$$- [\![\rho\dot{x}_i(\dot{x}_j n_j - w_n)]\!] = 0 \ ,$$

and

$$(2.22)\qquad [\![\dot{x}_i t_{ij} - q_j - e_{jk\ell} E_k H_\ell + (\varepsilon_o E_j E_k + \mu_o H_j H_k)\dot{x}_k + (P_j E_k + \mu_o M_j H_k)\dot{x}_k +$$

$$- \frac{1}{2}(\varepsilon_o E_k E_k + \mu_o H_k H_k)\dot{x}_j]\!] n_j - [\![\{\frac{1}{2}\rho\dot{x}_i\dot{x}_i + \rho U +$$

$$+ \frac{1}{2}(\varepsilon_o E_i E_i + \mu_o H_i H_i)\}(\dot{x}_j n_j - w_n)]\!] = 0 \ .$$

To summarize, we may state that the Maxwell equations together with the balance laws of mass, momentum (in the original not approximated version), angular momentum and energy presented here are Euclidean invariant. However, since these equations must be supplemented by the Maxwell-Lorentz aether relations (2.2), the complete set of electromagnetic and mechanical field equations is Euclidean invariant, only in the non-relativistic approximation.

With this digression on invariance properties, which in our opinion are crucial, we now continue with the constitutive theory, which in accordance with the preceding conclusions must also be Euclidean invariant. The conditions for these requirements to be satisfied are given in Section 1.7, and in equation (1.86) some possible constitutive relations are presented. In the Chu-formulation the $\underline{B}$-field does not occur and therefore case d) must be excluded here. As we are interested in a comparison of different formulations of thermoelastic, polarizable and magnetizable bodies, we should present the constitutive theory for each case in (1.86). We shall discuss the details for case a) only, but the results for the two other cases will also be listed.

Case a):

$$(2.23) \qquad C = \hat{C}(F_{i\alpha}, \frac{P_i}{\rho}, \frac{\mu_o M_i}{\rho}, \theta, \theta_{,i}, \mathcal{Q}) .$$

Following the usual procedure, we introduce the Helmholtz free energy ψ by

$$(2.24) \qquad \psi = U - \theta\eta ,$$

and eliminate with the aid of $(2.7)^3$ r^{ext} from (1.53). This yields

$$(2.25) \qquad -\rho\dot{\psi} - \rho\eta\dot{\theta} + t_{ij}F^{-1}_{\alpha j}\dot{F}_{i\alpha} + \rho E_i \frac{d}{dt}(\frac{P_i}{\rho}) + \rho H_i \frac{d}{dt}(\frac{\mu_o M_i}{\rho}) + $$
$$+ J_i E_i - q_{i,i} + \theta\phi_{i,i} \geq 0 .$$

Constitutive equations are established for the (dependent) variables

$$t_{ij}, E_i, H_i, \eta, \phi_i, q_i \text{ and } J_i ,$$

which we all assume to be functions of the form (2.23). Furthermore, we assume the classical form of the entropy flux, namely

$$(2.26) \qquad \phi_i = \frac{q_i}{\theta} .$$

Actually this need not be done. Instead one could also postulate a general constitutive equation for ϕ_i and then prove the relation (2.26). For this formulation such a proof has not been given yet. However, in view of the equivalence of this theory with the next one (Model II) in which the result (2.26) was established, we may regard (2.26) as a proven statement.

We now proceed in the way as described in Section 1.7. Substituting the constitutive equations for the dependent variables into the Maxwell equations (2.3) and into the balance laws of mass, momenta and energy (2.9), what results are the field equations for x_i, P_i, M_i, θ and $\mathcal{Q}$. Any solution to given initial data is called a thermodynamic process. Following Coleman and Noll [5], we request the entropy inequality (or any inequality derived from it) to hold for any smooth thermodynamic process. Since ρF_i^{ext} and ρr^{ext} may have any arbitrarily assigned values, this implies that we may freely choose $F_{i\alpha}$, P_i/ρ, $\mu_o M_i/\rho$, θ, $\mathcal{Q}$, their material time derivatives and $(\theta_{,i})^{\cdot}$ and do not violate the field equations. Using the constitutive relation (2.23) for ψ, substituting the expression (2.26) for the entropy flux ϕ_i into (2.25) and performing all indicated differentiations results in an inequality which is explicitly linear in

$$\dot{F}_{i\alpha},\ (P_i/\rho)^{\cdot},\ (\mu_o M_i/\rho)^{\cdot},\ \dot{\theta},\ (\theta_{,i})^{\cdot} \text{ and } \dot{\mathcal{Q}}\ .$$

Since all these variables may have arbitrary values, their coefficients must vanish, which implies the following relations

$$(2.27)\qquad \psi = \hat{\psi}(F_{i\alpha},\ P_i/\rho,\ \mu_o M_i/\rho,\ \theta)\ ,$$

$$\eta = -\frac{\partial\hat{\psi}}{\partial\theta}\ ,$$

$$(2.28)\qquad E_i = \frac{\partial\hat{\psi}}{\partial(P_i/\rho)}\ ,$$

$$H_i = \frac{\partial\hat{\psi}}{\partial(\mu_o M_i/\rho)}\ ,$$

and

$$t_{ij} = \rho\,\frac{\partial\hat{\psi}}{\partial F_{i\alpha}}\,F_{j\alpha}\ .$$

Hence, the Helmholtz free energy cannot depend on the free charge and the temperature gradient.

Of (2.25) there remains the residual inequality

$$(2.29)\qquad J_i E_i - \frac{\theta_{,i}q_i}{\theta} \geq 0\ ,$$

where J_i and q_i are still general functions of the type (2.23).

Since ψ, η, E_i, H_i and t_{ij} must be objective scalar, vector and tensor valued fields under the full Euclidean group, ψ should have the form

$$\text{(2.30)} \qquad \psi = \hat{\Psi}(C_{\alpha\beta}, P_\alpha, M_\alpha, \theta) ,$$

with $C_{\alpha\beta}$, P_α and M_α as given in (1.88) and (1.89).
With this choice, the constitutive equations (2.28) become

$$\text{(2.31)} \qquad \begin{aligned} \eta &= -\frac{\partial\hat{\Psi}}{\partial\theta} , \\ E_i &= \frac{\partial\hat{\Psi}}{\partial P_\alpha} F_{i\alpha} , \\ H_i &= \frac{\partial\hat{\Psi}}{\partial M_\alpha} F_{i\alpha} \operatorname{sgn} J , \\ t_{ij} &= 2\rho \frac{\partial\hat{\Psi}}{\partial C_{\alpha\beta}} F_{i\alpha}F_{j\beta} + P_i E_j + \mu_o M_i H_j = \\ &= \rho[2 \frac{\partial\hat{\Psi}}{\partial C_{\alpha\beta}} + \frac{\partial\hat{\Psi}}{\partial P_\beta} P_\gamma C^{-1}_{\alpha\gamma} + \frac{\partial\hat{\Psi}}{\partial M_\beta} M_\gamma C^{-1}_{\alpha\gamma}]F_{i\alpha}F_{j\beta} . \end{aligned}$$

Conversely, it is easily shown that with this choice η, E_i, H_i and t_{ij} indeed behave as objective scalars, vectors and tensors, respectively (note that they are in correspondence with (1.90), (1.91) and (1.92)). Moreover, $(2.31)^4$ implies that the balance law of angular momentum $(2.9)^3$ is identically satisfied.

The above derivation makes no direct use of the Gibbs relation, as is usually done in irreversible thermodynamics. This must be so, because quite contrary to irreversible thermodynamics, the Gibbs relation is a proven statement here, valid not only in thermostatic equilibrium but also in a general thermodynamic process. Indeed, (2.31) together with (2.24) imply

$$\text{(2.32)} \qquad \begin{aligned} \frac{\partial\eta}{\partial\theta} &= \frac{1}{\theta}\frac{\partial U}{\partial\theta} , \\ \frac{\partial\eta}{\partial P_\alpha} &= \frac{1}{\theta}\{\frac{\partial U}{\partial P_\alpha} - F^{-1}_{\alpha i}E_i\} , \\ \frac{\partial\eta}{\partial M_\alpha} &= \frac{1}{\theta}\{\frac{\partial U}{\partial M_\alpha} - F^{-1}_{\alpha i}H_i \operatorname{sgn} J\} , \\ \frac{\partial\eta}{\partial C_{\alpha\beta}} &= \frac{1}{\theta}\{\frac{\partial U}{\partial C_{\alpha\beta}} - \frac{1}{2\rho}(t_{ij} - P_i E_j - \mu_o M_i H_j)F^{-1}_{\alpha i}F^{-1}_{\beta j}\} = \\ &= \frac{1}{\theta}\{\frac{\partial U}{\partial C_{\alpha\beta}} - \frac{1}{2}[\frac{1}{\rho_o} T^P_{\beta\alpha} - (P_\gamma E_\delta + M_\gamma H_\delta)C^{-1}_{\alpha\gamma}C^{-1}_{\beta\delta}]\} , \end{aligned}$$

from which we easily obtain the so called Gibbs relation

(2.33) $$d\eta = \frac{1}{\theta}\{dU - \frac{1}{2}[\frac{1}{\rho_o} T^P_{\beta\alpha} - (P_\gamma E_\delta + M_\gamma H_\delta)C^{-1}_{\alpha\gamma}C^{-1}_{\beta\delta}]dC_{\alpha\beta} + - E_\beta C^{-1}_{\alpha\beta}dP_\alpha - H_\beta C^{-1}_{\alpha\beta}dM_\alpha\} ,$$

where $T^P_{\alpha\beta}$, E_α and H_α are defined by (1.40) and (1.89), respectively. Mathematically the terms in curly brackets are called a Pfaffian form. Equation (2.33) delivers in a well-known manner integrability conditions for the coefficient functions. These will not be derived here, because picking a particular functional representation for the Helmholtz free energy $\hat{\Psi}(\cdot)$ and evaluating the fields η, $\mathcal{E}_i$, $\mathcal{H}_i$ and t_{ij} according to (2.31) guarantees the Gibbs relation to be satisfied identically. We do not say that these integrability conditions must not be known in general. On the contrary, for experimentalists these relations are easier to be determined than the free energies themselves, which may then be obtained by integration. If the electromagnetic fields vanish, the above Pfaffian form reduces to

(2.34) $$d\eta = \frac{1}{\theta}\{dU - \frac{1}{2\rho_o} T^P_{\alpha\beta}dC_{\alpha\beta}\} ,$$

the Gibbs relation as proved correct by Carathéodory [6] for thermostatic processes of thermoelastic materials. Irreversible thermodynamics postulates (2.34) to be the Gibbs relation for all those thermodynamic processes of thermoelastic materials which deviate only slightly from thermostatic equilibrium. Here (2.33) holds for all thermodynamic processes in the presence of the electromagnetic fields. Needless to say that, starting with a given electromechanical interaction model, it would probably be rather difficult for an irreversible thermodynamicist to guess a Gibbs relation of the form (2.33).

Case b): We now proceed with case b) which only differs from case a) in that the independent variable P_i/ρ is replaced by $\mathcal{E}_i$. In this case the constitutive equations can easily be obtained from the preceding ones by the Legendre transformation

(2.35) $$\tilde{\psi} = \psi - \frac{1}{\rho} \mathcal{E}_i P_i = U - \theta\eta - \frac{1}{\rho} \mathcal{E}_i P_i ,$$

where

(2.36) $\tilde{\psi} = \tilde{\psi}(F_{i\alpha}, E_i, \mu_o M_i/\rho, \theta) = \tilde{\Psi}(C_{\alpha\beta}, E_\alpha, M_\alpha, \theta)$.

With (2.35) and (2.36) the constitutive equations (2.31) must now be replaced by

$$\eta = -\frac{\partial\tilde{\Psi}}{\partial\theta} ,$$

(2.37) $$P_i = -\rho \frac{\partial\tilde{\Psi}}{\partial E_\alpha} F_{i\alpha} ,$$

$$H_i = \frac{\partial\tilde{\Psi}}{\partial M_\alpha} F_{i\alpha} \operatorname{sgn} J ,$$

and

$$t_{ij} = 2\rho \frac{\partial\tilde{\Psi}}{\partial C_{\alpha\beta}} F_{i\alpha}F_{j\beta} - E_i P_j + \mu_o M_i H_j .$$

Defining

(2.38) $$\tilde{\varepsilon} := U - \frac{1}{\rho} E_i P_i = U - E_\alpha P_\beta C^{-1}_{\alpha\beta} ,$$

the Gibbs relation becomes

(2.39) $$d\eta = \frac{1}{\theta}\{d\tilde{\varepsilon} - \frac{1}{2}[\frac{1}{\rho_o} T^P_{\alpha\beta} + (E_\gamma P_\delta - M_\gamma H_\delta)C^{-1}_{\alpha\gamma}C^{-1}_{\beta\delta}]dC_{\alpha\beta} +$$

$$+ P_\beta C^{-1}_{\alpha\beta} dE_\alpha - H_\beta C^{-1}_{\alpha\beta} dM_\alpha\} .$$

In a similar way we can derive constitutive equations for

Case c): Here we define

(2.40) $$\bar{\psi} := U - \theta\eta - \frac{\mu_o}{\rho} H_i M_i = \bar{\psi}(F_{i\alpha}, E_i, H_i, \theta) = \bar{\Psi}(C_{\alpha\beta}, E_\alpha, H_\alpha, \theta) .$$

The constitutive equations pertinent to this case then become

$$\eta = -\frac{\partial\bar{\Psi}}{\partial\theta} ,$$

$$P_i = -\rho \frac{\partial\bar{\Psi}}{\partial E_\alpha} F_{i\alpha} ,$$

(2.41) $$M_i = -\rho \frac{\partial\bar{\Psi}}{\partial H_\alpha} F_{i\alpha} \operatorname{sgn} J ,$$

$$t_{ij} = 2\rho \frac{\partial\bar{\Psi}}{\partial C_{\alpha\beta}} F_{i\alpha}F_{j\beta} - E_i P_j - \mu_o H_i M_j .$$

Furthermore, after the introduction of

$$(2.42)\qquad \bar{\varepsilon} := U - \frac{1}{\rho} E_i P_i - \frac{\mu_o}{\rho} H_i M_i = U - (E_\alpha P_\beta + H_\alpha M_\beta) C^{-1}_{\alpha\beta} ,$$

the Gibbs relation may be transformed into

$$(2.43)\qquad d\eta = \frac{1}{\theta}\{d\bar{\varepsilon} - \frac{1}{2}[\frac{1}{\rho_o} T^P_{\alpha\beta} + (E_\gamma P_\delta + H_\gamma M_\delta) C^{-1}_{\alpha\gamma} C^{-1}_{\beta\delta}] dC_{\alpha\beta} +$$

$$+ P_\beta C^{-1}_{\alpha\beta} dE_\alpha + M_\beta C^{-1}_{\alpha\beta} dH_\alpha\} .$$

Before we proceed to other electromechanical interaction models, we would like to mention that all the above constitutive models are equally possible ones. Case a) is particularly fashionable among applied physicists and has for the case that the material under consideration is polarizable-only or magnetizable-only been used by, amongst others, Toupin [7] and Brown [8], respectively. The formulation of case c) is used by Pao and Hutter [4]. There are reasons for preference of case c). Indeed, since any electromagnetic problem must be solved in the entire space, including the vacuum, any theory must ultimately be expressed in terms of E_i and H_i. It follows that in the cases a) and b) the constitutive equations for E_i and/or H_i must be invertible in the sense that

$$(2.44)\qquad \begin{aligned} \frac{P_i}{\rho} &= f_P(F_{i\alpha}, E_i, H_i, \theta) , \\ \frac{\mu_o M_i}{\rho} &= f_M(F_{i\alpha}, E_i, H_i, \theta) . \end{aligned}$$

For linear constitutive relations such inversions are trivial; they become more difficult when dealing with nonlinear theories.

2. The Two-Dipole Model with a Symmetric Stress Tensor (Model II)

The above two-dipole model is not the only one that has been proposed. There is another one, leading to a symmetric stress tensor. This model has been investigated by Hutter [9] only recently, but it is already suggested by Fano, Chu and Adler [1]. It is the two-dipole counterpart of Model V, treated below.

The Maxwell equations are the same as stated for model I. Electromagnetic body force, couple and energy supply, however, are taken as follows

$$\rho F_i^e = \mathcal{Q}^e E_i^C + e_{ijk} \mathcal{J}_j^e \mu_o H_k^C + \mathcal{Q}^m H_i^C - e_{ijk} \mathcal{J}_j^m \varepsilon_o E_k^C \ ,$$

(2.45) $$\rho L_{ij}^e = 0 \ ,$$

$$\rho r^e = \mathcal{J}_i^e \mathcal{E}_i + \mathcal{J}_i^m \mathcal{H}_i \ .$$

Here,

(2.46) $$\mathcal{J}_i^e = J_i^e - \mathcal{Q}^e \dot{x}_i, \qquad \text{and} \qquad \mathcal{J}_i^m = J_i^m - \mathcal{Q}^m \dot{x}_i \ ,$$

and the superscripts e and m indicate that the corresponding current or charge densities are those due to electric and magnetic charges, respectively. The quantities can easily be read off from the Maxwell equations (2.3):

$$\mathcal{Q}^e = \mathcal{Q} - P_{i,i}^C, \qquad J_i^e = J_i + \rho \frac{d}{dt}\left(\frac{P_i^C}{\rho}\right) - (P_j^C \dot{x}_i)_{,j} \ ,$$

(2.47) $$\mathcal{Q}^m = -\mu_o M_{i,i}^C, \qquad J_i^m = \rho \frac{d}{dt}\left(\frac{\mu_o M_i^C}{\rho}\right) - (\mu_o M_j^C \dot{x}_i)_{,j} \ ,$$

and

$$\mathcal{J}_i^e = \mathcal{J}_i + \overset{*}{P}{}_i^C, \qquad \mathcal{J}_i^m = \mu_o \overset{*}{M}{}_i^C \ .$$

Substitution of (2.47) into (2.45) yields

$$\rho F_i^e = \mathcal{Q} E_i^C + \mu_o e_{ijk} J_j H_k^C + P_j^C E_{i,j}^C + \mu_o e_{ijk} \dot{x}_j H_{k,\ell}^C P_\ell^C +$$

$$+ \rho \mu_o e_{ijk} \frac{d}{dt}\left(\frac{P_j^C}{\rho}\right) H_k^C + \mu_o M_j^C H_{i,j}^C - \varepsilon_o e_{ijk} \dot{x}_j E_{k,\ell}^C \mu_o M_\ell^C +$$

(2.48) $$- \rho \varepsilon_o e_{ijk} \frac{d}{dt}\left(\frac{\mu_o M_j^C}{\rho}\right) E_k^C - (P_j \mathcal{E}_i + \mu_o \mathcal{M}_j \mathcal{H}_i)_{,j} \ ,$$

$$\rho L_{ij}^e = 0 \ ,$$

and

$$\rho r^e = \mathcal{J}_i \mathcal{E}_i + \overset{*}{P}_i \mathcal{E}_i + \mu_o \overset{*}{M}_i \mathcal{H}_i \ .$$

Before we proceed, let us look at the invariance properties of (2.48). To this end, recall that $\mathcal{E}_i$, $\mathcal{H}_i$, $P_i = P_i^C$, $\mu_o \mathcal{M}_i = \mu_o M_i^C$, $\mathcal{Q}$ and $\mathcal{J}_i$ are objective quantities with respect to the Euclidean transformation group. Hence, the electromagnetic energy supply is an objective scalar under this group. Incidentally, it should be noted that ρr^e is also an objective scalar under the extended Lorentz group, provided we restrict ourselves to the semi-relativistic approximation.

Regarding the electromagnetic body force (2.48)[1] we see that the above expression differs from (2.7)[1] only in a term which is objective under Euclidean and under Lorentz transformations in the semi-relativistic sense. Hence, all conclusions drawn in Model I do apply also here, and we refer the reader to the pertinent discussion in Model I. In conclusion we state the governing equations of this theory consistently to within the order of exactness of the non-relativistic approximation. Hence, all terms preceded by a c^{-2} factor must be dropped; as a result we arrive at the following balance laws for mass, linear and angular momentum and energy:

$$\dot{\rho} + \rho \dot{x}_{i,i} = 0 \; ,$$

$$\rho \ddot{x}_i = t_{ij,j} + \rho F_i^{ext} + \mathcal{Q} E_i + \mu_o e_{ijk} J_j H_k - P_{j,j} E_i + \\ + \mu_o e_{ijk} \overset{\star}{P}_j H_k - \mu_o M_{j,j} H_i \; ,$$

(2.49) $$t_{[ij]} = 0 \; ,$$

$$\rho \dot{U} = t_{ij} \dot{x}_{i,j} - q_{i,i} + J_i E_i + \overset{\star}{P}_i E_i + \mu_o \overset{\star}{M}_i H_i + \rho r^{ext} \; .$$

Here, we wish to point out that the fact that ρr^e is by itself an objective scalar is a corroboration of the assumption of symmetry of the stress tensor (compare the relevant discussion in Model I).

Furthermore, a long but straightforward calculation shows that, within the non-relativistic approximation (compare these expressions with model I)

$${}^{II}t^M_{ij} = {}^{II}t^M_{ji} = \varepsilon_o E_i E_j + \mu_o H_i H_j - \frac{1}{2} \delta_{ij} (\varepsilon_o E_k E_k + \mu_o H_k H_k) \; ,$$

$${}^{II}g_i = 0 \; ,$$

(2.50) $${}^{II}\pi_i = -e_{ijk} E^C_j H^C_k = -e_{ijk} E_j H_k - (\varepsilon_o E_k E_k + \mu_o H_k H_k) \dot{x}_i + (\varepsilon_o E_i E_j + \mu_o H_i H_j) \dot{x}_j$$

$${}^{II}\omega = -\frac{1}{2} (\varepsilon_o E_k E_k + \mu_o H_k H_k) \; .$$

Hence, the jump conditions for momentum and energy become

$$[\![t_{ij} + \varepsilon_o E_i E_j + \mu_o H_i H_j - \frac{1}{2} \delta_{ij} (\varepsilon_o E_k E_k + \mu_o H_k H_k)]\!] n_j + $$

(2.51) $$- [\![\rho \dot{x}_i (\dot{x}_j n_j - w_n)]\!] = 0 \; ,$$

and

(over)

$$[\![\dot{x}_i t_{ij} - q_j - e_{jk\ell}E_k H_\ell + (\varepsilon_o E_j E_k + \mu_o H_j H_k)\dot{x}_k - \tfrac{1}{2}(\varepsilon_o E_k E_k + \mu_o H_k H_k)\dot{x}_j]\!] n_j +$$

$$- [\![\{\tfrac{1}{2}\rho\dot{x}_i\dot{x}_i + \rho U + \tfrac{1}{2}(\varepsilon_o E_i E_i + \mu_o H_i H_i)\}(\dot{x}_j n_j - w_n)]\!] = 0 .$$

The jump conditions of electromagnetic fields and mass are the same as those for Model I (see (2.11)) and will not be repeated here.

We now turn to the constitutive theory. We shall not present all the details here, but list the results only. We start by eliminating ρr^{ext} from the entropy inequality (1.53) and the energy balance (2.49)[4]. The result reads

$$(2.52) \qquad -\rho\dot{U} + \rho\theta\dot{\eta} + \rho E_i \frac{d}{dt}\left(\frac{P_i}{\rho}\right) + \rho H_i \frac{d}{dt}\left(\frac{\mu_o M_i}{\rho}\right) +$$

$$+ (t_{ij} - E_i P_j - \mu_o H_i M_j)\dot{x}_{i,j} + J_i E_i - \frac{\theta_{,i} q_i}{\theta} \geq 0 ,$$

where we have also set

$$(2.53) \qquad \phi_i = \frac{q_i}{\theta} .$$

In a recent article by Hutter [9], dealing with a theory of this complexity, the form of the entropy flux vector introduced here as a postulate has been proved correct.

As before, the constitutive theory depends on the selection of the independent variables. We may again choose constitutive relations of the form (1.86), but we shall give them for case a) only. The equations for other cases can be derived in the same way as was done for Model I.

When the constitutive relations are of the form

$$(2.54) \qquad C = \hat{C}\left(F_{i\alpha}, \frac{P_i}{\rho}, \frac{\mu_o M_i}{\rho}, \theta, \theta_{,i}, \mathcal{Q}\right) ,$$

the Helmholtz free energy

$$(2.55) \qquad \psi := U - \eta\theta = \hat{\Psi}(C_{\alpha\beta}, P_\alpha, M_\alpha, \theta) ,$$

does not depend on $\mathcal{Q}$ and $\theta_{,i}$ and serves as potential for η, E_i, H_i and t_{ij}, viz.:

$$\eta = -\frac{\partial\hat{\Psi}}{\partial\theta} ,$$

$$E_i = \frac{\partial\hat{\Psi}}{\partial P_\alpha} F_{i\alpha} ,$$

(2.56)

$$H_i = \frac{\partial\hat{\Psi}}{\partial M_\alpha} F_{i\alpha} \operatorname{sgn} J ,$$

$$t_{ij} = 2\{\rho \frac{\partial\hat{\Psi}}{\partial C_{\alpha\beta}} F_{i\alpha}F_{j\beta} + E_{(i}P_{j)} + \mu_o H_{(i}M_{j)}\} .$$

There remains

(2.57) $$J_i E_i - \frac{\theta_{,i} q_i}{\theta} \geq 0 ,$$

as the reduced entropy inequality.
Finally, the Gibbs relation takes the form

(2.58) $$d\eta = \frac{1}{\theta}\{dU - [\frac{1}{2\rho_o} T^P_{\alpha\beta} - C^{-1}_{\alpha\gamma}C^{-1}_{\beta\delta}(E_\gamma P_\delta + E_\delta P_\gamma + H_\gamma M_\delta + H_\delta M_\gamma)]dC_{\alpha\beta} + - C^{-1}_{\alpha\beta}E_\beta dP_\alpha - C^{-1}_{\alpha\beta}H_\beta dM_\alpha\} .$$

For the reader's sake of curiosity, we mention that this theory will be shown to be fully equivalent to the theory of model I.
This completes the description of the two-dipole models. Both descriptions outlined above use a dipole model not only for polarization but also for magnetization. There are no others as far as we know and all the remaining models bear to a lesser or larger extent the notion of dipole structure for polarization and the structure of electric circuits for magnetization.
In conclusion we wish to mention a paper by Alblas, [10], who also used a Chu-formulation and who derived a theory which completely agrees with our model I. The derivation is on the basis of a global energy balance law (see also Sections 2.3 and 2.8) and for a more general problem, including spin- and dissipation effects.
In what follows, we shall discuss theories presented by van de Ven, [11], De Groot & Suttorp, [12], Müller, [13] and Hutter, [14].

2.3 THE MAXWELL-MINKOWSKI FORMULATION (Model III)

The foregoing two interaction models were constructed by postulating body force, body couple and energy supply terms due to the electromagnetic fields. The approach in this Section is different. Basically, the balance laws of mass and momenta are derived from a global energy balance law by postulating certain invariance properties. In [11] van de Ven follows this method, first formulated by Green and Rivlin [15], and developed for a ferromagnetic solid by Alblas [16]. By postulating that the energy balance is invariant under rigid-body motions and by making some à priori assumptions concerning the invariance properties of the quantities involved, the equations of mass and momenta are derived. In this derivation all terms preceded by a factor c^{-2} are neglected.

Clearly, the outcome of this approach essentially depends on what à priori assumptions regarding the invariance properties of the various quantities are made. Moreover, special care must be observed when neglecting terms which contain a c^{-2}-factor. Indeed, there is an essential difference between the approximations performed in the above mentioned references and the non-relativistic approximation in Section 1.6. While we apply Giorgi-units (MKSA), Gaussian units are used in [11], and it is a well-known fact that a c^{-2}-term in one system of units is not necessarily a c^{-2}-term in the other system of units as well. For instance

$$\frac{1}{c}\,\underline{E} \times \underline{H}$$

in Gaussian units becomes

$$\frac{1}{c^2}\,\underline{E} \times \underline{H}$$

in Giorgi units. As a consequence of both, the à priori assumptions on invariance that are not in conformity with the results of Section 1.6 and the discrepancy in the neglect of terms like the one demonstrated above, the equations derived in [11] are not Euclidean invariant in the sense as defined in Section 1.6. However, it is possible to derive from an analogous energy balance as was postulated in [11] local balance equations of mass, linear and angular momentum and energy that are invariant under Euclidean transformations by simply applying in a consistent way non-relativistic invariance requirements. This then fully justifies the à priori assumptions we will impose.

Before proceeding the following remark seems to be in order: It is conceivable that one might object to associate the names of Maxwell and Minkowski with the formulation presented below, and indeed neither Maxwell nor Minkowski ever presented a formulation of this kind. Yet, for the special case of a magnetostatic problem, the magnetoelastic stresses that will be obtained here, are identical with those used by W.F. Brown Jr. in his monograph, [8]; and Brown stated that these stresses could be derived by taking over from Maxwell a formula for the magnetic force. This is justification for us to associate this (dynamic) theory with the name of Maxwell. Moreover, since the Maxwell equations will be used in the Minkowski formulation, we shall refer to the formulation presented in this Section as the Maxwell-Minkowski formulation. Clearly, the association of names with certain theories or formulations of theories bears its well-known disadvantages and a reader not willing to accept our proposal may reject it and invent his own name for it. As an example the reader may recall that there is no unique version of the Maxwell-Minkowski stress tensor either. We shall come back to this subject at the end of this Section.

As said above, the Maxwell equations will be expressed in terms of the Minkowski field variables E_i^M, D_i^M, H_i^M, and B_i^M, which are related to the field variables introduced in (1.21) and (1.27) by

$$\text{(2.59)} \qquad \begin{aligned} \mathcal{E}_i &= E_i^M + e_{ijk}\dot{x}_j B_k^M, & D_i &= D_i^M , \\ \mathcal{H}_i &= H_i^M - e_{ijk}\dot{x}_j D_k^M, & B_i &= B_i^M . \end{aligned}$$

Accordingly, the following set of Maxwell equations is obtained:

$$\text{(2.60)} \qquad \begin{aligned} B_{i,i} &= 0, & e_{ijk}E^M_{k,j} + \frac{\partial B_i}{\partial t} &= 0 , \\ D_{i,i} &= \mathcal{Q}, & e_{ijk}H^M_{k,j} - \frac{\partial D_i}{\partial t} &= J_i . \end{aligned}$$

Here, and in the following, since no confusion is possible, the superscript M for D_i and B_i is omitted. In a purely formal way polarization per unit volume and magnetization per unit volume can be introduced:

$$\text{(2.61)} \qquad P_i^M := D_i - \varepsilon_o E_i^M, \qquad \mu_o M_i^M := B_i - \mu_o H_i^M ,$$

and it is convenient to introduce also the variable

$$\text{(2.62)} \qquad \mathcal{M}_i := M_i^M + e_{ijk}\dot{x}_j P_k^M .$$

It is now a routine matter (see for instance Appendix A) to prove that under the Euclidean transformation group

$$
(2.63)\qquad \begin{aligned}
&\mathcal{E}_i,\ D_i \text{ and } P_i^M \text{ transform as objective vectors, and}\\
&\mathcal{H}_i,\ B_i \text{ and } \mathcal{M}_i \text{ transform as objective axial vectors,}
\end{aligned}
$$

respectively.

In particular, and as an easy calculation also shows directly, under the rigid-body translation

$$
(2.64)\qquad x_i' = x_i - b_i(t), \qquad t' = t
$$

the following transformation laws must hold

$$
(2.65)\qquad \begin{aligned}
D_i' &= D_i, & E_i^{M'} &= E_i^M + e_{ijk}\dot{b}_j B_k ,\\
B_i' &= B_i, & H_i^{M'} &= H_i^M - e_{ijk}\dot{b}_j D_k ,\\
\mathcal{Q}' &= \mathcal{Q}, & J_i' &= J_i - \mathcal{Q}\dot{b}_i ,\\
P_i^{M'} &= P_i^M, & M_i^{M'} &= M_i^M + e_{ijk}\dot{b}_j P_k^M .
\end{aligned}
$$

On the other hand, the Minkowski fields can also be subjected to the special Lorentz transformation (1.69), which when written in the semi-relativistic approximation, becomes

$$
(2.66)\qquad x_i^* = x_i - V_i t, \qquad t^* = t - \frac{1}{c^2} x_i V_i .
$$

Then, the transformation rules are as follows

$$
(2.67)\qquad \begin{aligned}
D_i^* &= D_i + \frac{1}{c^2} e_{ijk} V_j H_k^M, & E_i^{M^*} &= E_i^M + e_{ijk} V_j B_k ,\\
B_i^* &= B_i - \frac{1}{c^2} e_{ijk} V_j E_k^M, & H_i^{M^*} &= H_i^M - e_{ijk} V_j D_k ,\\
\mathcal{Q}^* &= \mathcal{Q} - \frac{1}{c^2} J_i V_i, & J_i^* &= J_i - \mathcal{Q} V_i ,\\
P_i^{M^*} &= P_i^M - \frac{1}{c^2} e_{ijk} V_j M_k^M, & M_i^{M^*} &= M_i^M + e_{ijk} V_j P_k^M ,
\end{aligned}
$$

A comparison of (2.65) and (2.67), which constitute the transformation rules for special Euclidean and Lorentz transformations, shows that the primed and stared quantities equal only in the non-relativistic approximation. This conclusion is also correct if the Minkowski-fields are subject to general Euclidean or Lorentz transformations. In view of the results of Section 1.6 this must be so expected.

A nice application of equations (2.65) and (2.67) is obtained if the invariance properties of the Maxwell equations are investigated. To this end they are best expressed in terms of the objective fields $\mathcal{E}_i$, D_i, $\mathcal{H}_i$, B_i, $\mathcal{J}_i$ and $\mathcal{Q}$ and read then (see (1.23) and (1.28))

$$(2.68)\qquad \begin{aligned} B_{i,i} &= 0, & e_{ijk}\mathcal{E}_{k,j} + \overset{*}{B}_i &= 0\,, \\ D_{i,i} &= \mathcal{Q}, & e_{ijk}\mathcal{H}_{k,j} - \overset{*}{D}_i &= \mathcal{J}_i\,. \end{aligned}$$

If all the fields in these equations were independent it would follow at once from the above listed transformation properties, that the equations (2.68) would be invariant under Euclidean transformations. The relations (2.61), which replace the Maxwell-Lorentz aether relations in the previous discussions, are not invariant under Euclidean transformations, however, for according to (2.65) and (2.61) we have

$$(2.69)\qquad \begin{aligned} P_i^{M'} &= P_i^M = D_i - \varepsilon_o E_i^M = D_i' - \varepsilon_o E_i^{M'} + \varepsilon_o e_{ijk}\dot{b}_j B_k = \\ &= D_i' - \varepsilon_o E_i^{M'} + \frac{1}{c^2}\, e_{ijk}\dot{b}_j (H_k + M_k) \\ M_i^{M'} &= \mu_o M_i^M + \mu_o e_{ijk}\dot{b}_j P_k^M = B_i - \mu_o H_i^M + \mu_o e_{ijk}\dot{b}_j P_k^M = \\ &= B_i' - \mu_o H_i^{M'} - \mu_o e_{ijk}\dot{b}_j (D_k - P_k) \\ &= B_i' - \mu_o H_i^{M'} - \frac{1}{c^2}\, e_{ijk}\dot{b}_j E_k\,. \end{aligned}$$

Because equation (2.64) is a special Euclidean transformation, we conclude that the relations (2.61) are not invariant under Euclidean transformations in general except in the non-relativistic sensè.

A similar calculation can also be performed with (2.67). The result is that (2.61) is a Lorentz invariant equation. We leave this proof to the reader. One further particular point in (2.69) must be mentioned. In reaching the non-relativistic invariance property the relation

$$(2.70)\qquad B_i = \mu_o (H_i^M + M_i^M)$$

must be used and only afterwards terms containing a c^{-2}-factor can be dropped. This procedure is tantamount to assuming B_i to be proportional to μ_o. Apparently this should be done consistently and hence we make the following statements that will be observed as basic rules henceforth:

i) B_i must be considered to be proportional to μ_o.
ii) Under this provision terms containing a c^{-2}-factor should be dropped.

When these rules are observed the relations (2.61) and (2.62) may be replaced by the very suggestive formulas

(2.71) $$D_i = \varepsilon_o E_i + P_i, \quad \text{and} \quad B_i = \mu_o H_i + \mu_o M_i$$

in which we have set

(2.72) $$P_i = P_i^M .$$

These preliminary remarks may suffice, and so we proceed to derive the balance laws of mass, linear and angular momentum and energy starting with a global balance law of energy that is subjected to superimposed rigid-body motions (Euclidean transformations). Although our approach is different in detail we shall in the following derivation essentially follow van de Ven [11].

Starting equation is a global energy balance law of the following form

$$\frac{d}{dt} \int_V \{\rho U + \frac{1}{2}(\varepsilon_o E_i E_i + \mu_o H_i H_i) + \frac{1}{2} \rho \dot{x}_i \dot{x}_i + \rho T\} dv =$$

(2.73) $$= \int_V \{\rho r^{ext} + \rho F_i^{ext} \dot{x}_i\} dv + \int_{\partial V} \{t_{ij} \dot{x}_i - q_j - e_{jk\ell} E_k^M H_\ell^M +$$

$$+ \frac{1}{2}(\varepsilon_o E_k E_k + \mu_o H_k H_k) \dot{x}_j + R_j\} da_j .$$

This equation, in which ρT and R_i are still to be determined in terms of the electromagnetic field quantities and the motion, is a generalization of the purely mechanical energy balance law. Indeed, when all electromagnetic quantities are set to zero, what results is the classical non-relativistic energy balance law. It is also easy to interpret in (2.73) the terms of electromagnetic origin. Firstly,

$$\frac{1}{2}(\varepsilon_o E_i E_i + \mu_o H_i H_i)$$

is an electromagnetic energy density, and, secondly

$$\underline{E}^M \times \underline{H}^M$$

is the Poynting vector. As is well-known, both are not unique and we shall see that other electromagnetic models correspond to different postulates of the electromagnetic energy density and the Poynting vector. This does not mean, however, that these models will also yield different results for physically measurable quantities.

A clue as to what should be chosen for the as yet undetermined quantities ρT and R_i is obtained, if (2.73) is written for a body whose mass density is vanishingly small. In this case we expect (2.73) to become the energy equations for the electromagnetic fields in vacuo. It is well-known that this energy equation is a consequence of the Maxwell equations (2.68). This energy equation can be brought into the form (2.73) (with vanishing t_{ij}, q_i, $\dot{x}_i$, U, r^{ext}, F_i^{ext}), and it is not hard to show that in this case and to within terms of order c^{-2}

$$(2.74) \qquad \rho T = 0 \qquad \text{and} \qquad R_i = 0 \ .$$

In other words, in a vacuum ρT and $\underline{R}$ must vanish.

Next we determine ρT and $\underline{R}$ in a ponderable body, and for that purpose (2.73) must be written in local form. Assuming sufficient smoothness of the fields involved this yields

$$(2.75) \qquad (\dot{\rho} + \rho\dot{x}_{i,i})(U + \tfrac{1}{2}\dot{x}_i\dot{x}_i + T) + \rho\dot{T} - R_{i,i} + (e_{ijk}E_j^M H_k^M)_{,i} +$$
$$+ \{\rho\ddot{x}_i - \rho F_i^{ext} - t_{ij,j} - \tfrac{1}{2}(\varepsilon_o E_k E_k + \mu_o H_k H_k)_{,i}\}\dot{x}_i +$$
$$+ \rho\dot{U} - t_{ij}\dot{x}_{i,j} + q_{i,i} - \rho r^{ext} + \tfrac{1}{2}\tfrac{d}{dt}(\varepsilon_o E_k E_k + \mu_o H_k H_k) = 0 \ .$$

This equation is now subjected to Euclidean transformations and it is postulated that it is invariant under these transformations. The details of these calculations are somewhat tedious and are presented in full detail in Appendix B. They deliver expressions for ρT and $\underline{R}$ as well as local balance laws of mass, linear and angular momentum and energy. The results for ρT and $\underline{R}$ in Appendix B are

$$(2.76) \qquad \rho T = 0, \qquad \text{and} \qquad R_i = (P_j E_j + \mu_o M_j H_j)\dot{x}_i + e_{jk\ell}P_k B_\ell \dot{x}_j \dot{x}_i \ .$$

We note that only the second term in the expression for R_i follows from invariance requirements whereas the first term is chosen arbitrarily to simplify certain formulas. This seems to be a disadvantage of this method of derivation, however a change in R only results in a different stress tensor t_{ij} (and, eventually, a different energy flux q_i) and it can be shown that this has no effect in the ultimate form of the balance laws.

If (2.76) is substituted into (2.73) we obtain as global energy balance

$$\frac{d}{dt}\int_V \{\rho U + \tfrac{1}{2}(\varepsilon_o E_i E_i + \mu_o H_i H_i) + \tfrac{1}{2}\rho \dot{x}_i \dot{x}_i\}dv =$$

$$(2.77)\qquad = \int_V \{\rho r^{ext} + \rho F_i^{ext}\dot{x}_i\}dv + \int_{\partial V} \{t_{ij}\dot{x}_i - q_j - e_{jk\ell}E_k H_\ell +$$

$$+ (D_j E_k + B_j H_k)\dot{x}_k - \tfrac{1}{2}(\varepsilon_o E_k E_k + \mu_o H_k H_k)\dot{x}_j\}da_j ,$$

from which a local balance law can be derived, which with the use of (2.68) and (2.71) may be written in the form

$$(\dot{\rho} + \rho \dot{x}_{i,i})(U + \tfrac{1}{2}\dot{x}_i\dot{x}_i) + \rho\dot{U} - J_i E_i - E_i \overset{*}{P}_i - \mu_o H_i \overset{*}{M}_i +$$

$$(2.78)\qquad + q_{i,i} - \rho r^{ext} - [t_{ij} + E_i P_j + \mu_o H_i M_j]\dot{x}_{i,j} +$$

$$+ [\rho\ddot{x}_i - t_{ij,j} - \rho F_i^{ext} - \mathcal{Q}E_i - e_{ijk}J_j B_k - P_j E_{j,i} +$$

$$- \mu_o M_j H_{j,i} - e_{ijk}(D_j \overset{*}{B}_k + \overset{*}{D}_j B_k)]\dot{x}_i = 0 .$$

The invariance requirements under which

$$\rho,\ U,\ t_{ij},\ (\ddot{x}_i - F_i^{ext}),\ q_i,\ \rho r^{ext},\ \mathcal{Q},\ J_i,\ E_i,\ P_i,\ H_i,\ M_i,\ D_i,\ B_i$$

are assumed to transform as objective quantities then yield the local balance laws of

mass

$$(2.79)\qquad \dot{\rho} + \rho\dot{x}_{i,i} = 0$$

momentum

$$(2.80)\qquad \rho\ddot{x}_i - t_{ij,j} - \rho F_i^{ext} = \rho F_i^{e} = \mathcal{Q}E_i + e_{ijk}J_j B_k + $$

$$+ P_j E_{j,i} + \mu_o M_j H_{j,i} + e_{ijk}(D_j \overset{*}{B}_k + \overset{*}{D}_j B_k) ,$$

and angular momentum

$$(2.81)\qquad t_{[ij]} = \rho L_{ij}^{e} = P_{[i}E_{j]} + \mu_o M_{[i}H_{j]} ,$$

while the local energy balance reduces to

$$\rho\dot{U} - t_{ij}\dot{x}_{i,j} + q_{i,i} - \rho r^{ext} = \rho r^{e} = J_i E_i + \tag{2.82}$$
$$+ \rho E_i \frac{d}{dt}\left(\frac{P_i}{\rho}\right) + \rho\mu_o H_i \frac{d}{dt}\left(\frac{M_i}{\rho}\right) .$$

The source terms in (2.80) and (2.82) can be transformed into the form (1.62) whereby to within terms of order c^{-2}

$$\begin{aligned} {}^{III}t^{M}_{ij} &= E_i D_j + H_i B_j - \frac{1}{2}\delta_{ij}(\varepsilon_o E_k E_k + \mu_o H_k H_k) , \\ {}^{III}g_i &= 0 , \\ {}^{III}\pi_i &= -e_{ijk}E_j H_k + (D_i E_j + B_i H_j)\dot{x}_j - (\varepsilon_o E_j E_j + \mu_o H_j H_j)\dot{x}_i , \\ {}^{III}\omega &= -\frac{1}{2}(\varepsilon_o E_i E_i + \mu_o H_i H_i) . \end{aligned} \tag{2.83}$$

is obtained.

Substitution of (2.59) into (1.58) and of (2.83) into (1.66) yields the following set of jump conditions

$$[\![e_{ijk}\overset{M}{E}_j n_k + B_i w_n]\!] = 0, \qquad [\![D_i]\!]n_i = 0 ,$$

$$[\![e_{ijk}\overset{M}{H}_j n_k - D_i w_n]\!] = 0, \qquad [\![B_i]\!]n_i = 0 ,$$

$$[\![\rho(\dot{x}_i n_i - w_n)]\!] = 0 ,$$

$$\begin{aligned} &[\![t_{ij} + E_i D_j + H_i B_j - \frac{1}{2}\delta_{ij}(\varepsilon_o E_k E_k + \mu_o H_k H_k)]\!]n_j + \\ &- [\![\rho\dot{x}_i(\dot{x}_j n_j - w_n)]\!] = 0 , \end{aligned} \tag{2.84}$$

$$\begin{aligned} &[\![\dot{x}_i t_{ij} - q_j - e_{jk\ell}E_k H_\ell + (D_j E_k + B_j H_k)\dot{x}_k + \\ &- \frac{1}{2}(\varepsilon_o E_k E_k + \mu_o H_k H_k)\dot{x}_j]\!]n_j - [\![\{\frac{1}{2}\rho\dot{x}_i\dot{x}_i + \rho U + \\ &+ \frac{1}{2}(\varepsilon_o E_i E_i + \mu_o H_i H_i)\}(\dot{x}_j n_j - w_n)]\!] = 0 . \end{aligned}$$

We note that $(2.84)^{6,7}$ can still be written in a somewhat different from. As described in [11] (see pp. 65 and 66) it can be shown that for a material <u>surface</u> the relations $(2.84)^{6,7}$ are equivalent to

$$(2.85)\quad \begin{aligned} [\![t_{ij}]\!]n_j &= \tfrac{1}{2}[\![(\mu_o \mathcal{M}_j n_j)^2 + (P_j n_j)^2]\!]n_i\ , \\ [\![q_j]\!]n_j &= [\![\dot{x}_i t_{ij} - \tfrac{1}{2}\{(\mu_o \mathcal{M}_k n_k)^2 + (P_k n_k)^2\}\dot{x}_j]\!]n_j\ . \end{aligned}$$

If the singular surface is a boundary separating or ponderable body from a vacuum, it is easily shown that equation $(2.85)^2$ reduces to $[\![q_j]\!]n_j = 0$. In this special case, therefore, the normal component of the energy flux vector must vanish.

In order to make the theory complete, the balance equations must be supplemented by constitutive equations. Our presentation will be brief as it follows exactly the approach in the preceding Sections. As independent variables we choose

$$(2.86)\qquad C_{\alpha\beta}, E_\alpha, M_\alpha, \theta, \theta_{,\alpha} \text{ and } \mathcal{Q}\ ,$$

where $C_{\alpha\beta}$, E_α and M_α have already been introduced in (1.88) and (1.89).

Defining the energy functional $\tilde{\psi}$ by (see (2.35))

$$(2.87)\qquad \tilde{\psi} := U - \eta\theta - \frac{P_i}{\rho} \mathcal{E}_i\ ,$$

and using, as before, for the entropy flux the classical relation

$$\phi_i = \frac{q_i}{\theta}\ ,$$

we deduce from the reduced entropy inequality that $\tilde{\psi}$ can neither depend on $\theta_{,\alpha}$ nor $\mathcal{Q}$; hence

$$(2.88)\qquad \tilde{\psi} = \tilde{\Psi}(C_{\alpha\beta}, E_\alpha, M_\alpha, \theta)\ .$$

Moreover,

$$(2.89)\quad \begin{aligned} \eta &= -\frac{\partial\tilde{\Psi}}{\partial\theta}\ , & \mathcal{H}_i &= \frac{\partial\tilde{\Psi}}{\partial M_\alpha} F_{i\alpha} \operatorname{sgn} J\ , \\ P_i &= -\rho \frac{\partial\tilde{\Psi}}{\partial E_\alpha} F_{i\alpha}\ , & t_{ij} &= 2\rho \frac{\partial\tilde{\Psi}}{\partial C_{\alpha\beta}} F_{i\alpha}F_{j\beta} - \mathcal{E}_i P_j + \mu_o \mathcal{M}_i \mathcal{H}_j\ . \end{aligned}$$

Of the reduced entropy inequality there remains the residual inequality

$$(2.90)\qquad \mathcal{J}_i \mathcal{E}_i - \frac{\theta_{,i} q_i}{\theta} \geq 0\ ,$$

and the Gibbs relation becomes

$$d\eta = \frac{1}{\theta}\{d[U - P_\alpha E_\beta C^{-1}_{\alpha\beta}] - \frac{1}{2}[\frac{1}{\rho_o} T^P_{\alpha\beta} - (P_\gamma E_\delta + M_\gamma H_\delta) C^{-1}_{\alpha\gamma} C^{-1}_{\beta\delta}] dC_{\alpha\beta} + \quad (2.91)$$
$$+ P_\alpha dE_\alpha - H_\alpha dM_\alpha\} ,$$

with definitions of E_α, H_α, P_α and M_α as given in (1.89).

We still wish to point out that with the constitutive equation (2.89)[4] the angular momentum equation (2.81) is satisfied identically.

At this point we have set up a complete theory, consisting of balance laws, constitutive equations and jump conditions, for the interactions of electromagnetic and thermoelastic fields in solids based on a Minkowskian formulation of electrodynamics. Yet, there is a certain arbitrariness in the basic postulate of the theory, as manifested by the energy balance (2.77). As a result, the so called Maxwell stress tensor (here denoted by t^M_{ij}) does not appear in a unique form. Therefore, we shall give here some other formulations for this global energy balance which all lead to seemingly different theories, which, however, at the end all turn out to be equivalent.
First, we wish to consider a Minkowskian formulation as given by Penfield and Haus ([2], Section 7.3).

Taking as energy balance the expression

$$\frac{d}{dt}\int_V \{\rho U + \frac{1}{2}(\varepsilon_o \mathcal{E}_i \mathcal{E}_i + \mu_o \mathcal{H}_i \mathcal{H}_i) + \frac{1}{2}\rho \dot{x}_i \dot{x}_i\} dv =$$
$$(2.92) \quad = \int_V \{\rho r^{ext} + \rho F^{ext}_i \dot{x}_i\} dv + \int_{\partial V} \{t_{ij}\dot{x}_i - q_j - e_{jk\ell}\mathcal{E}_k \mathcal{H}_\ell +$$
$$+ (D_j \mathcal{E}_k + B_j \mathcal{H}_k)\dot{x}_k - (\mathcal{E}_k D_k + \mathcal{H}_k B_k)\dot{x}_j\} da_j ,$$

which differs from (2.73) or (2.77) only in the choice of R_i, the following local balance equations can be derived:

$$\rho\ddot{x}_i - t_{ij,j} - \rho F^{ext}_i = \rho F^e_i = \mathcal{Q}\mathcal{E}_i + e_{ijk}\mathcal{J}_j B_k - B_{j,i}\mathcal{H}_j - D_{j,i}\mathcal{E}_j + \quad (2.93)$$
$$+ e_{ijk}(D_j \overset{\star}{B}_k + \overset{\star}{D}_j B_k) = [\mathcal{E}_i D_j + \mathcal{H}_i B_j - \delta_{ij}(\mathcal{E}_k D_k + \mathcal{H}_k B_k)]_{,j} = t^M_{ij,j} ,$$

(2.94) $$t_{[ij]} = \rho L^{e}_{ij} = P_{[i}E_{j]} + \mu_o M_{[i}H_{j]} ,$$

(2.95) $$\frac{\partial W_M}{\partial t} + (W_M\dot{x}_i)_{,i} - t_{ij}\dot{x}_{i,j} + q_{i,i} - \rho r^{ext} = \rho r^{e} = J_iE_i + E_i\dot{D}_i + H_i\dot{B}_i ,$$

where

(2.96) $$W_M := \rho U + \frac{1}{2}(\varepsilon_o E_kE_k + \mu_o H_kH_k) .$$

We note that these balance laws equal those of Penfield and Haus, if the latter are taken in the non-relativistic approximation (cf. eqs. (7.39)-(7.43) of [2]). Furthermore, one can easily show that the only difference with the previous formulation lies in the stress tensor. Indeed, when we require the momentum equations in the two formulations to be the same, what results is as follows:

(2.97) $$t_{ij}(2.93) - t_{ij}(2.80) = \delta_{ij}[D_kE_k + B_kH_k - \frac{1}{2}(\varepsilon_o E_kE_k + \mu_o H_kH_k)] .$$

Once this difference is taken into account the two systems based on (2.77) and (2.92), respectively, are completely equivalent.

Still another possible form of the energy balance is

(2.98) $$\begin{aligned}\frac{d}{dt}\int_V \{\rho U + \frac{1}{2}(E_iD_i + H_iB_i) + \frac{1}{2}\rho\dot{x}_i\dot{x}_i\}dv &= \int_V \{\rho r^{ext} + \rho F^{ext}_i\dot{x}_i\}dv + \\ &+ \int_{\partial V}\{t_{ij}\dot{x}_i - q_j - e_{jk\ell}E_kH_\ell + (D_jE_k + B_jH_k)\dot{x}_k) + \\ &- \frac{1}{2}(E_kD_k + H_kB_k)\dot{x}_j\}da_j ,\end{aligned}$$

and it leads to the following balance equations of momentum and energy

(2.99) $$\rho\ddot{x}_i - t_{ij,j} - \rho F^{ext}_i = \rho F^{e}_i = [E_iD_j + H_iB_j - \frac{1}{2}\delta_{ij}(E_kD_k + H_kB_k)]_{,j} ,$$

and

(2.100) $$\begin{aligned}\rho\dot{U} - t_{ij}\dot{x}_{i,j} + q_{j,j} - \rho r^{ext} &= \rho r^{e} = \\ &= J_iE_i + \frac{1}{2}(\dot{E}_iD_i - E_i\dot{D}_i + \dot{H}_iB_i - H_i\dot{B}_i) .\end{aligned}$$

These equations resemble in a way a formulation in which the expressions for the electromagnetic body force and energy supply are derived from a four-dimensional formulation of the Maxwell-equations, as outlined by Møller ([17],

Ch. 7) and in the report of Pao, ([18], Section 6). However, there is one essential difference, as in (2.99) and (2.100) the rest-frame fields $\mathcal{E}_i$ and $\mathcal{H}_i$ are used, whereas in [17] and [18] the laboratory fields E_i and H_i are employed. Since the balance laws as presented in [17] or [18] are not invariant in the non-relativistic sense, they can never be deduced from an energy balance in the way described above. This makes it very questionable whether the formulation, discussed in [17], describes the interactions between matter and field in any meaningful way (see also [2], p. 202 and [18], pp. 121-122). It is not too difficult to show that the results corresponding to (2.98) and those based upon (2.77) are in correspondence provided that

$$(2.101)\qquad \begin{aligned} t_{ij}(2.98) - t_{ij}(2.77) &= \frac{1}{2}\,\delta_{ij}(\mathcal{E}_k P_k + \mathcal{H}_k \mathcal{M}_k) , \\ \rho U(2.98) - \rho U(2.77) &= -\frac{1}{2}(\mathcal{E}_i P_i + \mathcal{H}_i \mathcal{M}_i) . \end{aligned}$$

So far we have given three possible Minkowski formulations of electromagnetoelastic interactions. Whenever in the sequel reference is made to the Maxwell-Minkowski formulation (model III), the first theory of this Section is meant (i.e. system (2.79)-(2.82)).

We conclude by stating that the two theories outlined in the preceding Section could, at least for the consistent non-relativistic part, also be based upon a global energy balance. The underlying energy balance for model I is (see also Alblas [10])

$$\frac{d}{dt}\int_{V}\{\rho U + \frac{1}{2}\rho \dot{x}_i \dot{x}_i + \frac{1}{2}(\varepsilon_o \mathcal{E}_k \mathcal{E}_k + \mu_o \mathcal{H}_k \mathcal{H}_k)\}dv = \int_{V}\{\rho r^{ext} + \rho F_i^{ext}\dot{x}_i\}dv +$$

$$(2.102)\qquad + \int_{\partial V}\{t_{ij}\dot{x}_i - q_j - e_{jk\ell}\mathcal{E}_k \mathcal{H}_\ell - \frac{1}{2}(\varepsilon_o \mathcal{E}_k \mathcal{E}_k + \mu_o \mathcal{H}_k \mathcal{H}_k)\dot{x}_j +$$

$$+ (\varepsilon_o \mathcal{E}_j \mathcal{E}_k + \mu_o \mathcal{H}_j \mathcal{H}_k)\dot{x}_k + (P_j^C \mathcal{E}_k + \mu_o M_j^C \mathcal{H}_k)\dot{x}_k\}da_j .$$

In the usual way this relation leads to expressions for ρF_i^e, ρL_{ij}^e and ρr^e as given in (2.16), $(2.7)^2$ and $(2.7)^3$, respectively, except for the terms preceded by a c^{-2}-factor in (2.16).

In an analogous way, model II can be based upon the balance law

$$\frac{d}{dt}\int_V \{\rho U + \frac{1}{2}\rho\dot{x}_i\dot{x}_i + \frac{1}{2}(\varepsilon_o E_k E_k + \mu_o H_k H_k)\}dv =$$

$$(2.103)\quad = \int_V \{\rho r^{ext} + \rho F_i^{ext}\dot{x}_i\}dv + \int_{\partial V} \{t_{ij}\dot{x}_i - q_j - e_{jk\ell}E_k H_\ell +$$

$$- \frac{1}{2}(\varepsilon_o E_k E_k + \mu_o H_k H_k)\dot{x}_j + (\varepsilon_o E_j E_k + \mu_o H_j H_k)\dot{x}_k\}da_j \; .$$

In the following sections we shall show that energy balances of this kind can also serve as bases for the theories outlined there.

2.4 THE STATISTICAL FORMULATION (Model IV)

The proper physical approach toward a formulation of polarizable and magnetizable continua is through methods of statistical mechanics. Early attempts go back to Rosenfeldt [19]. A comprehensive treatment - in the light of relativistically covariant statistical mechanics - is given by De Groot and Suttorp [12]. This book may also serve as guideline for a historical account on the subject. In this theory, matter consists of stable groups of electrically charged particles, such as electrons, ions etc. The field effect of these particles within each stable group is represented by electric and magnetic multipoles, the statistical averages of which give rise to the definition of electric polarization P_i and magnetization M_i.
With regard to the statistical derivation of the Maxwell equations, De Groot and Suttorp are entirely general and they do not introduce any non-relativistic approximations. As a result, the macroscopic Maxwell equations are <u>proved</u> to be Lorentz invariant, a property hitherto assumed to hold.
We introduce the statistical formulation again formally by transformation rules, viz.

$$(2.104)\quad \begin{aligned} B_i &= B_i^S \, , \\ E_i &= E_i^S + e_{ijk}\dot{x}_j B_k^S \, , \\ D_i &= \varepsilon_o E_i^S + P_i^S \, , \\ H_i &= \frac{1}{\mu_o} B_i^S - e_{ijk}\dot{x}_j \varepsilon_o E_k^S - M_i^S - e_{ijk}\dot{x}_j P_i^S \, . \end{aligned}$$

Accordingly, the variables of the statistical formulation bear the superscript S. (B_i^S, E_i^S, P_i^S and M_i^S are called magnetic flux density, electric field strength, polarization and magnetization, respectively.)
Substituting (2.104) into the Maxwell equations (1.23) and (1.28), what results reads as follows

$$B_{i,i}^S = 0 ,$$

$$e_{ijk}E_{k,j}^S + \frac{\partial B_i^S}{\partial t} = 0 ,$$

(2.105)

$$\varepsilon_o E_{i,i}^S = Q - P_{i,i}^S ,$$

$$\frac{1}{\mu_o} e_{ijk}B_{k,j}^S - \varepsilon_o \frac{\partial E_i^S}{\partial t} = J_i + \frac{\partial P_i^S}{\partial t} + e_{ijk}M_{k,j}^S .$$

Incidentally, Penfield and Haus [2] call this formulation the Boffi formulation and, as a comparison with (2.60) and (2.61) shows, these variables agree with the ones introduced in the preceding section as the Minkowski variables. The equations (2.105) are obtained also when the action of matter upon the electromagnetic fields is derived from very crude physical models of stationary rigid bodies. Indeed they can be found in almost any physics book treating electromagnetism (see e.g. Feynman [20]). They hold, however, for a much broader class of physical processes than anticipated in these books and embrace all those for which the action of matter on the electromagnetic fields has been discussed above.
The expressions for the electromagnetic force, couple and energy supply have been derived by de Groot and Suttorp in full generality, including relativistic effects. However, they also present two approximated versions, differing in the degree of approximation, which in our terminology are called the non-relativistic and the semi-relativistic approximations. Here, we shall discuss the non-relativistic case only.
In their second Chapter, De Groot and Suttorp arrive at the following non-relativistic expressions for body force, body couple and energy supply (cf. [12], pp. 47,63)

$$\rho F_i^e = QE_i^S + e_{ijk}J_jB_k^S + P_j^S E_{j,i}^S + M_j^S B_{j,i}^S + \rho e_{ijk}\frac{d}{dt}(\frac{1}{\rho} P_j^S B_k^S) =$$

$$= \mathcal{Q}\mathcal{E}_i + e_{ijk}\mathcal{J}_jB_k + P_j\mathcal{E}_{j,i} + \mathcal{M}_jB_{j,i} + e_{ijk}(\overset{*}{P}_jB_k + P_j\overset{*}{B}_k) ,$$

(over)

$$\rho L^e_{ij} = P_{[i}E_{j]} + M_{[i}B_{j]} \ ,$$

(2.106)

$$\rho r^e = J_i E_i + \rho E_i \frac{d}{dt}\left(\frac{P_i}{\rho}\right) - M_i \dot{B}_i \ ,$$

where

(2.107) $B_i = B^S_i, \qquad P_i = P^S_i \qquad \text{and} \qquad M_i = M^S_i + e_{ijk}\dot{x}_j P^S_k \ .$

Thus, the conservation of mass, the balance laws of linear and angular momentum and the balance of energy become

$$\dot{\rho} + \rho\dot{x}_{i,i} = 0 \ ,$$

$$\rho\ddot{x}_i = t_{ij,j} + \rho F^{ext}_i + Q E_i + e_{ijk}J_j B_k + P_j E_{j,i} +$$

(2.108)
$$+ M_j B_{j,i} + e_{ijk}(\overset{*}{P}_j B_k + P_j \overset{*}{B}_k) \ ,$$

$$t_{[ij]} = P_{[i}E_{j]} + M_{[i}B_{j]} \ ,$$

and

$$\rho\dot{U} = t_{ij}\dot{x}_{i,j} - q_{i,i} + J_i E_i + \rho E_i \frac{d}{dt}\left(\frac{P_i}{\rho}\right) - M_i \dot{B}_i + \rho r^{ext} \ .$$

Since, as already said before, the statistical fields are identical to the Minkowski fields, we note that

$$Q,\ J_i,\ E_i,\ P_i,\ B_i \text{ and } M_i$$

transform under Euclidean transformations as an objective scalar, as objective vectors and as objective axial vectors, respectively. Consequently, the electromagnetic body force and body couple as given by the above expressions are an objective vector and skew-symmetric tensor, respectively. Of course, this holds true also in the non-relativistic sense as defined in Section 1.6. As far as the electromagnetic energy supply is concerned, we should note, as we already have seen in Section 2.2.1, that ρr^e cannot be an objective scalar, because the stress tensor is not symmetric in this formulation. We should rather look at the expression

$$\rho r^e + t_{[ij]}\dot{x}_{i,j} \ ,$$

which according to $(2.106)^3$, $(2.108)^3$ and (2.105) equals

$$\rho r^{e} + t_{[ij]}\dot{x}_{i,j} = J_i E_i + E_i(\overset{*}{P}_i + e_{ijk}M_{k,j}) + (e_{ijk}E_j M_k)_{,i} + \tag{2.109}$$
$$+ (E_i P_j - M_i B_j + \delta_{ij} M_k B_k)\dot{x}_{(i,j)}$$

Each term on the right-hand side of (2.109) is obviously an objective scalar. Hence, the balance equations are invariant under the Euclidean group. However, these laws are not invariant under Lorentz transformations, not even in the semi-relativistic sense.

As the momentum equation and energy equation are invariant under Euclidean transformations it may be expected that the results given above are also derivable from a global energy balance as derived in Section 2.3. Indeed this is true and the underlying energy balance reads

$$\frac{d}{dt}\int_V \{\rho U + \tfrac{1}{2}(\varepsilon_o E_i E_i + \frac{1}{\mu_o} B_i B_i) + \tfrac{1}{2}\rho \dot{x}_i \dot{x}_i\} dv =$$
$$= \int_V \{\rho r^{ext} + \rho F_i^{ext}\dot{x}_i\} dv + \int_{\partial V} \{t_{ij}\dot{x}_i - q_j - e_{jk\ell}E_k(\frac{1}{\mu_o} B_\ell - M_\ell) + \tag{2.110}$$
$$- \tfrac{1}{2}(\varepsilon_o E_k E_k + \frac{1}{\mu_o} B_k B_k)\dot{x}_j + (\varepsilon_o E_j E_k + \frac{1}{\mu_o} B_j B_k)\dot{x}_k +$$
$$+ (P_j E_k - B_j M_k + \delta_{jk} M_\ell B_\ell)\dot{x}_k\} da_j \ .$$

Using an approach entirely analogous to the one outlined in Section 2.3 and invoking the invariance requirement that (2.110) is Euclidean invariant, it may be shown that the above energy balance law implies the local balance equations (2.108).

The electromagnetic body force and energy supply can also be expressed in the form (1.62). Indeed, straightforward calculations show that

$$^{IV}t^{M}_{ij} = \varepsilon_o E^S_i E^S_j + \frac{1}{\mu_o} B_i B_j + E^S_i P_j - M^S_i B_j +$$
$$- \frac{1}{2}\delta_{ij}(\varepsilon_o E^S_k E^S_k + \frac{1}{\mu_o} B_k B_k - 2B_k M^S_k) + e_{ik\ell}P_k B_\ell \dot{x}_j =$$
$$= \varepsilon_o E_i E_j + \frac{1}{\mu_o} B_i B_j + E_i P_j - M_i B_j +$$
$$- \frac{1}{2}\delta_{ij}(\varepsilon_o E_k E_k + \frac{1}{\mu_o} B_k B_k - 2M_k B_k) \ ,$$

(over)

$$(2.111) \qquad {}^{IV}g_i = 0 \ ,$$

$${}^{IV}\pi_i = -e_{ijk}E^S_j\left(\frac{1}{\mu_o}B_k - M^S_k\right) + E^S_j P_j \dot{x}_i =$$

$$= -e_{ijk}E_j\left(\frac{1}{\mu_o}B_k - M_k\right) - \left(\varepsilon_o E_j E_j + \frac{1}{\mu_o}B_j B_j\right)\dot{x}_i +$$

$$+ \left(\varepsilon_o E_i E_j + \frac{1}{\mu_o}B_i B_j\right)\dot{x}_j + \left(P_i E_j - B_i M_j + \delta_{ij}M_k B_k\right)\dot{x}_j \ .$$

$${}^{IV}\omega = -\frac{1}{2}\left(\varepsilon_o E^S_k E^S_k + \frac{1}{\mu_o}B_k B_k\right) = -\frac{1}{2}\left(\varepsilon_o E_k E_k + \frac{1}{\mu_o}B_k B_k\right) \ .$$

We note that in the above calculations $(\varepsilon_o \underset{\sim}{E} \times \underset{\sim}{B})$-terms are neglected, which according to the statement made in the preceding Section on page 55 is consistent with our requirements.

The jump conditions for the fields, mass, momentum and energy of matter and fields become now

$$[\![B_i]\!]n_i = 0, \qquad [\![\varepsilon_o E^S_i + P_i]\!]n_i = 0 \ ,$$

$$[\![e_{ijk}E^S_j]\!]n_k + [\![B_i]\!]w_n = 0 \ ,$$

$$[\![e_{ijk}\left(\frac{1}{\mu_o}B_j - M^S_j\right)]\!]n_k - [\![\varepsilon_o E^S_i + P^S_i]\!]w_n = 0 \ ,$$

$$[\![\rho(\dot{x}_i n_i - w_n)]\!] = 0 \ ,$$

$$(2.112) \qquad [\![t_{ij} + \varepsilon_o E_i E_j + \frac{1}{\mu_o}B_i B_j + E_i P_j - M_i B_j +$$

$$- \frac{1}{2}\delta_{ij}\left(\varepsilon_o E_k E_k + \frac{1}{\mu_o}B_k B_k - 2M_k B_k\right)]\!]n_j - [\![\rho\dot{x}_i(\dot{x}_j n_j - w_n)]\!] = 0 \ ,$$

$$[\![\dot{x}_i t_{ij} - q_j - e_{jk\ell}E_k\left(\frac{1}{\mu_o}B_\ell - M_\ell\right) - \frac{1}{2}\left(\varepsilon_o E_k E_k + \frac{1}{\mu_o}B_k B_k\right)\dot{x}_j +$$

$$+ \left(\varepsilon_o E_i E_j + \frac{1}{\mu_o}B_i B_j\right)\dot{x}_i + \left(E_i P_j - M_i B_j + \delta_{ij}M_k B_k\right)\dot{x}_i]\!]n_j +$$

$$- [\![\left\{\frac{1}{2}\rho\dot{x}_i\dot{x}_i + \rho U + \frac{1}{2}\left(\varepsilon_o E_i E_i + \frac{1}{\mu_o}B_i B_i\right)\right\}(\dot{x}_j n_j - w_n)]\!] = 0 \ .$$

It remains to develop the constitutive theory. For this purpose, we notice that, in a non-relativistic theory, the invariance group which the principle of material frame indifference relies upon, is the Galilean, or more generally the Euclidean, group. Hence, we may choose the constitutive relations of the form derived in Section 1.6. In this Section we restrict ourselves to the case

(2.113) $\quad C = \overset{+}{C}(C_{\alpha\beta}, P_\alpha, B_\alpha, \theta, \theta_{,\alpha}, \mathcal{Q})$,

where P_α and B_α are defined in (1.89).

The application of the entropy principle is again a routine matter and we simply state the results. With the Helmholtz free energy

(2.114) $\quad \psi = U - \eta\theta$,

and with (2.26), one finds that

(2.115) $\quad \psi = \overset{+}{\Psi}(C_{\alpha\beta}, P_\alpha, B_\alpha, \theta)$,

while the entropy η, the electromotive intensity $\mathcal{E}_i$, the rest frame magnetization $\mathcal{M}_i$ and the stress t_{ij} are given by

(2.116)
$$\eta = -\frac{\partial\overset{+}{\Psi}}{\partial\theta} ,$$
$$\mathcal{E}_i = \frac{\partial\overset{+}{\Psi}}{\partial P_\alpha} F_{i\alpha} ,$$
$$\mu_o \mathcal{M}_i = -\rho \frac{\partial\overset{+}{\Psi}}{\partial B_\alpha} F_{i\alpha} \operatorname{sgn} J ,$$
$$t_{ij} = 2\rho \frac{\partial\overset{+}{\Psi}}{\partial C_{\alpha\beta}} F_{i\alpha} F_{j\beta} + P_i \mathcal{E}_j - B_i \mathcal{M}_j .$$

These results imply that the balance law of moment of momentum is satisfied identically. Furthermore, the reduced entropy inequality is

(2.117) $$\mathcal{J}_i \mathcal{E}_i - \frac{\theta_{,i} q_i}{\theta} \geq 0$$

and the Gibbs relation becomes

(2.118) $$d\eta = \frac{1}{\theta}\{dU - \frac{1}{2}[\frac{1}{\rho_o} T^P_{\alpha\beta} + C^{-1}_{\alpha\gamma} C^{-1}_{\beta\delta}(P_\gamma E_\delta - B_\gamma M_\delta)]dC_{\alpha\beta} + $$
$$- C^{-1}_{\alpha\beta} E_\alpha dP_\beta - C^{-1}_{\alpha\beta} M_\alpha dB_\beta\} ,$$

where E_α and M_α are given in (1.89).

This completes the non-relativistic theory of magnetizable and polarizable solids in the statistical description.

2.5 THE LORENTZ FORMULATION (Model V)

The Lorentz description of electromagnetism is founded on his theory of electrons, originally formulated for dielectric materials only, see [3]. According to this description, the body is supposed to consist of a set of electrically interfering charged particles. These particles respond to their own fields (as well as to possible external fields) and may move rapidly, thereby producing highly fluctuating (microscopic) electromagnetic fields as well. The pertinent field equations are the Maxwell equations in a vacuum and when averaged, the Maxwell equations in the Lorentz formulation emerge (see for instance the booklet by L. Rosenfeldt [19]). In the light of De Groot and Suttorp's statistical description this average is a crude statistical model. This might lead the reader to the conclusion that the Lorentz formulation is only approximate. This is not so. On the contrary, this theory has been put on a sound, relativistically correct, axiomatics by Truesdell and Toupin [3]. (For a non-relativistic presentation, though nevertheless relativistically correct formulation, see the book by Müller [13].) The fact that it may be defined also from any set of Maxwell equations by simply performing a variable transformation should be a corroboration of its correctness.

This is exactly what we shall do here. Indeed, we introduce the Lorentz formulation using the definitions

$$D_i = \varepsilon_o E_i^L + P_i^L, \qquad \mathcal{E}_i = E_i^L + e_{ijk}\dot{x}_j B_k^L,$$
$$B_i = B_i^L, \qquad \mathcal{H}_i = \frac{1}{\mu_o} B_i^L - M_i^L - \varepsilon_o e_{ijk}\dot{x}_j E_k^L. \tag{2.119}$$

Upon substitution of (2.119) into (1.23) and (1.28), the Maxwell equations in the Lorentz formulation are obtained as follows:

$$B_{i,i}^L = 0,$$
$$e_{ijk}E_{k,j}^L + \frac{\partial B_i^L}{\partial t} = 0,$$
$$\varepsilon_o E_{i,i}^L = \mathcal{Q} - P_{i,i}^L, \tag{2.120}$$
$$\frac{1}{\mu_o} e_{ijk}B_{k,j}^L - \varepsilon_o \frac{\partial E_i^L}{\partial t} = J_i + \frac{\partial P_i^L}{\partial t} + e_{ijk}(e_{k\ell m}P_\ell^L \dot{x}_m)_{,j} + e_{ijk}M_{k,j}^L.$$

These equations, introduced here formally, are often applied in the modern literature, as for instance by Toupin [7], Liu and Müller [21], Benach and Müller [22] and Hutter [14]. The equations are also mentioned by Pao [18], but he states "that for magnetizable materials their validity is questionable".

Comparing (2.120) with the Maxwell equations of the statistical model (2.105), we see that

$$(2.121)\qquad \begin{aligned} &E^L_{\cdot i} = E^S_{\cdot i}, \qquad B^L_{\cdot i} = B^S_{\cdot i} = B_{\cdot i}\ , \\ &P^L_{\cdot i} = P^S_{\cdot i} = P_{\cdot i}, \qquad \text{and} \qquad M^L_{\cdot i} = M^S_{\cdot i} + e_{\cdot ijk}\dot{x}_{\cdot j}P^L_{\cdot k} = \mathcal{M}_{\cdot i}\ . \end{aligned}$$

The Maxwell equations in the Lorentz description can also be deduced from the assumption that every particle is equipped with a number of electric dipoles and with an electric circuit. If this dipole-circuit model is treated non-relativistically again the Maxwell equations in the Lorentz description are obtained. Of course, such a derivation bears inherently the notion of approximation. What is approximated thereby is not the theory as such, however, but the use of the model. That the theory is indeed correct from a relativistic point of view cannot be seen from such a derivation, and this must be regarded as a major disadvantage of the model. This fact might also be the reason why Penfield and Haus use a model in which each particle is equipped with a number of dipoles and an electric circuit both of which are treated relativistically. What they obtain is different from the Lorentz description. They call these Maxwell equations the Ampèrean model. The difference between the Ampèrean and the Lorentzean variables is a small difference in the polarization vector of the amount

$$c^{-2}(\underline{\dot{x}} \times \underline{M})\ .$$

In a non-relativistic theory this term is negligible, and it emerges from the fact that in the Lorentz description

$$P^L_{\cdot i} = P^M_{\cdot i}\ ,$$

while in the Ampèrean description

$$P^A_{\cdot i} = P^C_{\cdot i}\ .$$

Nevertheless, both the Ampèrean and the Lorentz formulation are relativistically correct. This should make it clear that the approximations must be sought in the model rather than in the basic theory.

With this digression on approximations we proceed and mention that in the Lorentz formulation

$$-P^L_{i,i}$$

is the charge density due to polarization,

$$e_{ijk}M^L_{k,j}$$

the current due to magnetization and

$$\frac{\partial P^L_i}{\partial t} + e_{ijk}(e_{k\ell m}P^L_\ell \dot{x}_m)_{,j}$$

the polarization current, all as shown on the right-hand side of (2.120). This suggests to postulate as electromagnetic body force, body couple and energy supply the expressions

$$\rho F^e_i = \mathcal{Q}^{tot}E^L_i + e_{ijk}J^{tot}_j B_k = \mathcal{Q}^{tot}\mathcal{E}_i + e_{ijk}\mathcal{J}^{tot}_j B_k \ ,$$

$$(2.122) \qquad \rho L^e_{ij} = 0 \ ,$$

$$\rho r^e = \mathcal{J}^{tot}_i \mathcal{E}_i = J^{tot}_i E^L_i - \rho F^e_i \dot{x}_i \ ,$$

where $\mathcal{Q}^{tot}$ and J^{tot}_i are the total charge and total current as given on the right-hand side of $(2.120)^{3,4}$. (Compare this postulate with the one for model II, (2.45)). The postulates (2.122) have, in this generality, first been introduced by Toupin [7]. They were then applied by Liu and Müller [21], Benach and Müller [22], and Hutter [14] in theories of polarizable and magnetizable fluids, fluid mixtures and solids, respectively. Substituting $\mathcal{Q}^{tot}$ and J^{tot}_i, as obtained from $(2.120)^{3,4}$, into (2.122) yields

$$\rho F^e_i = (\mathcal{Q} - P_{j,j})\mathcal{E}_i + e_{ijk}(\mathcal{J}_j + \overset{*}{P}_j + e_{jmn}\mathcal{M}_{n,m})B_k \ ,$$

$$(2.123) \qquad \rho L^e_{ij} = 0 \ ,$$

$$\rho r^e = \mathcal{J}_i\mathcal{E}_i + \mathcal{E}_i(\overset{*}{P}_i + e_{ijk}\mathcal{M}_{k,j}) \ ,$$

where the convective derivative denoted by a superimposed star is defined in (1.24).

With (2.123) the balance laws of mass, momenta and energy assume the form

$$\dot\rho + \rho\dot x_{i,i} = 0 \ ,$$

(2.124)
$$\rho\ddot x_i = t_{ij,j} + \rho F_i^{ext} + (\mathcal{Q} - P_{j,j})\mathcal{E}_i + e_{ijk}(\mathcal{J}_j + \overset{*}{P}_j + e_{jmn}\mathcal{M}_{n,m})B_k \ ,$$

$$t_{[ij]} = 0 \ ,$$

$$\rho\dot U = t_{ij}\dot x_{i,j} - q_{i,i} + \mathcal{J}_i\mathcal{E}_i + \mathcal{E}_i(\overset{*}{P}_i + e_{ijk}\mathcal{M}_{k,j}) + \rho r^{ext} \ .$$

The expressions (2.123) immediately show that ρr^e and ρF_i^e are an objective scalar and objective vector under the Euclidean transformation group. For this to be true, we have used the fact that $\mathcal{Q}$, $\mathcal{J}_i$, $\mathcal{E}_i$, B_i, P_i and $\mathcal{M}_i$ are objective quantities, as we have already seen in the preceding Sections. A proof of this can also be found in Truesdell and Toupin [3], Chapter 7. Hence, the momentum and energy equations enjoy the classical (non-relativistic) invariance requirements. The energy equation is, however, only invariant under Euclidean transformations, because we have also taken the stress tensor to be symmetric. This may serve as a justification for the rather unmotivated choice of $\rho L_{ij}^e = 0$.
Just as in the preceding Sections, this formulation can also be derived from a global energy balance which in this case reads

(2.125)
$$\frac{d}{dt}\int_{V}\{\rho U + \frac{1}{2}\rho\dot x_i\dot x_i + \frac{1}{2}(\varepsilon_o\mathcal{E}_k\mathcal{E}_k + \frac{1}{\mu_o}B_kB_k)\}dv = \int_{V}\{\rho r^{ext} + \rho F_i^{ext}\dot x_i\}dv +$$
$$+ \int_{\partial V}\{t_{ij}\dot x_i - q_j - e_{jk\ell}\mathcal{E}_k\frac{B_\ell}{\mu_o} + \frac{1}{2}(\varepsilon_o\mathcal{E}_k\mathcal{E}_k + \frac{1}{\mu_o}B_kB_k)\dot x_j\}da_j \ .$$

Furthermore, it may be shown that the expressions (2.123) can be written in the form (1.62) (with the non-relativistic approximations), yielding

$$^{V}t_{ij}^{M} = \varepsilon_o\mathcal{E}_i\mathcal{E}_j + \frac{1}{\mu_o}B_iB_j - \frac{1}{2}\delta_{ij}(\varepsilon_o\mathcal{E}_k\mathcal{E}_k + \frac{1}{\mu_o}B_kB_k) \ ,$$

$$^{V}g_i = 0 \ ,$$

(2.126)
$$^{V}\pi_i = -e_{ijk}E_j^L\frac{B_k}{\mu_o} = -e_{ijk}\mathcal{E}_j(\mathcal{H}_k + \mathcal{M}_k) - (\varepsilon_o\mathcal{E}_k\mathcal{E}_k + \frac{1}{\mu_o}B_kB_k)\dot x_i +$$
$$+ (\varepsilon_o\mathcal{E}_i\mathcal{E}_j + \frac{1}{\mu_o}B_iB_j)\dot x_j \ ,$$

$$^{V}\omega = -\frac{1}{2}(\varepsilon_o\mathcal{E}_k\mathcal{E}_k + \frac{1}{\mu_o}B_kB_k) \ .$$

With these relations, the jump conditions for momentum and energy of matter and fields can be derived. We list them below, together with the jump conditions for the electromagnetic fields and the density

$$[\![B_i]\!]n_i = 0, \qquad [\![\varepsilon_o E_i^L + P_i]\!]n_i = 0 ,$$

$$[\![e_{ijk}E_j^L]\!]n_k + [\![B_i]\!]w_n = 0 ,$$

$$[\![e_{ijk}(\frac{1}{\mu_o} B_j - M_j + e_{j\ell m}\dot{x}_\ell P_m)]\!]n_k - [\![\varepsilon_o E_i^L + P_i]\!]w_n = 0 ,$$

$$[\![\rho(\dot{x}_i n_i - w_n)]\!] = 0 ,$$

(2.127)

$$[\![t_{ij} + \varepsilon_o E_i E_j + \frac{1}{\mu_o} B_i B_j - \frac{1}{2}\delta_{ij}(\varepsilon_o E_k E_k + \frac{1}{\mu_o} B_k B_k)]\!]n_j +$$

$$- [\![\rho\dot{x}_i(\dot{x}_j n_j - w_n)]\!] = 0 ,$$

$$[\![t_{ij}\dot{x}_i - q_j - e_{jk\ell}E_k \frac{B_\ell}{\mu_o} - \frac{1}{2}(\varepsilon_o E_k E_k + \frac{1}{\mu_o} B_k B_k)\dot{x}_j + (\varepsilon_o E_i E_j + \frac{1}{\mu_o} B_i B_j)\dot{x}_i]\!]n_j +$$

$$- [\![\{\frac{1}{2}\rho\dot{x}_i\dot{x}_i + \rho U + \frac{1}{2}(\varepsilon_o E_k E_k + \frac{1}{\mu_o} B_k B_k)\}(\dot{x}_j n_j - w_n)]\!] = 0 .$$

We wish to note that the jump conditions $(2.127)^{6,7}$, and all corresponding ones of the preceding Sections, could have been derived from the respective global energy balance laws. To illustrate this, note that from the global energy balance (2.125) a jump condition can be derived as demonstrated in Section 1.5. This jump condition (given by $(2.127)^7$) is then subjected to a rigid-body translation of the form

$$x_i \Rightarrow x_i - b_i(t) ,$$

under which $(\dot{x}_j n_j - w_n)$ remains invariant. Since under such transformations

$$-e_{ijk}E_j^L B_k/\mu_o$$

changes into (remember that although B_i^L is invariant, B_i^L/μ_o is not (see Section 1.6))

$$-e_{ijk}E_j^L \frac{B_k}{\mu_o} - \{(\varepsilon_o E_i E_j + \frac{1}{\mu_o} B_i B_j) - \delta_{ij}(\varepsilon_o E_k E_k + \frac{1}{\mu_o} B_k B_k)\}b_i ,$$

and then $(2.127)^7$ immediately leads to $(2.127)^6$.

We now turn our attention to the constitutive theory and for that purpose it is advantageous to transform the energy balance equation (2.124)[4] into the form

$$\begin{aligned}\rho\dot{U} &= t_{ij}\dot{x}_{i,j} - q_{i,i} + J_i E_i - (e_{ijk}E_j M_k)_{,i} + \\ &+ \rho E_i \frac{d}{dt}\left(\frac{P_i}{\rho}\right) - M_i \dot{B}_i - (E_i P_j - M_i B_j + M_k B_k \delta_{ij})\dot{x}_{i,j} + \rho r^{ext} .\end{aligned} \tag{2.128}$$

Then, eliminating ρr^{ext} from (1.53) and introducing the Helmholtz free energy by

$$\psi = U - \eta\theta ,$$

reveals the inequality

$$\begin{aligned}&-\rho\dot{\psi} - \rho\theta\dot{\eta} + \rho E_i \frac{d}{dt}\left(\frac{P_i}{\rho}\right) - M_i\dot{B}_i + \{t_{ij} - E_i P_j + M_i B_j - M_k B_k \delta_{ij}\}\dot{x}_{i,j} + \\ &+ J_i E_i + [\theta\phi_{i,i} - q_{i,i} - (e_{ijk}E_j M_k)_{,i}] \geq 0 .\end{aligned} \tag{2.129}$$

Before we proceed, a comment seems to be in order regarding the last, bracketed, term of this inequality. Evidently, there is no unique definition of energy flux and entropy flux in terms of thermodynamic and electromagnetic variables. To see this, let us introduce a new energy flux vector q_i^S by

$$q_i^S := q_i + e_{ijk}E_j M_k . \tag{2.130}$$

(As will be shown in the next Chapter, q_i^S agrees with the energy flux vector of the statistical model IV). With the choice

$$\phi_i = q_i^S/\theta , \tag{2.131}$$

the bracketed term in (2.129) then becomes

$$[-\theta_{,i} q_i^S/\theta] ,$$

and this is what one would expect classically, if one insisted to call q_i^S the heat flux vector. However, this is a purely formal definition, and no argument whatsoever justifies such an identification. Moreover, if we substitute (2.130) into the classical relation for the entropy flux (2.26), we find

$$\phi_i = \frac{q_i}{\theta} + \frac{1}{\theta} e_{ijk}E_j M_k , \tag{2.132}$$

and, if we insist to call q_i the heat flux, this equation teaches us that in this case the entropy flux does not obey the classical relation, namely heat flux divided by absolute temperature. This is the reason why we prefer to call q_i the energy flux vector. Moreover, this shows that it is imperative to refrain from setting the entropy flux equal to the heat flux divided by absolute temperature from the outset, because it is not evident what should be understood under heat flux. Instead, one must prescribe ϕ_i as a general constitutive variable, whose form should be determined in due course with the exploitation of the entropy inequality. For one particular theory of the complexity treated here, the relation

$$(2.133) \qquad \phi_i = \frac{q_i^S}{\theta} = \frac{q_i + e_{ijk}\mathcal{E}_j\mathcal{M}_k}{\theta} ,$$

was proved by Hutter [14], and the same relation was also found earlier by Liu and Müller [21] for the case of a simple fluid. For the purpose of this monograph we shall regard (2.133) as a postulate.
As was done for the other models, we now establish constitutive relations, which in view of (2.129) are assumed in the form

$$(2.134) \qquad C = \overset{+}{C}(F_{i\alpha}, P_i/\rho, B_i, \theta, \theta_{,i}, \mathcal{Q}) ,$$

where C denotes the set $\{U, \psi, \eta, \mathcal{E}_i, \mu_o\mathcal{M}_i/\rho, q_i, \mathcal{J}_i, t_{ij}\}$. Moreover, we require the constitutive relations to be invariant under the Euclidean group of transformations. In view of the transformation rules listed before this is already ascertained.
In the usual way we can derive the following results:

$$(2.135) \qquad \psi = \overset{+}{\Psi}(C_{\alpha\beta}, P_\alpha, B_\alpha, \theta) ,$$

and

$$(2.136) \qquad \begin{aligned} \eta &= -\frac{\partial\overset{+}{\Psi}}{\partial\theta} , \\ \mathcal{E}_i &= \frac{\partial\overset{+}{\Psi}}{\partial P_\alpha} F_{i\alpha} , \\ \frac{\mu_o\mathcal{M}_i}{\rho} &= -\frac{\partial\overset{+}{\Psi}}{\partial B_\alpha} F_{i\alpha}\operatorname{sgn} J , \\ t_{ij} &= 2\rho\frac{\partial\overset{+}{\Psi}}{\partial C_{\alpha\beta}} F_{i\alpha}F_{j\beta} + (P_i\mathcal{E}_j + P_j\mathcal{E}_i) - (\mathcal{M}_iB_j + \mathcal{M}_jB_i) + \delta_{ij}\mathcal{M}_kB_k . \end{aligned}$$

The reduced entropy balance reads

$$(2.137)\qquad J_i E_i - \frac{\theta_{,i}}{\theta}(q_i + e_{ijk}E_j M_k) \geq 0 ,$$

and the Gibbs relation becomes

$$(2.138)\qquad d\eta = \frac{1}{\theta}\{dU - \frac{1}{2}[\frac{1}{\rho_o} T^P_{\alpha\beta} - C^{-1}_{\alpha\gamma}C^{-1}_{\beta\delta}(P_\gamma E_\delta + P_\delta E_\gamma) + - C^{-1}_{\alpha\gamma}C^{-1}_{\beta\delta}(M_\gamma B_\delta + M_\delta B_\gamma) - C^{-1}_{\alpha\beta}M_\gamma M_\gamma]dC_{\alpha\beta} + - C^{-1}_{\alpha\beta}E_\beta dP_\alpha + C^{-1}_{\alpha\beta}M_\beta dB_\alpha\} .$$

This completes the presentation of the Lorentz formulation.

2.6 THERMOSTATIC EQUILIBRIUM CONSTITUTIVE EQUATIONS FOR THE CONDUCTIVE CURRENT DENSITY AND THE ENERGY FLUX

In the preceding Sections we determined constitutive equations for entropy, two electromagnetic field vectors and the stress tensor. All of the above mentioned variables are derivable from a thermodynamic potential (the latter being the Helmholtz free energy or its Legendre transformations). Given such a thermodynamic potential as a function of its independent variables the constitutive relations for the remaining fields were derivable. However, it is not possible to construct in the same way constitutive relations for the conductive current density and for the energy flux. Constitutive relations for these quantities are merely restricted by the residual entropy inequality, which in all but the Lorentz formulation may be written as

$$(2.139)\qquad \gamma := J_i E_i - \frac{\theta_{,i} q_i}{\theta} \geq 0 .$$

In the Lorentz formulation q_i must be replaced by

$$(2.140)\qquad q^S_i := q_i + e_{ijk}E_j M_k .$$

To find the restrictions imposed by the residual entropy inequality we look at thermostatic processes, which will be defined as processes with uniform and time-independent temperature and vanishing conductive current. The results depend on whether we are dealing with an electrical insulator or conductor. Hence, we discuss these cases seperately.

a) In an <u>electrical insulator</u> the conductive current J_i vanishes identically. Hence, (2.139) becomes

$$\gamma = -\frac{\theta_{,i} q_i}{\theta} \geq 0 \;, \tag{2.141}$$

and in equilibrium

$$\gamma = \gamma|_E = 0 \;. \tag{2.142}$$

We shall characterize thermostatic equilibrium by the index $|_E$. It follows that γ, which is a function of all the independent variables q_i depends upon, assumes its minimum when $\theta_{,i} = 0$. Necessary conditions for this to be the case are

$$\left.\frac{\partial\gamma}{\partial\theta_{,i}}\right|_E = 0 \;, \text{ and} \qquad -\left.\frac{\partial^2\gamma}{\partial\theta_{,i}\partial\theta_{,j}}\right|_E \text{ is positive-semi definite ,} \tag{2.143}$$

or

$$q_i|_E = 0 \;, \text{ and} \qquad \left.\frac{\partial q_{(i}}{\partial\theta_{,j)}}\right|_E \text{ is negative-semi definite} \tag{2.144}$$

Hence, for an electrical insulator in thermostatic equilibrium the energy flux vector must necessarily vanish.

To see what consequences these restrictions impose, let us consider the case for which the constitutive relations are given in the form

$$C = \check{C}(F_{i\alpha}, E_i, B_i, \theta, \theta_{,i}, \mathcal{Q}) \;. \tag{2.145}$$

A constitutive relation for the energy flux vector that is objective under Euclidean transformations and depends on the above variables is of the form

$$q_i = F_{i\alpha}\check{q}_\alpha(C_{\beta\gamma}, E_\gamma, B_\gamma, \theta, \theta_{,\beta}, \mathcal{Q}) \;, \tag{2.146}$$

where E_γ, B_γ and $C_{\beta\gamma}$ have been defined in (1.89) and where

$$\theta_{,\alpha} := F_{i\alpha}\theta_{,i} \;. \tag{2.147}$$

A necessary and sufficient condition for the energy flux vector to vanish in thermostatic equilibrium is to write q_i as

(2.148) $$q_i = -F_{i\alpha}\check{\kappa}_{\alpha\beta}(C_{\gamma\delta},E_\gamma,B_\gamma,\theta,\theta_{,\gamma},\mathcal{Q})\theta_{,\beta}$$

This representation is suitable for solids to which we shall restrict ourselves henceforth. In an electrical insulator (2.148) is also the most general constitutive relation for the energy flux vector which vanishes in thermostatic equilibrium. For this form of the constitutive relation, the condition $(2.144)^2$ requires that the matrix

(2.149) $$A_{ij} := F_{(i\alpha}F_{j)\beta}\check{\kappa}_{\alpha\beta}\big|_E$$

is positive-semi definite.

In the theory of solids one often restricts oneself to small deformations and small deviations from thermostatic equilibrium. In such special cases, in the constitutive equations all terms that are quadratic or of higher order in the above mentioned small quantities are neglected. Since the constitutive relation for q_i is already explicitly linear in $\theta_{,i}$, this requires that $\check{\kappa}_{\alpha\beta}$ is independent of $C_{\alpha\beta}$ and $\theta_{,\alpha}$ and that $F_{i\alpha}$ may be replaced by $\delta_{i\alpha}$. For such a case (2.149) reduced to

(2.150) $$A_{ij} = \check{\kappa}_{(ij)}\big|_E \quad \text{is positive-semi definite .}$$

Nothing can be concluded from the equilibrium condition about the skew-symmetric part of $\check{\kappa}_{ij}\big|_E$. It is well-known that as a consequence of the Onsager relations, one usually requires that

(2.151) $$\check{\kappa}_{[ij]}\big|_E = 0 \; .$$

Finally, in this linear approximation it is not difficult to show that

(2.152) $$q_i = -\check{\kappa}_{ij}(E_k,B_k,\theta,\mathcal{Q})\theta_{,j} \; .$$

Here, as a consequence of the Onsager relations, we require

(2.153) $$\check{\kappa}_{[ij]} = 0 \; ,$$

and, furthermore,

(2.154) $$\check{\kappa}_{ij}\theta_{,i}\theta_{,j} \geq 0 \; .$$

In this drastically simplified version the constitutive relation for the energy flux vector is known as the Fourier law of heat conduction.

b) In an electrical conductor one may write

$$
(2.155)\qquad
\begin{aligned}
q_i &= \overset{\vee}{q}_i(F_{j\alpha},E_j,B_j,\theta,\theta_{,j},\mathcal{Q}) \, ,\\
J_i &= \overset{\vee}{J}_i(F_{j\alpha},E_j,B_j,\theta,\theta_{,j},\mathcal{Q}) \, ,
\end{aligned}
$$

and then γ is a function of

$$F_{i\alpha},E_i,B_i,\theta,\theta_{,i} \text{ and } \mathcal{Q} \, .$$

Since the electrical insulator as a special case of an electrical conductor with zero conductivity was considered on the previous pages already, we exclude it here. For such a case, the second of the constitutive relations (2.155) may be inverted in the sense that

$$
(2.156)\qquad E_i = \overset{\nabla}{E}_i(F_{j\alpha},J_j,B_j,\theta,\theta_{,j},\mathcal{Q}) \, .
$$

If this relation is substituted into (2.155)[1], we may regard γ to be a function of

$$F_{i\alpha},J_i,B_i,\theta,\theta_{,i} \text{ and } \mathcal{Q} \, ,$$

so that (2.139) becomes

$$
(2.157)\qquad \gamma = J_i\overset{\nabla}{E}_i - \frac{\theta_{,i}}{\theta}\overset{\nabla}{q}_i = \overset{\nabla}{\gamma}(F_{i\alpha},J_i,B_i,\theta,\theta_{,i},\mathcal{Q}) \geq 0 \, .
$$

Since then

$$
(2.158)\qquad \gamma\big|_E = \overset{\nabla}{\gamma}(F_{i\alpha}\big|_E,0,B_i\big|_E,\theta\big|_E,0,\mathcal{Q}\big|_E) = 0 \, ,
$$

we must necessarily have

$$
(2.159)\qquad \frac{\partial\overset{\nabla}{\gamma}}{\partial J_i}\bigg|_E = 0, \qquad \frac{\partial\overset{\nabla}{\gamma}}{\partial\theta_{,i}}\bigg|_E = 0 \, ,
$$

and

$$
(2.160)\qquad \left.\begin{pmatrix}
\dfrac{\partial^2\overset{\nabla}{\gamma}}{\partial J_i\partial J_j} & \dfrac{\partial^2\overset{\nabla}{\gamma}}{\partial J_i\partial\theta_{,j}}\\[2ex]
\dfrac{\partial^2\overset{\nabla}{\gamma}}{\partial J_i\partial\theta_{,j}} & \dfrac{\partial^2\overset{\nabla}{\gamma}}{\partial\theta_{,i}\partial\theta_{,j}}
\end{pmatrix}\right|_E \quad \text{is positive-semi definite} \, .
$$

Of necessity then

$$(2.161)\qquad \overset{\nabla}{E}_i\Big|_E = 0, \qquad \text{and} \qquad \overset{\nabla}{q}_i\Big|_E = 0 .$$

Therefore, the electromotive intensity and the heat flux vector vanish in thermostatic equilibrium. Hence, we could have defined thermostatic equilibrium also as a time-independent process for which E_i and $\theta_{,i}$ vanish. Then, the above relations would read

$$(2.162)\qquad \frac{\partial\check{\gamma}}{\partial E_i}\Big|_E = 0, \qquad \frac{\partial\check{\gamma}}{\partial\theta_{,i}}\Big|_E = 0 ,$$

and

$$(2.163)\qquad \begin{pmatrix} \dfrac{\partial^2\check{\gamma}}{\partial E_i\partial E_j} & \dfrac{\partial^2\check{\gamma}}{\partial E_i\partial\theta_{,j}} \\ \dfrac{\partial^2\check{\gamma}}{\partial E_i\partial\theta_{,j}} & \dfrac{\partial^2\check{\gamma}}{\partial\theta_{,i}\partial\theta_{,j}} \end{pmatrix}\Bigg|_E \quad \text{is positive-semi definite ,}$$

and they yield

$$(2.164)\qquad \check{J}_i\Big|_E = 0, \qquad \check{q}_i\Big|_E = 0 ,$$

as well as

$$(2.165)\qquad \begin{pmatrix} \dfrac{\partial\check{J}_{(i}}{\partial E_{j)}} & \left(\dfrac{\partial\check{J}_i}{\partial\theta_{,j}} - \dfrac{1}{\theta}\dfrac{\partial\check{q}_j}{\partial E_i}\right) \\ \left(\dfrac{\partial\check{J}_i}{\partial\theta_{,j}} - \dfrac{1}{\theta}\dfrac{\partial\check{q}_j}{\partial E_i}\right) & -\dfrac{1}{\theta}\dfrac{\partial\check{q}_{(i}}{\partial\theta_{,j)}} \end{pmatrix}\Bigg|_E \quad \text{is positive-semi definite.}$$

The most general form of the constitutive equations (2.155) for a solid satisfying the principle of material frame indifference is

$$(2.166)\qquad \begin{aligned} q_i &= F_{i\alpha}\check{q}_\alpha(C_{\beta\gamma},E_\beta,B_\beta,\theta,\theta_{,\beta},\mathcal{Q}) , \\ J_i &= F_{i\alpha}\check{J}_\alpha(C_{\beta\gamma},E_\beta,B_\beta,\theta,\theta_{,\beta},\mathcal{Q}) \end{aligned}$$

When they are written in the form

$$(2.167)\qquad \begin{aligned} q_i &= F_{i\alpha}\{-\check{\kappa}_{\alpha\beta}\theta_{,\beta} + \check{\beta}^{(q)}_{\alpha\beta}E_\beta\} , \\ J_i &= F_{i\alpha}\{\check{\beta}^{(J)}_{\alpha\beta}\theta_{,\beta} + \check{\sigma}_{\alpha\beta}E_\beta\} , \end{aligned}$$

they automatically satisfy the equilibrium conditions (2.164). Moreover, in order to satisfy (2.165),

$$\text{(2.168)}\qquad A := \begin{pmatrix} \overset{\vee}{\sigma}_{\alpha\beta}F_{(i\alpha}F_{j)\beta} & (\overset{\vee}{\beta}{}^{(J)}_{\alpha\beta} - \frac{1}{\theta}\overset{\vee}{\beta}{}^{(q)}_{\beta\alpha})F_{i\alpha}F_{j\beta} \\ (\overset{\vee}{\beta}{}^{(J)}_{\alpha\beta} - \frac{1}{\theta}\overset{\vee}{\beta}{}^{(q)}_{\beta\alpha})F_{i\alpha}F_{j\beta} & \frac{1}{\theta}\overset{\vee}{\kappa}_{\alpha\beta}F_{(i\alpha}F_{j)\beta} \end{pmatrix}$$

must be positive-semi definite. Necessary conditions (that are not sufficient however) for this to be satisfied are

$$\text{(2.169)}\qquad \overset{\vee}{\sigma}_{\alpha\beta}F_{(i\alpha}F_{j)\beta} \text{ and } \frac{1}{\theta}\overset{\vee}{\kappa}_{\alpha\beta}F_{(i\alpha}F_{j)\beta} \text{ are positive-semi definite .}$$

More interesting than the general case is again the linearized version of the theory. Assuming small deformations and small deviations from thermostatic equilibrium, it may be justified to postulate the constitutive relations for q_i and J_i, (2.167), to be linear in $\theta_{,i}$ and E_i and independent of the deformations. In this case, the coefficient functions $\overset{\vee}{\kappa}_{\alpha\beta}$, $\overset{\vee}{\beta}{}^{(J)}_{\alpha\beta}$, $\overset{\vee}{\beta}{}^{(q)}_{\alpha\beta}$ and $\overset{\vee}{\sigma}_{\alpha\beta}$ are independent of $C_{\alpha\beta}$, E_α and $\theta_{,\alpha}$, and $F_{i\alpha}$ may be replaced by $\delta_{i\alpha}$. Under these simplified restrictions

$$\text{(2.170)}\qquad A \approx \begin{pmatrix} \overset{\vee}{\sigma}_{(ij)} & (\overset{\vee}{\beta}{}^{(J)}_{ij} - \frac{1}{\theta}\overset{\vee}{\beta}{}^{(q)}_{ji}) \\ (\overset{\vee}{\beta}{}^{(J)}_{ij} - \frac{1}{\theta}\overset{\vee}{\beta}{}^{(q)}_{ji}) & \frac{1}{\theta}\overset{\vee}{\kappa}_{(ij)} \end{pmatrix} ,$$

must be positive-semi definite. Of necessity, then $\overset{\vee}{\sigma}_{(ij)}$ is positive definite (because $\overset{\vee}{\sigma} \equiv 0$ is excluded) and $\overset{\vee}{\kappa}_{(ij)}$ is positive semi definite. The Onsager relations require here that

$$\text{(2.171)}\qquad \begin{aligned} &\overset{\vee}{\sigma}_{[ij]} = 0, \qquad \overset{\vee}{\kappa}_{[ij]} = 0 , \\ &\overset{\vee}{\beta}{}^{(J)}_{ij} = \frac{1}{\theta}\,\beta^{(q)}_{ji} =: \frac{1}{\theta}\,\beta_{ji} . \end{aligned}$$

Of course, $\overset{\vee}{\sigma}_{ij}$, $\overset{\vee}{\kappa}_{ij}$ and $\overset{\vee}{\beta}_{ij}$ are still functions of B_k, θ and $\mathcal{Q}$. When linearized and with the use of (2.171), the constitutive equations (2.167) become

$$\text{(2.172)}\qquad \begin{aligned} q_i &= -\overset{\vee}{\kappa}_{ij}(B_k,\theta,\mathcal{Q})\theta_{,j} + \overset{\vee}{\beta}_{ij}(B_k,\theta,\mathcal{Q})E_j , \\ J_i &= \overset{\vee}{\beta}_{ji}(B_k,\theta,\mathcal{Q})\frac{\theta_{,j}}{\theta} + \overset{\vee}{\sigma}_{ij}(B_k,\theta,\mathcal{Q})E_j . \end{aligned}$$

These equations are known as Fourier's law of heat conduction and Ohm's law of electrical conduction and in this form they appear in usual treatises on crystallography and solid state physics (e.g. [23], [24]). The coefficients $\check{\kappa}_{\alpha\beta}$ and $\check{\sigma}_{\alpha\beta}$ are termed the thermal and electrical conductivity, respectively. The coefficients $\check{\beta}_{\alpha\beta}$ are responsible for thermoelectric effects (e.g. Seebeck effect, Peltier heat, Thomson heat, cf. [23], Chapter 12 or [25], Chapter 23), and in magnetizable materials they also cause galvanomagnetic or thermomagnetic effects (e.g. Ettinghausen or Nernst effect). These effects are investigated in more detail than is possible within the scope of this monograph, in a series of papers by Pipkin and Rivlin, [26] and [27], and Borghesani and Morro, [28] and [29]. Pipkin and Rivlin derive general, nonlinear conduction laws for isotropic materials considering separately the effects of deformation on the electric current and the influences of the electric field, the magnetic induction and the temperature gradient on the constitutive equations for the electric current, the heat flux and the magnetic field strength, whereas Borghesani and Morro derive the fourth-order versions of Fourier's law and Ohm's law.

Finally we remark that for most purposes of this monograph the consequences implied by the Onsager relations will be of no relevance. In particular, the results of Chapter 3 are independent of such relations. When using the Onsager relations we shall in the following therefore explicitly state it.

In summary, we have given an exposure of five different descriptions of deformable polarizable and magnetizable continua. Each model describes in its own way the interactions between the electromagnetic fields and the thermoelastic body. The field equations consist of the following set of equations:

i) The Maxwell equations, which must be counted as seven independent equations (see the remark in Section 1.3.2).
ii) Five balance laws of mass, momentum and energy.
iii) Sixteen constitutive relations for entropy, two electromagnetic field vectors and the stress tensor (the latter given in a form such that the balance law of moment of momentum is satisfied identically).
iv) Six constitutive relations for the electric current and the energy flux vector.

These are 34 equations for the following unknowns: four electromagnetic field vectors (12), electric current and electric charge density (4), mass density (1), motion χ_i (3), stress (9), temperature (1), energy flux (3) and entropy (1).

These field equations are supplemented by jump (or boundary) conditions for the electromagnetic field variables, mass, momentum and energy of matter and field.

2.7 DISCUSSION

In the preceding Sections five different descriptions of deformable polarizable and magnetizable continua were treated. The interaction of the electromagnetic fields with matter was achieved by introducing two additional electromagnetic field vectors to the two basic fields occurring in vacuo. In the five formulations the choice of these field vectors was not unique, however. We did not emphasize the models which lie behind these descriptions, although each of the formulations can be based upon well-defined models. The Chu variables of electrodynamics for instance can be founded on a two-dipole model, and as long as one restricts oneself to the derivation of the Maxwell equations, such a model leads to unique answers. The model may bear its disadvantages insofar as it is handled non-relativistically, while the resulting equations are postulated to be relativistically correct equations. This does not change the basic fact that the resulting Maxwell equations are unique. When electromagnetic body force and energy supply are derived, however, the dipole model is no longer unique and two theories of magnetizable and polarizable bodies can be developed. We have seen this when presenting the models I and II.

One can also develop a theory on the basis that polarization is modeled by a dipole while magnetization is treated as an electric circuit. Dependent on the degree of complexity of derivation, different Maxwell equations emerge (statistical and Lorentz formulation), and also the expressions for the electromagnetic body force, body couple and energy supply are different in these derivations. The reader could therefore be misled by these derivations as he might conclude that some of the theories are superior to others, because the derivation of the equations resembles a more profound approach. This is not so, and our point of view is different, as we regard all formulations as equally sound as long as none has been proved to be superior to any other one. Of course this requires that we deemphasize the models behind the equations. This is why we do not even share the viewpoint of many physicists, who would reject the two-dipole description on the basis that magnetic monopoles have never been observed experimentally and that therefore magnetization must be electric circuits. Corroboration to all the above statements will be found in the next Chapter of this monograph. For instance, the statistical model and the Lorentz model (models IV and V) can be derived by methods of statistical physics, but dependent on which author's book one opens, either one appears to be an approximation of the other. The reason that in the most rigorous treatment the Lorentz formulation is not obtained is a matter of definition of the macroscopic electric and magnetic dipole moments in terms of statistical averages of microscopic quantities. Nontheless, Pao concludes in his reviewing article [18] that for magnetizable materials the validity of the Lorentz model is questionable (page 24 of [18]). Yet, we shall prove that the non-relativistic versions of polarizable or magnetizable continua in the models IV and V are equivalent. A similar situation also exists for a body force expression as suggested by Fano, Chu and Adler that agrees with our model II. Again according to Pao (page 82 of [18]), Penfield and Haus [2] have considered these force expressions incomplete. We shall prove that the models I and II are entirely equivalent. Needless to say that Penfield and Haus [2] and Pao and Hutter [4] have advocated for model I.

The above statements should make it clear then that the derivation of the models is not important and that the formulations should be unified by demonstrating their equivalence, rather than emphazising their differences. For this reason we have presented the five models per se and have studied their invariance properties quite extensively. This led to the approach of model III, in which electromagnetic body force, body couple and energy supply were derived from a general energy expression that is subjected to specific invariance requirements under the general Euclidean transformation group. In order to illustrate the above mentioned unification and because some authors seem to have objections against the approach of Section 2.3 (as they do not believe that this approach will lead to unique results) we have also given the energy laws which could serve as basis for the other models. For model I this was already done by Alblas [10]. This, together with the results of Chapter 3, in which all models are shown to be equivalent, demonstrates that there is a large amount of freedom in the choice of the specific terms occurring in such an energy law, all leading to equivalent results.

We found further that in formulations in which the stress tensor was symmetric the electromagnetic body force and energy supply are objective quantities with respect to the Euclidean transformation group. In formulations with non-symmetric stress body force and body couple turned out to be objective, while the energy supply term first had to be corrected by the power of working of the skew symmetric part of the stress. As a result, the energy equation remained invariant.

These transformation properties make it possible that the various formulations of electromagnetic interactions with thermoelastic bodies have a chance to be equivalent. Indeed, full equivalence cannot be achieved when corresponding quantities in different formulations do not transform alike under the Euclidean transformation group. Similarly, if approximations in the non-relativistic sense are performed, equivalence can only be attempted to be proved within such approximations.

This discussion would not be complete if we would not point out that other descriptions for field matter interactions often agree with our models I-V. We find it important to discuss these descriptions also. In order to narrow the number of formulations down to a reasonable size, we shall not discuss any quasistatic formulation, such as the one by Toupin [30], Alblas [16], Tiersten [31], [32], Pao and Yeh [34] and other (see e.g. Penfield and Haus [2]), with the exception of the monograph of Brown [8], however. The reason for this exception is that Brown already was aware of the non-uniqueness of the magnetoelastic stresses, and he stated explicitly that these stresses are only then completely determined once the total system of momentum and moment of momentum equations, boundary conditions and constitutive equations are given. For a magnetostatic theory, Brown introduced four different stress tensors, namely: (see Section 5.6 of [8])

i) The pole-model $\bar{t}_{ij}$ (eqs. (5.21)-(5.24)).
ii) The Ampèrian current model $\bar{t}'_{ij}$ (eqs. (5.21')-(5.24')).
iii) The Maxwell model I t_{ij} (eqs. (5.29)-(5.32)).
iv) The Maxwell model II t'_{ij} (eqs. (5.29')-(5.32')).

Comparing these stress models with the ones introduced in this Chapter, we see that when the latter are simplified to the magnetostatic case, the following relations between our stresses and those of Brown hold:

$$(2.173)\qquad \begin{aligned} {}^{I}t_{ij} &= {}^{III}t_{ij} = t_{ij}, \qquad & {}^{II}t_{ij} &= \bar{t}_{ij}, \\ {}^{IV}t_{ij} &= t'_{ij}, \qquad & {}^{V}t_{ij} &= \bar{t}'_{ij}. \end{aligned}$$

In comparing other author's work, any complications due to spin interactions will be left aside. This does not mean that spin interaction theories will not be compared here, but any contribution due to magnetic spin will be set to zero without further mentioning. Furthermore, fully relativistic theories will only be mentioned in connection with non-relativistic approximations.

First when the material is polarizable-only, model V is in full agreement with a description of a dynamical theory of elastic dielectrics as presented by Toupin [7], and indeed it must be so as our force and energy supply expressions agree with those of Toupin. On the other hand, Dixon and Eringen attempted to derive a theory of field-matter interaction using a statistical description (which we believe to be incomplete) [35]. They include in their derivation also macroscopic electric quadrupoles, but when these are neglected their Maxwell equations and body force and energy supply expressions are in agreement with model V. However, the jump condition of energy derived by Dixon and Eringen (their equation (5.16)) is not correct. Indeed, their global energy balance contains a volume source which becomes indefinitely large when a surface of discontinuity is approached. Hence, apart from these shortcomings in the derivation, their results are non-relativistically correct. Moreover, Eringen in collaboration with Grot [36], and Grot [37] present relativistic formulations of solids and fluids in the electromagnetic fields, in which they postulate an energy-momentum tensor of field-matter interaction which reduces to the force and energy expressions of model V when terms of $0(V^2/c^2)$ are neglected.

Of a quite different nature is Tiersten's approach to describe the interaction of the electromagnetic fields with deformable continua. He uses a mixture concept and describes the field-matter interactions by coupling a so called lattice-continuum with charge-, electronic-, ionic- and spin-continua, each bearing the notion of a particular physical effect. All work of Tiersten is based on such a mixture concept. In Ref [38] he and Tsai describe the interaction of the electromagnetic fields with heat conducting deformable electric insulators. In a simplified version of this theory, where only a charge continuum is interacting with the lattice continuum their body force, body couple and energy supply (see equations (3.48), (8.4) and (8.7) in Ref. [38]) agree with model IV. An explicit proof for this equivalence can also be found in Pao's review article, [18]. The same comments hold for the paper of Lorenzi and Tiersten, [39].

Still other methods of derivation are those in which the governing equations derive from an over-all energy balance law which is postulated to be invariant under Euclidean transformations. The technique, first introduced by Green and Rivlin [15] for non-relativistic multipolar theories, was applied to describe deformable polarizable and magnetizable continua by Alblas [10], Parkus [40] and van de Ven [11]. We have shown in this monograph that all formulations could have been derived from such an over-all energy balance law. When we compare the work of Alblas [10] (who uses the Chu variables) with our model I complete agreement is found. As concerns the work of Parkus [40], he treats magnetizable and polarizable materials separately and when dealing with magnetizable materials restricts himself to quasistatic processes. His treatment of dielectrics on the other hand agrees with that of Toupin. Parkus uses Hamilton's principle and the same is done by Vlasov and Ishmukhametov [41], who arrive at results which agree with our model III. We could have added other electromagnetic models, if we had wished to do so, and for reasons of completeness these models should also be mentioned. Fano, Chu and Adler [1] and Penfield and Haus [2] also present the so called Ampèrean description, as do Hutter and Pao [42] in a paper dealing with magnetizable elastic solids with thermal and electrical conduction. The Ampèrean formulation (in which magnetization is modeled by a relativistic electric circuit) can easily be transformed into the Lorentz- or the statistical formulation and when this is done, it is found that the body force expression agrees with the one presented in model IV. The expressions for the energy supply and the body couple, when restricted to the non-relativistic approximation, fully agree with the ones of model IV.

Maugin on the other hand, partly in collaboration with Eringen and Collet, [43]-[48] uses the principle of virtual power to derive the basic equations for several different theories of magnetoelastic interactions. The interest of these authors is generally more limited, as they treat specialized subjects, such as spin relaxation and surface effects, all in the quasi-static approximation. In [47] (Collet and Maugin), and in [48] (Maugin) they treat dynamic processes in a non-relativistic approximation. Although there is a slight difference between our non-relativistic approximation and the one of them, which can be traced back to the use of Gaussian units as opposed to MKSA-units, the results of [47] and [48] when brought to our non-relativistic approximation completely agree with model IV. Maugin and Eringen have also presented a fully relativistic treatment in [49].

Finally, starting from a four-dimensional relativistic formulation, Boulanger and Mayné [50] derive expressions for the electromagnetic body force and energy supply. In cooporation with van Geen, they apply their results in [51] for the investigation of magnetooptical, electrooptical and photoelastic effects in elastic polarizable and magnetizable isotropic media. When comparing the relations for momenta and energy following from Eq. (14) of [50] or the balance laws (4)-(6) of [51], no immediate equivalence with one of our models is found. However, their balance laws can easily be related to those of, for instance, model IV by taking in the equations (4)-(6) of [51]

$$\rho\varepsilon = {}^{IV}(\rho U) + B_i M_i \ ,$$

and

$$\sigma_{ij} = {}^{IV}t_{ij} + E_i P_j + B_i M_j - \delta_{ij} B_k M_k \ .$$

(In this context it should be noted that in Eq. (16) of [50] two terms are missing, which should follow from (14), and which reappear in Eq. (4) of [51].)

In conclusion we might justly state that the models I to V embrace within the non-relativistic approximation the description of dynamic theories of polarizable and magnetizable materials known to date. There are formulations simpler than the ones presented above, but these formulations aim at describing more restrictive situations such as static and quasistatic processes. There are also more complicated descriptions, but those include additional phenomena, as for instance spin interaction, polarization gradient effects and the like. These phenomena are outside the scope of this monograph.

Since the completion of this Chapter a series of articles did appear which we would have taken into account if being aware of them. To mention are works by Alblas [52], Maugin and Eringen [53], Prechtl [54] and Romano [55]. Alblas presents a general exposition of electro- and magnetoelasticity with special topics such as electro- and magnetostatics including constitutive equations and linearization procedures and magnetoelastic stability. His formulation is that of Chu. Maugin and Eringen formulate a theory of magnetoelastic interaction including electric quadrupoles. Their approach is similar to that of Dixon and Eringen [35] and they seem to correct the erroneous jump condition in this work. When discarding the electric quadrupoles they arrive at body force expressions identical to those of our model IV. Prechtl, on the other hand presents a relativistic Chu-formulation; he reduces it to a three dimensional form, which in the non-relativistic approximation is equivalent to our model II. Unfortunately, Prechtl does not present constitutive relations, so that a full comparison is not achieved. Romano gives a quasi static magnetoelastic Chu-formulation.

REFERENCES

[1] Fano, R.M., L.C. Chu and R.B. Adler, *Electromagnetic Fields, Energy and Forces*, John Wiley & Sons, Inc., New York, 1960 (Reprinted by the M.I.T. Press).

[2] Penfield, P. and H.A. Haus, *Electrodynamics of Moving Media*, The M.I.T. Press, Cambridge, Massachusettes, 1967.

[3] Truesdell, C. and R.A. Toupin, *The Classical Field Theories*, Encyclopedia of Physics, Vol. III/1, ed. S. Flügge, Springer-Verlag, Berlin, 1960.

[4] Pao, Y.H. and K. Hutter, *Electrodynamics of Moving Elastic Solids and Viscous Fluids*, Proc. I.E.E.E. 63 (1975), 1011-1021.

[5] Coleman, B.D. and W. Noll, *The Thermodynamics of Elastic Materials with Heat Conduction and Viscosity*, Arch. Rat. Mech. Anal. 13 (1963), 167-178.

[6] Carathéodory, C., *Untersuchungen über die Grundlagen der Thermodynamik*. Math. Annalen 67 (1909), 355-386.

[7] Toupin, R.A., *A Dynamical Theory of Elastic Dielectrics*, Int. J. Eng. Sc. 1 (1963), 101-126.

[8] Brown, Jr., W.F., *Magnetoelastic Interactions*, Springer Tracts in Natural Philosophy, Vol. 9, Springer-Verlag, Berlin, 1966.

[9] Hutter, K., *A Thermodynamic Theory of Fluids and Solids in the Electromagnetic Fields*, Arch. Rat. Mech. Anal., 64 (1977), 269-198.

[10] Alblas, J.B., *Electro-Magneto-Elasticity*, Topics in Applied Continuum Mechanics, eds. J.L. Zeman and F. Ziegler, 71-114, Springer-Verlag, Wien, 1974.

[11] Ven, A.A.F. van de, *Interaction of Electromagnetic and Elastic Fields in Solids*, Dr. of Science Thesis, University of Technology Eindhoven, the Netherlands, 1975.

[12] De Groot, S.R. and L.G. Suttorp, *Foundations of Electrodynamics*, North-Holland Publishing Co., Amsterdam, 1972.

[13] Müller, I., *Thermodynamik*, Bertelsmann Universitätsverlag, Düsseldorf, Germany, 1973.

[14] Hutter, K., *On Thermodynamics and Thermostatics of Vivcous Thermoelastic Solids in the Electromagnetic Fields. A Lagrangian Formulation*, Arch. Rat. Mech. Anal. 58 (1975), 339-368.

[15] Green, A.E. and R.S. Rivlin, *Multipolar Continuum Mechanics*, Arch. Rat. Mech. Anal. 17 (1964), 113-147.

[16] Alblas, J.B., *Continuum Mechanics of Media with Internal Structure*, Instituto Nazionale di Alta Matematica, Symposia Mathematica, Vol. I, 229-251, London, Academic Press, Oderisi, Gubbio, 1969.

[17] Møller, C., *The Theory of Relativity*, Oxford University Press, London, 1972.

[18] Pao, Y.H., *Electromagnetic Forces in Deformable Media*, Mechanics Today, Vol. 4, ed. S. Nemat-Nasser, Pergamon Press Inc., New York, 1978.

[19] Rosenfeldt, L., *Theory of Electrons*, North-Holland, Amsterdam, 1951.

[20] Feynman, R.P., R.B. Leighton and M. Sands, *The Feynman Lectures on Physics*, Vol. 2, Addison-Wesley, Reading, Massachusettes, 1964.

[21] Liu, I.S. and I. Müller, *On the Thermodynamics and Thermostatics of Fluids in Electromagnetic Fields*, Arch. Rat. Mech. Anal. 46 (1972), 149-176.

[22] Benach, R. and I. Müller, *Thermodynamics and the Description of Magnetizable Dielectric Mixtures of Fluids*, Arch. Rat. Mech. Anal. 53 (1974), 312-346.

[23] Nye, J.F., *Physical Properties of Crystals*, Oxford University Press, Oxford, England, 1957.

[24] Born, M. and K. Huang, *Dynamical Theory of Crystal Lattices*, Oxford Press, 1954.

[25] Wannier, G.H., *Statistical Physics*, John Wiley & Sons, Inc., New York, London, Sydney, 1966.

[26] Pipkin, A.S. and R.S. Rivlin, *Electrical Conduction in Deformed Isotropic Materials*, J. Math. Phys. 1 (1960), 127-130.

[27] Pipkin, A.S. and R.S. Rivlin, *Galvanomagnetic and Thermomagnetic Effects in Isotropic Materials*, J. Math. Phys. 1 (1960), 542-546.

[28] Borghesani, R. and A. Morro, *Thermodynamics and Isotropy in Thermal and Electrical Conduction*, Meccanica, 9 (1974), 63-69.

[29] Borghesani, R. and A. Morro, *Thermodynamic Restrictions on Thermoelastic, Thermomagnetic and Galvanomagnetic Coefficients*, Meccanica, 9 (1974), 157-161.

[30] Toupin, R.A., *The Elastic Dielectric*, J. Rational Mechanics and Analysis 5 (1956), 850-915.

[31] Tiersten, H.F., *Coupled Magnetomechanical Equations for Magnetically Saturated Insulators*, J. Math. Phys. 5 (1964), 1298-1318.

[32] Tiersten, H.F., *On the Nonlinear Equations of Thermoelectroelasticity*, Int. J. Eng. Sc. 9 (1971), 587-603.

[33] Jordan, N.F. and A.C. Eringen, *On the Static Nonlinear Theory of Electromagnetic Thermoelastic Solids*, Int. J. Eng. Sc. 2 (1964), 59-114.

[34] Pao, Y.H. and C.S. Yeh, *A Linear Theory of Soft Ferromagnetic Elastic Solids*, Int. J. Eng. Sc. 11 (1973), 415-436.

[35] Dixon, R.C. and A.C. Eringen, *A Dynamical Theory of Polar Elastic Dielectrics*, Int. J. Eng. Sc. 3 (1965), 359-398.

[36] Grot, R.A. and A.C. Eringen, *Relativistic Continuum Mechanics*, Part I, Mechanics and Thermodynamics; Part II, Electromagnetic Interactions with Matter, Int. J. Eng. Sc. 4 (1966), 611-638, 639-670.

[37] Grot, R.A., *Relativistic Continuum Theory for the Interaction of Electromagnetic Fields with Deformable Bodies*, J. Math. Phys. 11 (1970), 109-113.

[38] Tiersten, H.F. and C.F. Tsai, *On the Interaction of the Electromagnetic Field with Heat Conducting Deformable Insulators*, J. Math. Phys. 13 (1972), 361-378.

[39] De Lorenzi, H.G. and H.F. Tiersten, *On the Interaction of the Electromagnetic Field with HeatConducting Deformable Semiconductors*, J. Math. Phys. 16 (1975), 938-957.

[40] Parkus, H., *Variational Principles in Thermo- and Magneto-Elasticity*, CISM, Courses and Lectures-No. 58, Udine 1970, Springer-Verlag, Wien, 1972.

[41] Vlasov, K.B. and B.Kh. Ishmukhametov, *Equations of Motion and State for Magnetoelastic Media*, Soviet Phys. JETP 19 (1964), 142-148.

[42] Hutter, K. and Y.H. Pao, *A Dynamic Theory for Magnetizable Elastic Solids with Thermal and Electrical Conduction*, J. of Elasticity, 4 (1974), 89-114.

[43] Maugin, G.A. and C.A. Eringen, *Deformable Magnetizable Saturated Media*, Part I, Field Equations; Part II, Constitutive Theory, J. Math. Phys. 13 (1972), 143-155, 1334-1347.

[44] Maugin, G.A., *A Continuum Theory of Deformable Ferrimagnetic Bodies*, Part I, Field Equations; Part II, Thermodynamics, Constitutive Theory, J. Math. Phys. 17 (1976), 1727-1738, 1739-1751.

[45] Collet, B. and G.A. Maugin, *Couplage Magnetoelastique de Surface dans les Materiaux Ferromagnétiques*, C.R. Acad. Sc. Paris, Série A, 280 (1975), 1641-1644.

[46] Maugin, G.A., *On the Spin Relaxation in Deformable Ferromagnets*, Physica A81 (1975), 454-468.

[47] Collet, B. and G.A. Maugin, Part I, *Sur l'Electrodynamique des Milieux Continus avec Interactions*; Part II, *Thermodynamique des Milieux Continus Electromagnétique avec Interactions*, C.R. Acad. Sc. Paris, Série B, 279 (1974), 379-382, 439-442.

[48] Maugin, G.A., *Deformable Dielectrics, Field Equations for a Dielectric Made of Several Molecular Species*, Arch. of Mech., 28 (1976), 679-692.

[49] Maugin, G.A. and C.A. Eringen, *Polarized Elastic Materials with Electronic Spin- A Relativistic Approach*, J. Math. Phys. 13 (1972), 1777-1788.

[50] Boulanger, P. and G. Mayné, *Etude Théorique de l'Interaction d'un Champ Electromagnétique et d'un Continu non Conducteur Polarisable et Magnétisable*, C.R. Acad. Sc. Paris, Série A, 274 (1972), 591-594.

[51] Boulanger, P., G. Mayne, and R. van Geen, *Magnetooptical, Electrooptical and Photoelastic Effects in an Elastic Polarizable and Magnetizable Isotropic Continuum*, Int. J. Solids Structures, 9 (1973), 1439-1464.

[52] Alblas, J.B., *General Theory of Electro- and Magneto-Elasticity*, in Electromagnetic Interactions in Elastic Solids, ed. by H. Parkus, Springer, Wien, 1978.

[53] Maugin, G.A. & A.C. Eringen, *On the Equations of the Electrodynamics of Deformable Bodies of Finite Extent*, J. de Mécanique 16 (1977), 101-147.

[54] Prechtl, A., *On the Electrodynamics of Deformable Media*, Acta Mechanica 28 (1977), 255-294.

[55] Romano, A., *A Macroscopic Nonlinear Theory of Magnetothermoelastic Continua*, Arch. Rat. Mech. Anal. 65 (1977), 1-24.

3. EQUIVALENCE OF THE MODELS

3.1 PRELIMINARY REMARKS

In Chapter 2 various descriptions of electromechanical interaction models were laid down, but no attempt was made to search for interrelations amongst these models. In this Chapter we perform the first steps toward a proof or disproof of the equivalence of these models. Specifically, we are going to present the conditions that need to be satisfied in order to render two formulations equivalent.

To begin with, we should point out that two models are called equivalent provided they deliver the same results for physically observable quantities in any arbitrary initial-boundary-value problem. By observable (measurable) quantities we mean hereby all those physical quantities that can be measured uniquely by two different independent observers. All kinematical quantities that are derivable from the motion $\chi_i(\underline{X},t)$ are measurable in principle. Regarding electromagnetic field quantities we take the viewpoint that they are not measurable except in vacuo, where they can be observed by measuring the force excerted on a test charge. Finally, (empirical) temperature is also regarded as being measurable, because, as is known from thermodynamics, it is a measure for the hotness of a body. The instruments to measure temperature are thermometers. They relate our sensation of the hotness of a body to a physical quantity, say pressure or volume, and the latter are measurable.

Differences in the various electromagnetic models appear in the body force, body couple and energy supply of electromagnetic origin, but as is well-known, such differences may be compensated in the stress tensor, the internal energy and the energy flux. Except for the latter, these quantities are determined by a thermodynamic potential (the Helmholtz free energy or its Legendre transformations). In other words, equivalence of two electromechanical interaction models depends to a large extent on the fact that the free energies of the two formulations can be determined so as to render stress, entropy and electromagnetic field variables compatible with the above mentioned compensation between stress and body force, etc. Equivalence of two theories is therefore a thermodynamic requirement.

Practically, two formulations are equivalent if upon transforming one into the other, not only the field equations but also the initial conditions and jump conditions are alike. Here in this Chapter we investigate the equivalence of the models described in the preceding Chapter. In particular, the conditions under which this equivalence is achieved will be formulated. Needless to say that all comparison will be made to within the order of the non-relativistic approximation.

3.2 COMPARISON OF THE MODELS I AND II

Here we discuss the two-dipole models, and we shall for this purpose heavily rely upon the results presented in Section 2.2.
To begin with, note that both models are based on the same set of electromagnetic field variables and therefore obey the same set of Maxwell equations (2.3). Differences do occur in the electromagnetic body force, body couple and energy supply, which can most easily be identified by comparing the expressions according to (2.7) and (2.48). From $(2.7)^1$ and $(2.48)^1$ we read off that

$$(3.1)\qquad {}^{II}(\rho F_i^e) = {}^{I}(\rho F_i^e) - (P_j E_i + \mu_o M_j H_i)_{,j} \ .$$

We conclude that the balance laws of momentum $(2.9)^2$ and $(2.49)^2$ are identical provided that

$$(3.2)\qquad {}^{II}t_{ij} = {}^{I}t_{ij} + E_i P_j + \mu_o H_i M_j \ .$$

In order to compare the balance laws of moment of momentum and energy, we must compare

$$\rho L_{ij}^e - t_{[ij]}, \qquad \text{and} \qquad \rho r^e + t_{ij}\dot{x}_{i,j} \ ,$$

respectively.
Using ρL_{ij}^e as given in $(2.7)^2$ and $(2.48)^2$ and ρr^e according to (2.12) and $(2.48)^3$ it is immediately seen that

$$(3.3)\qquad {}^{I}(\rho L_{ij}^e - t_{[ij]}) = {}^{II}(\rho L_{ij}^e - t_{[ij]}) \ ,$$

and

$$(3.4)\qquad {}^{I}(\rho r^e + t_{ij}\dot{x}_{i,j}) = {}^{II}(\rho r^e + t_{ij}\dot{x}_{i,j}) \ ,$$

provided that the stresses are related by (3.2). Hence, with the conditions (3.2), the balance laws of model I and model II have been proved to be equivalent.

It remains to compare the jump conditions. Since the Maxwell equations are the same, there cannot be a difference in the jumps of the electromagnetic field variables. On the other hand, as can immediately be seen from (2.21), (2.22) and (2.51), the jump conditions for momentum and energy of matter and fields are identical provided that the stress tensors are related according to (3.2).

The above comparison makes no use of thermodynamic arguments. Otherwise stated, equivalence of the two theories is guaranteed only if the relations (3.2) hold. That such relations indeed can hold follows from thermodynamic arguments, and they are, although trivial in this case, not obvious in general. In order to invoke these conditions, we must compare theories based on the same constitutive postulates, and, of course, in this case we must choose in both models the same set of independent constitutive variables. Explicitly we have done this for case a):

$$C = \hat{C}(C_{\alpha\beta}, P_\alpha, M_\alpha, \theta, \theta_{,i}, \mathcal{Q}) \ ,$$

and the results are listed as (2.31) and (2.56). It follows from these that

$$^{I}t_{ij} = 2\rho \frac{\partial\, ^{I}\hat{\Psi}}{\partial C_{\alpha\beta}} F_{i\alpha} F_{j\beta} + P_i \mathcal{E}_j + \mu_o \mathcal{M}_i \mathcal{H}_j \ ,$$

and that

$$^{II}t_{ij} = 2\rho \frac{\partial\, ^{II}\hat{\Psi}}{\partial C_{\alpha\beta}} F_{i\alpha} F_{j\beta} + 2P_{(i}\mathcal{E}_{j)} + 2\mu_o \mathcal{M}_{(i}\mathcal{H}_{j)} \ ,$$

and it is a straightforward matter to show that (3.2) is satisfied if we choose

$$^{II}\hat{\Psi} = {}^{I}\hat{\Psi} \ , \tag{3.5}$$

or

$$^{II}U = {}^{I}U \ . \tag{3.6}$$

With this choice, all the other dependent constitutive quantities that derive from the free energy, e.g. entropy, electromotive intensity, and magnetomotive intensity are guaranteed to be the same. It thus only remains to mention, that in order to obtain full agreement the energy flux vectors must also be chosen to be the same, i.e.

$$^{II}q_i = {}^{I}q_i \ . \tag{3.7}$$

The reader can readily prove the above statements to be correct also for all other constitutive theories that are obtained from the above one by merely interchanging some of the dependent and independent variables.
We therefore have proved the

Proposition. *Within the constitutive class of thermoelastic polarizable and magnetizable materials the two-dipole models I and II are equivalent, provided that the free energies and the energy flux vectors are the same functions of their independent variables.*

3.3 COMPARISON OF THE MODELS I AND III

Before comparing the models I and III one should realize that in model I electromagnetic fields according to the Chu-model are used, whereas in model III the Minkowski-fields are used. As stated by Penfield and Haus, [1], Ch. 7, and as follows from the relations (2.1) and (2.59), these fields are related in the following way

$$E_i^M = E_i^C - \mu_o e_{ijk}\dot{x}_j M_k^C, \qquad D_i^M = \varepsilon_o E_i^C + P_i^C ,$$
(3.8)
$$H_i^M = H_i^C + e_{ijk}\dot{x}_j P_k^C, \qquad B_i^M = \mu_o H_i^C + \mu_o M_i^C .$$

According to the definitions (2.61) for P_i^M and M_i^M we also have

$$P_i^M = P_i^C + \frac{1}{c^2} e_{ijk}\dot{x}_j M_k^C ,$$
(3.9)
$$M_i^M = M_i^C - e_{ijk}\dot{x}_j P_k^C .$$

Hence, to within the non-relativistic approximation,

$$(3.10) \qquad P_i^C = P_i^M = P_i, \qquad \text{and} \qquad M_i^C = M_i^M + e_{ijk}\dot{x}_j P_k^M = \mathcal{M}_i ,$$

where both, P_i and $\mathcal{M}_i$ are known to be objective under Euclidean transformations.
From the above relations it follows immediately that the rest frame fields $\mathcal{E}_i$ and $\mathcal{H}_i$ are equal in both formulations, and, moreover, that polarization and magnetization in the Chu-formulation are equal to the corresponding rest frame fields in the Minkowski-formulation.

Then, by comparing $(2.7)^{2,3}$ with the corresponding quantities as given by (2.81) and (2.82), respectively, it is obvious that the expressions for the body couple and energy supply of model I and model III are completely identical.

Since, in the non-relativistic approximation, in both models the electromagnetic momentum vector g_i is zero, it is most convenient to relate the respective electromagnetic body forces by comparing the expressions for the Maxwell stresses t^M_{ij} according to $(2.10)^1$ and $(2.83)^1$ With the aid of $(3.8)^{2,4}$, we may in (2.83) replace the Minkowski fields B^M_i and D^M_i by the Chu fields and can then show that to within terms containing a c^{-2}-factor

$$(3.11) \qquad {}^{III}t^M_{ij} = E_i D^M_j + H_i B^M_j - \frac{1}{2}\delta_{ij}(\varepsilon_o E_k E_k + \mu_o H_k H_k) =$$

$$= \varepsilon_o E^C_i E^C_j + \mu_o H^C_i H^C_j + E_i P^C_j + \mu_o H_i M^C_j +$$

$$- \frac{1}{2}\delta_{ij}(\varepsilon_o E^C_k E^C_k + \mu_o H^C_k H^C_k) = {}^{I}t^M_{ij}$$

Hence, we have demonstrated that the balance laws (2.17), $(2.9)^2$, $(2.9)^3$ and (2.80), (2.81), (2.82), are equivalent, provided that

$$(3.12) \qquad {}^{III}t_{ij} = {}^{I}t_{ij}, \qquad {}^{III}q_i = {}^{I}q_i, \qquad \text{and} \qquad {}^{III}U = {}^{I}U .$$

Under these conditions, and to within the non-relativistic approximation, also the jump conditions for momentum and energy of matter and fields of model I and model III are equal (cf. (2.21), (2.22) and $(2.84)^{6,7}$).

All the above is true, of course, again with the provision that constitutive relations show the dependent constitutive variables to be the same in the two theories. This can be seen to be true immediately by comparing the constitutive equations (2.37) with those of (2.89) (thus ${}^{I}\Psi = {}^{III}\Psi$). The reader may also show this to be correct, if any one of the dependent and independent variables are interchanged.

Thus, we have proved the following

Proposition. *Within the constitutive class of thermoelastic polarizable and magnetizable materials the two-dipole model I and the Maxwell-Minkowski model III are non-relativistically equivalent provided that the constitutive relations for the internal energies (or the free energies) and the energy flux vectors are the same functions of their independent variables.*

3.4 COMPARISON OF THE MODELS III AND IV

We begin with the observation that the Maxwell equations in the statistical formulation IV are written in a form which differs from that in the Minkowski formulation. However, if in the equations (2.60) the $\underline{D}^M$ and $\underline{H}^M$ fields are eliminated by means of (2.61) the resulting equations equal (2.105). Hence, $\underline{E}$, $\underline{P}$, $\underline{B}$ and $\underline{M}$ are identical in both formulations, and we do not need superscripts to distinguish them. Of course the same is then true also for the fields $\underline{\mathcal{E}}$ and $\underline{\mathcal{M}}$.

When with the use of (2.71) in the expression for the body force, (2.80), $\mathcal{H}_i$ and D_i are eliminated and thereby terms of the form $\varepsilon_o \underline{E} \times \underline{B}$ are neglected (as is justified in view of the corresponding remarks made in Section 1.6) one obtains

$$ {}^{III}(\rho F_i^e) = \mathcal{Q}\mathcal{E}_i + e_{ijk}\mathcal{J}_j B_k + P_j \mathcal{E}_{j,i} + \mathcal{M}_j B_{j,i} - \mu_o \mathcal{M}_j \mathcal{M}_{j,i} + $$
$$ + e_{ijk}(\overset{*}{P}_j B_k + P_j \overset{*}{B}_k) , $$

or

$$ {}^{III}(\rho F_i^e) = {}^{IV}(\rho F_i^e) - (\tfrac{1}{2}\, \mu_o \mathcal{M}_j \mathcal{M}_j)_{,i} , \tag{3.13} $$

as follows from a comparison with (2.106)[1].

Hence, the electromagnetic body forces in model III and IV differ in the term

$$ (\tfrac{1}{2}\, \mu_o \mathcal{M}_j \mathcal{M}_j)_{,i} . $$

However, this term is easy to handle, because substitution of the expression for the body force into the balance laws of momentum (2.80) and (2.108) reveals that the above term leads to a difference in the stress tensors ${}^{III}t_{ij}$ and ${}^{IV}t_{ij}$ given by

$$ {}^{IV}t_{ij} = {}^{III}t_{ij} - \tfrac{1}{2}\, \delta_{ij} \mu_o \mathcal{M}_k \mathcal{M}_k . \tag{3.14} $$

This relation does not change the antisymmetric part of the stresses, and one can see at once that the expressions for the body couples (2.81) and (2.106)[2] are identical, if the relation (2.71) is invoked.

As concernes the jump conditions of momentum, it is not difficult to see that (2.84)[6] is equal to (2.112)[6], once D_i and $\mathcal{H}_i$ are eliminated and, furthermore, equation (3.14) is substituted.

Because of (3.14), it may then be expected that the electromagnetic energy supplies ρr^e in the models III and IV will also differ. This is indeed the case and as can easily be seen from (2.82) and (2.106)[3] they are related by

$$(3.15) \qquad {}^{IV}(\rho r^e) = {}^{III}(\rho r^e) - \rho \frac{d}{dt}(B_i \frac{M_i}{\rho}) + \rho\mu_o M_i \frac{d}{dt}(\frac{M_i}{\rho}) .$$

Substituting the expressions (3.14) and (3.15) into the energy balance law for model IV, (2.108)[4], and comparing the resulting equation with (2.82), we see that the two energy balance laws are equivalent, if the pertinent internal energies are related by

$$(3.16) \qquad {}^{IV}U = {}^{III}U - B_i \frac{M_i}{\rho} + \frac{1}{2}\rho\mu_o \frac{M_i}{\rho}\frac{M_i}{\rho} ,$$

and provided that the heat fluxes ${}^{III}q_i$ and ${}^{IV}q_i$ are identical, i.e.

$$(3.17) \qquad {}^{IV}q_i = {}^{III}q_i .$$

Furthermore, the same holds true for the jump conditions of energy of matter and fields (2.84)[7] and (2.112)[7] as can be proved in the usual way.
What remains, is to find the conditions for which the relations (3.14)-(3.16) can be made compatible with the constitutive equations. For that purpose we first consider the free energies. From the equations (3.16) and (2.114) we obtain

$$(3.18) \qquad \begin{aligned} {}^{IV}\psi &= {}^{IV}U - {}^{IV}\eta\theta = {}^{III}U - {}^{III}\eta\theta - B_i \frac{M_i}{\rho} + \frac{1}{2}\rho\mu_o \frac{M_i}{\rho}\frac{M_i}{\rho} = \\ &= {}^{III}\Psi + P_\alpha E_\beta C^{-1}_{\alpha\beta} - B_\alpha M_\beta C^{-1}_{\alpha\beta} + \frac{\rho}{2\mu_o} M_\alpha M_\beta C^{-1}_{\alpha\beta} , \end{aligned}$$

where according to (2.87) and (2.88)

$$(3.19) \qquad {}^{III}\Psi = \tilde{\Psi}(C_{\alpha\beta}, E_\alpha, M_\alpha, \theta) .$$

Here, an important point must be mentioned, namely the fact that the free energy function ${}^{IV}\Psi$ on the far left in (3.18) is a function of $C_{\alpha\beta}$, P_α, B_α and θ. That this must be so can immediately be seen from (2.108)[4] and the definition of ${}^{IV}\Psi$ in (2.114). On the other hand, ${}^{III}\Psi$ as given in (3.19) is a function of $C_{\alpha\beta}, E_\alpha, M_\alpha, \theta$, and, similarly, so must be the expression on the right hand side of (3.18). If $\tilde{\Psi}$ is used as free energy functional in a constitutive theory of model IV, then instead of (2.116) we obtain

$$(3.20) \qquad \eta = -\frac{\partial\tilde{\Psi}}{\partial\theta} , \quad P_i = -\rho \frac{\partial\tilde{\Psi}}{\partial E_\alpha} F_{i\alpha} , \quad H_i = \frac{\partial\tilde{\Psi}}{\partial M_\alpha} F_{i\alpha} \operatorname{sgn} J ,$$

(over)

and

$$ {}^{IV}t_{ij} = 2\rho \frac{\partial\tilde{\Psi}}{\partial C_{\alpha\beta}} F_{i\alpha}F_{j\beta} - E_iP_j + \mu_o M_i H_j - \frac{1}{2}\delta_{ij}\mu_o M_k M_k , $$

in the derivation of which use has also been made of $(2.71)^2$ and of the relations

$$ (3.21) \qquad 2\frac{\partial\rho}{\partial C_{\alpha\beta}} F_{i\alpha}F_{j\beta} = \frac{\partial\rho}{\partial F_{i\alpha}} F_{j\alpha} = -\rho\delta_{ij} , $$

and

$$ (3.22) \qquad \frac{\partial C^{-1}_{\gamma\delta}}{\partial C_{\alpha\beta}} F_{i\alpha}F_{j\beta} = -\frac{1}{2}(F^{-1}_{\gamma i}F^{-1}_{\delta j} + F^{-1}_{\delta i}F^{-1}_{\gamma j}) . $$

We note that the relations (3.20) are not only compatible with (2.89) but also with (3.14); needless to say once more that in all those comparisons the non-relativistic approximation is employed. Moreover, the above comparison has been made for one set of dependent and independent constitutive variables and can of course also be repeated for all other possibilities. In view of the relation (3.16) this can be done easily. We shall not repeat the details here and we conclude with the

<u>Proposition</u>. *Within the constitutive class of thermoelastic polarizable and magnetizable materials the Maxwell-Minkowski model III and the statistical model IV can be brought to a one-to-one correspondence provided that the internal energies ${}^{III}U$ and ${}^{IV}U$ are related by relation (3.16) and, furthermore, that the energy fluxes are the same.*

There still remains one practical question. Given a free energy function ${}^{III}\Psi$ and another one, ${}^{IV}\Psi$, each a function of its own variables, how can it be decided that the theories according to the models III and IV are equivalent? To this end, observe that ${}^{IV}\Psi$ as it occurs in (3.18) is a function of the variables $C_{\alpha\beta}$, P_α, B_α and θ, or

$$ (3.23) \qquad {}^{IV}\Psi = \overset{+}{\Psi}(C_{\alpha\beta}, P_\alpha, B_\alpha, \theta) , $$

whereas ${}^{III}\Psi$ as given by (3.19) is a function of $C_{\alpha\beta}$, E_α, M_α and θ instead. We now can express ${}^{IV}\Psi$ in terms of the same set of variables as ${}^{III}\Psi$ by substituting into (3.23) for P_α the relation

$$(3.24) \qquad P_\alpha = - \frac{\partial\tilde{\Psi}}{\partial E_\beta} C_{\alpha\beta} ,$$

which follows from $(3.20)^2$, (1.88) and $(1.89)^1$, and for B_α the expression

$$(3.25) \qquad B_\alpha = \frac{\partial\tilde{\Psi}}{\partial M_\beta} C_{\alpha\beta} + \frac{\rho}{\mu_o} M_\alpha ,$$

following from $(3.20)^3$, $(2.71)^2$ and $(1.89)^{2,5}$. In this way we obtain

$$(3.26) \qquad {}^{IV}\Psi = F(C_{\alpha\beta},E_\alpha,M_\alpha,\theta) := \overset{+}{\Psi}(C_{\alpha\beta},- \frac{\partial\tilde{\Psi}}{\partial E_\beta} C_{\alpha\beta}, \frac{\partial\tilde{\Psi}}{\partial M_\beta} C_{\alpha\beta} + \frac{\rho}{\mu_o} M_\alpha,\theta) .$$

If this expression is substituted into (3.18) and thereby an idendity is obtained, in other words if

$$(3.27) \qquad \begin{aligned} &\overset{+}{\Psi}(C_{\alpha\beta},- \frac{\partial\tilde{\Psi}}{\partial E_\beta} C_{\alpha\beta}, \frac{\partial\tilde{\Psi}}{\partial M_\beta} C_{\alpha\beta} + \frac{\rho}{\mu_o} M_\alpha,\theta) = \\ &= \tilde{\Psi}(C_{\alpha\beta},E_\alpha,M_\alpha,\theta) - \frac{\partial\tilde{\Psi}}{\partial E_\alpha} E_\alpha - \frac{\partial\tilde{\Psi}}{\partial M_\alpha} M_\alpha - \frac{\rho}{2\mu_o} M_\alpha M_\beta C^{-1}_{\alpha\beta} , \end{aligned}$$

then one condition that model III is equivalent to model IV is satisfied. Of course, the energy flux vectors must also be expressed in the same variables and must also be the same functions of these variables.

The above procedure illustrates how, in practice, two theories can be decided to be equivalent. The procedure may in reality be very elaborate, but it shows explicitly that equivalence of theories amounts to a comparison of thermodynamic potentials. We have done this here for the sets (P_α,B_α) in theory IV and (E_α,M_α) in theory III. In the table below we show which variable sets of these two formulations naturally correspond to each other. The above investigation can be made for each of them, of course.

Model IV	Model III
P_α,B_α	E_α,M_α
P_α,M_α	E_α,H_α
E_α,M_α	P_α,H_α
E_α,B_α	P_α,M_α

3.5 COMPARISON OF THE MODELS IV AND V

In this Section we compare the statistical model IV and the Lorentz model V. With regard to the electromagnetic variables these models differ only in the definitions of the magnetization vectors, which are related by

$$(3.28)\qquad M_i^S = M_i^L - e_{ijk}\dot{x}_j P_k^L = \mathcal{M}_i - e_{ijk}\dot{x}_j P_k \,.$$

Here, as before we have set

$$(3.29)\qquad M_i^L = \mathcal{M}_i, \qquad \text{and} \qquad P_i^L = P_i \,.$$

If this relation is used, the Maxwell equations of one formulation transform into those of the other (see (2.105) and (2.120)).
With the transformation rule (3.28) it is also straightforward to relate the body forces. This is most easily achieved by comparing the Maxwell stresses $(2.111)^1$ and $(2.126)^1$, since in both formulations the electromagnetic momenta are equal to zero. This gives

$$(3.30)\qquad \begin{aligned} {}^{V}t^M_{ij} &= {}^{IV}t^M_{ij} - E_i^S P_j + M_i^S B_j - \delta_{ij} B_k M_k^S - e_{ik\ell} P_k B_\ell \dot{x}_j = \\ &= {}^{IV}t^M_{ij} - \mathcal{E}_i P_j + \mathcal{M}_i B_j - \delta_{ij} B_k \mathcal{M}_k \,. \end{aligned}$$

In order for the balance laws of momentum of the two formulations to be equivalent, it is thus necessary that the stress tensors be related by

$$(3.31)\qquad {}^{V}t_{ij} = {}^{IV}t_{ij} + \mathcal{E}_i P_j - \mathcal{M}_i B_j + \delta_{ij} B_k \mathcal{M}_k$$

This also implies that, if balance of moment of momentum is satisfied in model V, in which the stress tensor is symmetric, so it is in model IV.
For a comparison of the energy equations, we first write $(2.108)^4$ in the form

$$(3.32)\qquad \begin{aligned} {}^{IV}(\rho\dot{U}) &= {}^{IV}t_{(ij)}\dot{x}_{i,j} - {}^{IV}q_{i,i} + \mathcal{J}_i \mathcal{E}_i + \mathcal{E}_i(\overset{*}{P}_i + e_{ijk}\mathcal{M}_{k,j}) + \\ &+ (e_{ijk}\mathcal{E}_j \mathcal{M}_k)_{,i} + (\mathcal{E}_i P_j - \mathcal{M}_i B_j + \delta_{ij}\mathcal{M}_k B_k)\dot{x}_{(i,j)} \,, \end{aligned}$$

for the derivation of which use has also been made of (2.109).
Comparing (3.32) with the balance law of energy of model V, $(2.124)^4$, using thereby the relations (3.30) and (3.31), reveals that the two balance laws are equivalent if

$$(3.33)\qquad {}^{V}(\rho U) = {}^{IV}(\rho U) \,,$$

and

$$(3.34)\qquad {}^{V}q_i = {}^{IV}q_i - e_{ijk}\mathcal{E}_j \mathcal{M}_k \,.$$

Note that this relation between the energy fluxes was introduced already in Section 2.5 (eq. (2.130)), where it was used to bring the entropy inequality into its expected classical form. With the relation (3.34) the reduced entropy inequalities (2.117) and (2.137) are then also identical.

Bearing the transformation rules (3.30), (3.31) and (3.34) in mind, it is further recognized that the jump conditions (2.112) and (2.127) are also the same.

In order to obtain complete equivalence, there remains to consider the constitutive equations. In Chapter 2 the constitutive theories for both models were developed using constitutive assumptions of the form

$$C = \overset{+}{C}(F_{i\alpha}, P_i/\rho, B_i, \theta, \theta_{,i}, \mathcal{Q}) \ .$$

The results are listed in the two sets of equations (2.116) and (2.136), and they reveal that the relations (3.30) and (3.31) are satisfied, provided that

$$(3.35) \qquad {}^{V}\overset{+}{\Psi}(C_{\alpha\beta}, P_\alpha, B_\alpha, \theta) = {}^{IV}\overset{+}{\Psi}(C_{\alpha\beta}, P_\alpha, B_\alpha, \theta) \ ,$$

which is in accordance with (3.33). Furthermore, this simultaneously guarantees that entropy and electromotive intensity are the same in both formulations.

The reader may show himself that these results remain also correct when some of the dependent and independent variables are interchanged.

We have thus proved

Proposition. *Within the constitutive class of thermoelastic polarizable and magnetizable materials the non-relativistic statistical model (IV) and the Lorentz model (V) are equivalent provided that the energy flux vectors of the two formulations are related according to*

$$(3.36) \qquad {}^{V}q_i = {}^{IV}q_i - e_{ijk}\mathcal{E}_j\mathcal{M}_k \ ,$$

and, furthermore, provided that the two free energies are the same functions of their variables.

3.6 CONCLUSIONS

In summary we have shown that all field interaction models presented in Chapter 2 and describing polarizable and magnetizable thermoelastic materials are equivalent to each other and differ only in terms which in the context of the non-relativistic approximation have been considered to be negligibly small anyhow. We may thus justly call these theories to be equivalent. On the other hand, it is true that results obtained for these various models could be different in exactly these neglected terms. They have been assumed to be unimportant in all of the above theories, because each of them neglects terms that are preceded by a c^{-2}-factor. Hence, if experiments for a material obeying our constitutive assumptions should deviate from what any of these formulations predicts, then the non-relativistic approximation must be replaced by a semi-relativistic one.

There exists a semi-relativistic version of the statistical model (cf. [2]), whereas a semi-relativistic counterpart of the Lorentz model may be derived from a fully relativistic theory of Grot and Eringen [3] by merely neglecting terms of the order of v^2/c^2. In the same way, still other semi-relativistic formulations can be found in [1], Ch. 7. For all these semi-relativistic formulations the constitutive treatment differs from what we have presented here. Therefore, the equivalence proof of all these models - although claimed to be established by Penfield and Haus, [1] - still remains to be done. (In [1], neither a constitutive theory nor jump conditions are presented.) We regard this as one of the important future research topics to be attacked.

On the other hand, if the semi-relativistic formulation is considered to be too complicated, still a small improvement of our non-relativistic approximation can be obtained by changing from MKSA-units to Gaussian units. In that case, as already said several times before, some c^{-2}-terms that are neglected in our non-relativistic approximation become c^{-1}-terms and must be retained. The most striking effect caused by this change of units, but not the only one, is the fact that the electromagnetic momentum vector $\underline{g}$ is retained, which becomes (in Gaussian units)

$$\frac{1}{c}(\underline{E} \times \underline{H}) ,$$

for the models I, II and III, and

$$\frac{1}{c}(\underline{E} \times \underline{B}) ,$$

for the models IV and V. If all transformations are executed in a consistent way, all models remain equivalent also in a non-relativistic approximation based on Gaussian units.

We have performed the equivalence proof for a polarizable and magnetizable solid only, which must be regarded as a severe restriction. There are more complex material behaviors than the one dealt with by us. For all those this proof is not established yet. One immediate generalization is the inclusion of viscosity in the sence that apart from $F_{i\alpha}$ its time rate $\dot{F}_{i\alpha}$

may be included amongst the independent constitutive variables (cf. [4], [5] or [6]). When this is done, it turns out that the stress tensor may be decomposed into two parts:

$$(3.37)\qquad t_{ij} = t_{ij}^{th} + t_{ij}^{e}$$

One part, namely t_{ij}^{th}, is then given by a thermodynamic potential and is independent of $\dot{F}_{i\alpha}$. The second, dissipative, part t_{ij}^{e} is called extra stress and is given by an independent constitutive relation involving $\dot{F}_{i\alpha}$. Since our equivalence proofs have been performed for $t_{ij}^{e} \equiv 0$, it immediately follows that the models I-V are also equivalent for a viscous polarizable and magnetizable thermoelastic solid if, in addition to the conditions stated in this Chapter, the extra stress remains the same in all formulations.
Another possible extension is obtained when for ferromagnetic materials spin interactions are taken into account. These effects can be reckoned with by including in the constitutive theory the magnetization gradients and by introducing an extra moment of momentum density emanating from the magnetic spin action (see e.g. [4], [7], [8]). This results in two extra terms in the angular momentum law, namely the magnetic spin and the gradient of a magnetic couple stress tensor. The latter is determined by the derivative of the thermodynamic potential (free energy) with respect to magnetization gradients. Moreover, the energy equation must also be supplemented by two terms, namely the kinetic energy of the magnetic spin and the energy flux due to the magnetic couple stress. Since all these terms always seems to be introduced in a unique way, they do not lead to differences in the respective formulations.
Analogous remarks hold for constitutive theories in which polarization gradients are taken into account (see e.g. [9], [10] or [11]).
We conclude this Chapter with a Table* that lists all the pertinent interrelations between the various models. In this Table, the relations between the four basic electromagnetic fields, between the stresses t_{ij}, the internal energies U and the energy fluxes q_i are listed. The differences between the latter three quantities are expressed in the fields

$$\mathcal{E}_i,\ P_i,\ B_i \text{ and } \mathcal{M}_i\ ,$$

which are identical in all formulations. We dit not list the remaining variables, which are also the same in all formulations. They are

$$\mathcal{Q},\ J_i,\ \rho,\ x_i = \chi_i(\underline{X},t),\ \theta \text{ and } \eta\ .$$

* *The Table is placed at the end of the book.*

REFERENCES

[1] Penfield, P. and H.A. Haus, *Electrodynamics of Moving Media*, The M.I.T. Press, Cambridge, Massachusettes, 1967.

[2] De Groot, S.R. and L.G. Suttorp, *Foundations of Electrodynamics*, North-Holland Publishing Co., Amsterdam, 1972.

[3] Grot, R.A. and A.C. Eringen, *Relativistic Continuum Mechanics*, Part I, Mechanics and Thermodynamics; Part II, Electromagnetic Interactions with Matter, Int. J. Eng. Sc., 4 (1966), 611-638, 639-670.

[4] Alblas, J.B., *Electro-Magneto-Elasticity*, Topics in Applied Continuum Mechanics, eds. J.L. Zeman and F. Ziegler, 71-114, Springer-Verlag, Wien, 1974.

[5] Maugin, G.A., *A Continuum Theory of Deformable Ferromagnetic Bodies*, Part II, Thermodynamics, Constitutive Theory, J. Math. Phys. 17 (1976), 1739-1751.

[6] Hutter, K., *On Thermodynamics and Thermostatics of Viscous Thermoelastic Solids in the Electromagnetic Fields*, Arch. Rat. Mech. Anal. 58 (1975), 339-368.

[7] Ven, A.A.F. van de, *Interaction of Electromagnetic and Elastic Fields in Solids*, Dr. of Science Thesis, University of Technology Eindhoven, the Netherlands, 1975.

[8] Tiersten, H.F., *Coupled Magnetomechanical Equations for Magnetically Saturated Insulators*, J. Math. Phys. 5 (1964), 1298-1318.

[9] Mindlin, R.D., *Polarization Gradient in Elastic Dielectrics*, Int. J. Solids Struct. 4 (1968), 637-642.

[10] Suhubi, E.S., *Elastic Dielectrics with Polarization Gradient*, Int. J. Eng. Sc. 7 (1969), 993-997.

[11] Chowdhury, K.L. and P.G. Glockner, *Constitutive Equations for Elastic Dielectrics*, Int. J. Non-Linear Mech. 11 (1976), 315-324.

4. MATERIAL DESCRIPTION

4.1 MOTIVATION

In the last two Chapters the governing equations of field matter interaction were chiefly based on the spatial or Eulerian description. Only very briefly have we given the material or Lagrangian formulation. Such a formulation is of advantage in describing the deformation of solids, because the boundary conditions for solids are usually prescribed on the undeformed body, which is generally the body in its reference configuration. As a consequence, any theory describing deformable solids in the electromagnetic fields should from the outset be given in the material description. This is not done in general. On the contrary, in almost all theories the spatial description is applied. The Lagrangian formulation is introduced only afterwards, and if so, only by introducing some approximations, e.g. linearizations. These linearization procedures, although being straightforward are nevertheless quite cumbersome. They become an almost trivial matter when the material description of the governing field equations is used from the outset.

It thus should be apparent that the Lagrangian formulation is a necessity. The reason for presenting it this late is that it is relatively unknown, so that the equations do at first sight not look familiar. They are not new, however, and have been derived before (see Hutter [1], [2]). Use of Lagrangian variables is also made by Alblas, [3], for the linearization of some specialized topics as quasi-electroelastostatics and quasi-magnetoelastostatics. In fact, this approach can be seen as an improvement of the method used by Toupin, [4]. In his paper on the non-relativistic electrodynamics of deformable media, Prechtl, [5], devoted one Section to a Lagrangian formulation of the balance equations and the jump conditions (no constitutive equations are used here). Both works of Alblas and Prechtl are based on the Chu-model. In the following, we shall describe the two-dipole models first and shall then pass on to the description of the statistical and the Lorentz formulation; we shall conclude with the Maxwell-Minkowski model.

4.2 THE MATERIAL DESCRIPTION OF THE TWO - DIPOLE MODELS (MODELS I AND II)

In Chapter 1, Section 3.3, we briefly presented the Lagrangian form of the Maxwell equations. Those equations appeared in one particular formulation. We could, as was done for the Eulerian description, introduce the various models by simply performing variable transformations. Little insight into the meaning of the new variables would be gained thereby, however, so that a fresh derivation is enlightening.

As is well-known, the Chu-formulation is based upon the postulations that

i) only two vector quantities are necessary to describe the electromagnetic fields in free space, and
ii) that material bodies contribute toward these fields by acting as sources for these fields.

With regard to postulate i) it is advantageous for the derivation of the Lagrangian Maxwell equations to work formally with four field vectors and to relate the remaining two to the former ones. Concerning postulate ii) we assume it to be known that magnetization and polarization act as charge and current distributions. Based on such a conception one arrives at the equations (see Hutter [2])

$$\int_{\partial V} B_i^a dv = \int_V \mathcal{Q}^m dv ,$$

$$\int_{\partial S} \mathcal{E}_i d\ell_i + \frac{d}{dt} \int_S B_i^a da_i = - \int_S \mathcal{J}_i^m da_i ,$$

$$(4.1) \qquad \int_{\partial V} D_i^a da_i = \int_V (\mathcal{Q} + \mathcal{Q}^P) dv ,$$

$$\int_{\partial S} \mathcal{H}_i d\ell_i - \frac{d}{dt} \int_S D_i^a da_i = \int_S (\mathcal{J}_i + \mathcal{J}_i^P) dv ,$$

$$\frac{d}{dt} \int_V \mathcal{Q}\, dv + \int_{\partial V} \mathcal{J}_i da_i = 0 .$$

Here, $\mathcal{E}_i$, $\mathcal{H}_i$, $\mathcal{Q}$ and $\mathcal{J}_i$ are as introduced before; they are the electromotive intensity, the effective magnetic field strength and the charge and conductive current densities due to free charges. $\mathcal{Q}^P$, $\mathcal{J}_i^P$, $\mathcal{Q}^m$ and $\mathcal{J}_i^m$ are the charge and conductive current densities due to polarization and magnetization and it is well-known that in the Chu-formulation (compare also (2.47))

$$(4.2) \qquad \int_V \mathcal{Q}^P dv = - \int_{\partial V} P_i^C da_i , \qquad \int_V \mathcal{Q}^m dv = - \int_{\partial V} \mu_o M_i^C da_i ,$$

$$\int_S \mathcal{J}_i^P da_i = \frac{d}{dt} \int_S P_i^C da_i , \qquad \int_S \mathcal{J}_i^m da_i = \frac{d}{dt} \int_S \mu_o M_i^C da_i .$$

Finally, we have introduced in (4.1) the auxiliary fields B^a_i and D^a_i which are related to the electric and magnetic field strength according to

$$B^a_i = \mu_o H^C_i , \qquad D^a_i = \varepsilon_o E^C_i . \tag{4.3}$$

These equations are sometimes called the Maxwell-Lorentz aether relations. To derive a material description from (4.1) and (4.2) one must simply transform the integrals over spatial volume and spatial surface into integrals over reference volume and reference surface, using thereby the relations (1.32). How this is done was explained in Section 1.3.3, and hence we only give the results:

$$\begin{aligned}
&\int_{\partial V_R} \mathbb{B}^a_\alpha dA_\alpha = \int_{V_R} \mathbb{Q}^m dV , \\
&\int_{\partial S_R} \mathbb{E}_\alpha dL_\alpha + \frac{d}{dt} \int_{S_R} \mathbb{B}^a_\alpha dA_\alpha = - \int_{S_R} \mathbb{J}^m_\alpha dA_\alpha , \\
&\int_{\partial V_R} \mathbb{D}^a_\alpha dA_\alpha = \int_{V_R} (\mathbb{Q} + \mathbb{Q}^P) dV , \\
&\int_{\partial S_R} \mathbb{H}_\alpha dL_\alpha - \frac{d}{dt} \int_{S_R} \mathbb{D}_\alpha dA_\alpha = \int_{S_R} (\mathbb{J}_\alpha + \mathbb{J}^P_\alpha) dA_\alpha , \\
&\frac{d}{dt} \int_{V_R} \mathbb{Q}\, dV + \int_{\partial V_R} \mathbb{J}_\alpha dA_\alpha = 0 ,
\end{aligned} \tag{4.4}$$

with

$$\begin{aligned}
&\int_{V_R} \mathbb{Q}^P dV = - \int_{\partial V_R} \mathbb{P}^C_\alpha dA_\alpha , \qquad \int_{V_R} \mathbb{Q}^m dV = - \int_{\partial V_R} \mu_o \mathbb{M}^C_\alpha dA_\alpha , \\
&\int_{S_R} \mathbb{J}^P_\alpha dA_\alpha = \frac{d}{dt} \int_{S_R} \mathbb{P}^C_\alpha dA_\alpha , \qquad \int_{S_R} \mathbb{J}^m_\alpha dA_\alpha = \frac{d}{dt} \int_{S_R} \mu_o \mathbb{M}^C_\alpha dA_\alpha .
\end{aligned} \tag{4.5}$$

The following definitions have been used

$$\begin{aligned}
&\mathbb{B}^a_\alpha = J F^{-1}_{\alpha i} B^a_i , \qquad && B^a_i = J^{-1} F_{i\alpha} \mathbb{B}^a_\alpha , \\
&\mathbb{J}_\alpha = |J| F^{-1}_{\alpha i} J_i , \qquad && J_i = |J^{-1}| F_{i\alpha} \mathbb{J}_\alpha , \\
&\mathbb{P}^C_\alpha = |J| F^{-1}_{\alpha i} P^C_i , \qquad && P^C_i = |J^{-1}| F_{i\alpha} \mathbb{P}^C_\alpha , \\
&\mathbb{Q} = |J| Q , \qquad && Q = |J^{-1}| \mathbb{Q} ,
\end{aligned}$$

(over)

$$(4.6)\qquad \begin{aligned} \mathbb{E}_\alpha &= F_{i\alpha}E_i\ , & E_i &= F^{-1}_{\alpha i}\mathbb{E}_\alpha\ ,\\ \mathbb{D}^a_\alpha &= |J|F^{-1}_{\alpha i}D^a_i\ , & D^a_i &= |J^{-1}|F_{i\alpha}\mathbb{D}^a_\alpha\ ,\\ \mathbb{H}_\alpha &= F_{i\alpha}H_i\,\mathrm{sgn}\,J\ , & H_i &= F^{-1}_{\alpha i}\mathbb{H}_\alpha\,\mathrm{sgn}\,J\ ,\\ \mu_o\mathbb{M}^C_\alpha &= JF^{-1}_{\alpha i}\mu_o M^C_i\ , & \mu_o M^C_i &= J^{-1}F_{i\alpha}\mu_o\mathbb{M}^C_\alpha\ , \end{aligned}$$

some of which were already defined in (1.46) but are repeated here for ease of reference. In the above equations integration is over material parts V_R, ∂V_R, S_R and ∂S_R, respectively, and all variables are thought to be functions of X_α and t so that, as usual, d/dt denotes time derivative at a fixed particle.

The equations (4.4) and (4.5) may be combined, and for sufficiently smooth fields they yield

$$(4.7)\qquad \begin{aligned} &\mathbb{B}^a_{\alpha,\alpha} + \mu_o\mathbb{M}^C_{\alpha,\alpha} = 0\ ,\\ &\dot{\mathbb{B}}^a_\alpha + \mu_o\dot{\mathbb{M}}^C_\alpha + e_{\alpha\beta\gamma}\mathbb{E}_{\gamma,\beta} = 0\ ,\\ &\mathbb{D}^a_{\alpha,\alpha} + \mathbb{P}_{\alpha,\alpha} = \mathbb{Q}\ ,\\ &-\dot{\mathbb{D}}^a_\alpha - \dot{\mathbb{P}}_\alpha + e_{\alpha\beta\gamma}\mathbb{H}_{\gamma,\beta} = \mathbb{J}_\alpha\ ,\\ &\mathbb{J}_{\alpha,\alpha} + \dot{\mathbb{Q}} = 0\ . \end{aligned}$$

The equations (4.7) are the Maxwell equations in the material description. We shall call $\mathbb{E}_\alpha$ and $\mathbb{H}_\alpha$ the Lagrangian electric and magnetic field strengths, $\mathbb{P}_\alpha$ and $\mathbb{M}^C_\alpha$ the Lagrangian polarization and magnetization of the Chu-formulation and $\mathbb{Q}$ and $\mathbb{J}_\alpha$ the Lagrangian free charge and current densities. Since, in a non-relativistic approximation, the meaning of $\mathbb{P}_\alpha$ is unique for all formulations the upper index C for $\mathbb{P}_\alpha$ has been dropped. This is not the case for $\mathbb{M}_\alpha$ and, therefore, the index C must be retained there.

Before we proceed a few comments seem to be in order: Firstly, all quantities listed on the right column of (4.6) are objective vectors (J_i, P^C_i, E_i, D^a_i), objective axial vectors (B^a_i, M^C_i, H_i) and an objective scalar (Q) under the Euclidean transformation group. As a consequence, all Lagrangian variables (listed on the left column of (4.6)) are scalars under this group, as they must.

Secondly, we could easily relate the variables occurring in (4.7) to those listed in (1.46) and indeed this transformation is achieved by setting

$$(4.8) \qquad \mathbb{B}_\alpha := \mathbb{B}^a_\alpha + \mu_o \mathbb{M}^C_\alpha , \qquad \mathbb{D}_\alpha := \mathbb{D}^a_\alpha + \mathbb{P}_\alpha ,$$

where $\mathbb{D}_\alpha$ and $\mathbb{B}_\alpha$ may be defined also in terms of B_i and D_i as was done in (1.46).

Thirdly, in order to determine the auxiliary fields $\mathbb{B}^a_\alpha$ and $\mathbb{D}^a_\alpha$, we simply must use the Maxwell-Lorentz aether relations and must invoke the transformations (4.6). When this is done and when terms proportional to c^{-2} are discarded, a straightforward calculation shows that

$$(4.9) \qquad \mathbb{D}^a_\alpha = \varepsilon_o |J| C^{-1}_{\alpha\beta} \mathbb{E}_\beta , \qquad \mathbb{B}^a_\alpha = \mu_o |J| C^{-1}_{\alpha\beta} \mathbb{H}_\beta .$$

These relations thus hold in a non-relativistic theory. When relativistic terms are retained they become much more complicated as can be seen in [2]. Of course, as was the case in the Eulerian description, the global equations (4.4) and (4.5) also imply jump conditions which follow from (1.60) by invoking the definitions (4.8). They then read

$$[\![\mathbb{B}^a_\alpha + \mu_o \mathbb{M}^C_\alpha]\!] N_\alpha = 0 ,$$

$$[\![\mathbb{D}^a_\alpha + \mathbb{P}_\alpha]\!] N_\alpha = 0 ,$$

$$(4.10) \qquad e_{\alpha\beta\gamma} [\![\mathbb{E}_\beta]\!] N_\gamma + [\![(\mathbb{B}^a_\alpha + \mu_o \mathbb{M}^C_\alpha) W_N]\!] = 0 ,$$

$$e_{\alpha\beta\gamma} [\![\mathbb{H}_\beta]\!] N_\gamma - [\![(\mathbb{D}^a_\alpha + \mathbb{P}_\alpha) W_N]\!] = 0 ,$$

$$[\![\mathbb{J}_\alpha]\!] N_\alpha + [\![\mathbb{Q} W_N]\!] = 0 .$$

To complete the description of the electro-mechanical interaction the balance laws of mass, momentum, moment of momentum and energy must be given. They are listed in (1.34)-(1.37), and their local forms appear in (1.39). It thus suffices to write down the material counterparts of the electromagnetic body force, body couple and energy supply. They are $\rho_o F^e_i$, $\rho_o L^e_{ij}$ and $\rho_o r^e$, respectively. To calculate them for the models I and II we only need to convert the expressions (2.15), $(2.7)^2$, (2.12) and (2.48) to Lagrangian form omitting thereby c^{-2}-terms. In this calculation it is also advantageous to use relation (3.1):

$$\rho\, {}^{I}F^e_i = \rho\, {}^{II}F^e_i + (E_i P_j + \mu_o H_i M^C_j)_{,j} .$$

The details are tedious even though they are straightforward, and what one obtains reads as follows:

For model I

$$
\begin{aligned}
{}^{I}(\rho_o F_i^e) &= F_{\alpha i}^{-1}\{(\mathbb{Q}-\mathbb{P}_{\beta,\beta})\mathbb{E}_\alpha + e_{\alpha\beta\gamma}(\mathbb{J}_\beta + \dot{\mathbb{P}}_\beta)\mathbb{B}_\gamma^a - \mu_o \mathbb{M}_{\beta,\beta}^C \mathbb{H}_\alpha\} + \\
&\quad + \{F_{\beta i}^{-1}(\mathbb{P}_\alpha \mathbb{E}_\beta + \mu_o \mathbb{M}_\alpha^C \mathbb{H}_\beta)\}_{,\alpha}\ , \\
{}^{I}(\rho_o L_{ij}^e) &= F_{[i\alpha}F_{\beta j]}^{-1}(\mathbb{P}_\alpha \mathbb{E}_\beta + \mu_o \mathbb{M}_\alpha^C \mathbb{H}_\beta)\ , \qquad (4.11)\\
{}^{I}(\rho_o r^e) &= \mathbb{J}_\alpha \mathbb{E}_\alpha + \dot{\mathbb{P}}_\alpha \mathbb{E}_\alpha + \mu_o \dot{\mathbb{M}}_\alpha^C \mathbb{H}_\alpha + F_{\alpha i}^{-1}(\mathbb{E}_\alpha \mathbb{P}_\beta + \mu_o \mathbb{H}_\alpha \mathbb{M}_\beta^C)\dot{F}_{i\beta}\ ,
\end{aligned}
$$

and for model II:

$$
\begin{aligned}
{}^{II}(\rho_o F_i^e) &= F_{\alpha i}^{-1}\{(\mathbb{Q}-\mathbb{P}_{\beta,\beta})\mathbb{E}_\alpha + e_{\alpha\beta\gamma}(\mathbb{J}_\beta + \dot{\mathbb{P}}_\beta)\mathbb{B}_\gamma^a - \mu_o \mathbb{M}_{\beta,\beta}^C \mathbb{H}_\alpha\}\ , \\
{}^{II}(\rho_o L_{ij}^e) &= 0\ , \qquad (4.12)\\
{}^{II}(\rho_o r^e) &= \mathbb{J}_\alpha \mathbb{E}_\alpha + \dot{\mathbb{P}}_\alpha \mathbb{E}_\alpha + \mu_o \dot{\mathbb{M}}_\alpha^C \mathbb{H}_\alpha\ ,
\end{aligned}
$$

in the derivation of which also the identities

$$
\begin{aligned}
&e_{ijk}F_{i\alpha}F_{j\beta}F_{k\gamma} = J e_{\alpha\beta\gamma}, \qquad (\tfrac{1}{J}F_{i\alpha})_{,i} = 0\ , \\
&\overset{*}{P}_i = |J^{-1}| F_{i\alpha}\dot{\mathbb{P}}_\alpha, \qquad P_{i,i} = |J^{-1}|\mathbb{P}_{\alpha,\alpha}\ , \qquad (4.13)\\
&\overset{*}{M}{}_i^C = J^{-1}F_{i\alpha}\dot{\mathbb{M}}_\alpha^C, \qquad M_{i,i}^C = J^{-1}\mathbb{M}_{\alpha,\alpha}^C\ ,
\end{aligned}
$$

were used.

Two conclusions are readily drawn from the above expressions. Firstly, mere inspection of (4.11) and (4.12) shows that the expressions for body force and body couple are (in a non-relativistic sense) an objective vector and an objective skew-symmetric tensor, respectively. It is also seen that the electromagnetic energy supply is an objective scalar only in model II. Needless to state which model seems to be simpler. Secondly, the body force expression (4.12)[1] is easily interpretable. It is composed of an electric and a magnetic Lorentz force (compare (2.45)[1])

$$
{}^{II}(\rho_o F_i^e) = F_{\alpha i}^{-1}\{(\mathbb{Q}^e\mathbb{E}_\alpha + e_{\alpha\beta\gamma}\mathbb{J}_\beta^e\mathbb{B}_\gamma^a) + (\mathbb{Q}^m\mathbb{H}_\alpha + e_{\alpha\beta\gamma}\mathbb{J}_\beta^m\mathbb{D}_\gamma^a)\}\ , \qquad (4.14)
$$

where

$$
\begin{aligned}
\mathbb{Q}^e &= \mathbb{Q} - \mathbb{P}_{\alpha,\alpha}, \qquad \mathbb{J}_\beta^e = \mathbb{J}_\beta + \dot{\mathbb{P}}_\beta\ , \qquad (4.15)\\
\mathbb{Q}^m &= -\mu_o \mathbb{M}_{\alpha,\alpha}^C, \qquad \mathbb{J}_\beta^m = \mu_o \dot{\mathbb{M}}_\beta^C\ ,
\end{aligned}
$$

as follows from (4.2).

As was done formally in (1.65) we can also derive the Maxwell stress tensor $\overset{M}{T}_{i\alpha}$, the electromagnetic momentum G_i, the energy flux Π_α and the energy density Ω. Formally, this is done by expressing the representations (2.10) and (2.50) in Lagrangian variables and using the transformations (see (1.65))

$$(4.16)\qquad \begin{aligned} \overset{M}{T}_{i\alpha} &= |J|(\overset{M}{t}_{ij} - g_i\dot{x}_j)F^{-1}_{\alpha j}, \qquad & G_i &= |J|g_i\ , \\ \Pi_\alpha &= |J|(\pi_i - \omega\dot{x}_i)F^{-1}_{\alpha i}, \qquad & \Omega &= |J|\omega\ . \end{aligned}$$

The calculations are again easy, though rather long, and what one obtains in a non-relativistic formulation can be written as follows:

For <u>model II</u>

$$(4.17)\qquad \begin{aligned} {}^{II}\overset{M}{T}_{i\alpha} &= F^{-1}_{\beta i}(\mathbb{D}^a_\alpha\mathbb{E}_\beta + \mathbb{B}^a_\alpha\mathbb{H}_\beta) - \frac{1}{2}F^{-1}_{\alpha i}(\mathbb{D}^a_\beta\mathbb{E}_\beta + \mathbb{B}^a_\beta\mathbb{H}_\beta)\ , \\ {}^{II}G_i &= 0\ , \\ {}^{II}\Pi_\alpha &= -e_{\alpha\beta\gamma}\mathbb{E}_\beta\mathbb{H}_\gamma - \frac{1}{2}(\mathbb{D}^a_\beta\mathbb{E}_\beta + \mathbb{B}^a_\beta\mathbb{H}_\beta)F^{-1}_{\alpha i}\dot{x}_i + (\mathbb{D}^a_\alpha\mathbb{E}_\beta + \mathbb{B}^a_\alpha\mathbb{H}_\beta)F^{-1}_{\beta i}\dot{x}_i = \\ &= -e_{\alpha\beta\gamma}\mathbb{E}_\beta\mathbb{H}_\gamma + {}^{II}\overset{M}{T}_{i\alpha}\dot{x}_i\ , \\ {}^{II}\Omega &= -\frac{1}{2}(\mathbb{E}_\alpha\mathbb{D}^a_\alpha + \mathbb{H}_\alpha\mathbb{B}^a_\alpha)\ . \end{aligned}$$

Alternatively, the corresponding expressions for model I follow from

$$(4.18)\qquad \begin{aligned} {}^{I}\overset{M}{T}_{i\alpha} &= {}^{II}\overset{M}{T}_{i\alpha} + F^{-1}_{\beta i}(\mathbb{P}_\alpha\mathbb{E}_\beta + \mu_o\overset{C}{\mathbb{M}}_\alpha\mathbb{H}_\beta), \qquad & {}^{I}G_i &= {}^{II}G_i = 0\ , \\ {}^{I}\Pi_\alpha &= {}^{II}\Pi_\alpha + (\mathbb{P}_\alpha\mathbb{E}_\beta + \mu_o\overset{C}{\mathbb{M}}_\alpha\mathbb{H}_\beta)F^{-1}_{\beta i}\dot{x}_i, \qquad & {}^{I}\Omega &= {}^{II}\Omega\ . \end{aligned}$$

The jump conditions for the two models can now be obtained by simply substituting (4.17) or (4.18) into the general jump conditions (1.67).

There remains the presentation of the constitutive theory. For that purpose, we first eliminate $\rho_o r^{ext}$ from (1.54) and $(1.39)^3$. When making use of $(4.14)^3$ thereby the following entropy inequality for model II is obtained:

$$(4.19)\qquad -\rho_o\dot{\psi} - \rho_o\eta\dot{\theta} + {}^{II}T_{i\alpha}\dot{F}_{i\alpha} + \dot{\mathbb{P}}_\alpha\mathbb{E}_\alpha + \mu_o\dot{\overset{C}{\mathbb{M}}}_\alpha\mathbb{H}_\alpha + \mathbb{J}_\alpha\mathbb{E}_\alpha - \frac{Q_\alpha\theta_{,\alpha}}{\theta} \geq 0\ .$$

Here we have introduced the Helmholtz free energy

$$(4.20)\qquad \psi = U - \eta\theta\ ,$$

and have also set

(4.21) $$\Phi_\alpha = \frac{Q_\alpha}{\theta} .$$

The entropy inequality for model I can be found from (4.19) by introducing into the latter the following relation between the Piola-Kirchhoff stress tensors of the two models (compare (3.2))

(4.22) $${}^{I}T_{i\alpha} = {}^{II}T_{i\alpha} - F^{-1}_{\beta i}(\mathbb{P}_\alpha \mathbb{E}_\beta + \mu_o \mathbb{M}^C_\alpha \mathbb{H}_\beta) .$$

Inequality (4.19) suggests to establish constitutive relations of the form

(4.23) $$C = \hat{C}(C_{\alpha\beta}, \frac{\mathbb{P}_\alpha}{\rho_o}, \frac{\mu_o \mathbb{M}^C_\alpha}{\rho_o}, \theta, \theta_{,\alpha}, \mathbb{Q}) .$$

All variables in the above list are objective <u>per se</u>, although they are different from the objective variables used in Chapters 2 and 3 (say P_α and M_α etc.). That the Lagrangian field variables are the <u>natural</u> objective combinations of deformation and electromagnetic fields will become apparent shortly. However, before deducing constitutive equations, we first replace in (4.19) the first Piola-Kirchhoff stress tensor $T_{i\alpha}$ by the second one, which according to (1.40) is defined as

(4.24) $$T^P_{\alpha\beta} := T_{i\alpha} F^{-1}_{\beta i} .$$

Once constitutive relations of the form (4.23) are postulated for ψ, η, $T^P_{\alpha\beta}$, $\mathbb{E}_\alpha$, $\mathbb{H}_\alpha$, $\mathbb{J}_\alpha$ and Q_α which all must be objective scalars, we may derive from (4.19) in the usual way that i) the free energy cannot depend on $\theta_{,\alpha}$ and $\mathbb{Q}$,

(4.25) $$\psi = \hat{\Psi}(C_{\alpha\beta}, \frac{\mathbb{P}_\alpha}{\rho_o}, \frac{\mu_o \mathbb{M}^C_\alpha}{\rho_o}, \theta) ,$$

and ii) that

(4.26)
$$\eta = -\frac{\partial\hat{\Psi}}{\partial\theta} ,$$
$$\mathbb{E}_\alpha = \frac{\partial\hat{\Psi}}{\partial\mathbb{P}_\alpha/\rho_o} ,$$
$$\mathbb{H}_\alpha = \frac{\partial\hat{\Psi}}{\partial\mu_o\mathbb{M}^C/\rho_o} ,$$
$${}^{II}T^P_{\alpha\beta} = 2\rho_o \frac{\partial\hat{\Psi}}{\partial C_{\alpha\beta}} ,$$

so that the reduced entropy inequality becomes

(4.27) $$\mathbb{J}_\alpha \mathbb{E}_\alpha - \frac{Q_\alpha \theta_{,\alpha}}{\theta} \geq 0 \ .$$

The second Piola-Kirchhoff stress tensor of model I on the other hand obeys the relation

(4.28) $$^{I}T^{P}_{\alpha\beta} = 2\rho_o \frac{\partial\hat{\Psi}}{\partial C_{\alpha\beta}} - C^{-1}_{\beta\gamma}(\mathbb{P}_\alpha \mathbb{E}_\gamma + \mu_o \mathbb{M}^C_\alpha \mathbb{H}_\gamma) \ ,$$

which with the aid of (4.24) and (4.26)[4] can be derived from (4.22). Hence, this stress tensor, in contrast to $^{II}T^P_{\alpha\beta}$, is not symmetric. However, in view of (4.28) and (4.11)[2] the balance law of moment of momentum which reads (see (1.39))

(4.29) $$^{I}(\rho_o L^e_{ij}) = {}^{I}T_{[i\alpha}F_{j]\alpha} = -F_{i\alpha}F_{j\beta}\,{}^{I}T^P_{[\alpha\beta]}$$

is satisfied identically.

When in the energy equation (1.39)[3] use is made of the relations (4.26) and (4.12)[3], the latter may be written as

(4.30) $$\rho_o \theta\dot{\eta} = \mathbb{J}_\alpha \mathbb{E}_\alpha - Q_{\alpha,\alpha} + \rho_o r^{ext} \ .$$

It should be noted that this relation is derived for model II, but holds for model I as well. It is this form of the energy equation which normally is used in applications.

One advantage of the Lagrangian variables introduced in (4.6) is the simplicity the constitutive equation for stress assumes. Corroboration is provided by a comparison of the formulas (2.56)[4] and (4.26)[4]. Another advantage is the form of the Gibbs relation when written in terms of these variables. Indeed, it follows from the definition of the Helmholtz free energy ψ, (4.20), and the results (4.26) that

$$\frac{\partial\eta}{\partial\theta} = \frac{1}{\theta}\frac{\partial U}{\partial\theta} \ ,$$

(4.31) $$\frac{\partial\eta}{\partial\mathbb{P}_\alpha/\rho_o} = \frac{1}{\theta}\{\frac{\partial U}{\partial\mathbb{P}_\alpha/\rho_o} - \mathbb{E}_\alpha\} \ ,$$

$$\frac{\partial\eta}{\partial\mu_o\mathbb{M}^C_\alpha/\rho_o} = \frac{1}{\theta}\{\frac{\partial U}{\partial\mu_o\mathbb{M}^C_\alpha/\rho_o} - \mathbb{H}_\alpha\} \ ,$$

(over)

$$\frac{\partial \eta}{\partial C_{\alpha\beta}} = \frac{1}{\theta}\{\frac{\partial U}{\partial C_{\alpha\beta}} - \frac{1}{2\rho_o}\, {}^{II}T^P_{\alpha\beta}\} ,$$

from which one readily deduces that

$$(4.32) \qquad d\eta = \frac{1}{\theta}\{dU - \frac{1}{2\rho_o}\, {}^{II}T^P_{\alpha\beta}dC_{\alpha\beta} - \mathbb{E}_\alpha d(\frac{\mathbb{P}_\alpha}{\rho_o}) - \mathbb{H}_\alpha d(\frac{\mu_o \mathbb{M}^C_\alpha}{\rho_o})\} .$$

Next we would like to explore the consequences, which follow from the Gibbs relation, but were not determined in the previous Chapter. The reasons for this postponement are formal ones, for the Gibbs relation assumes a particularly simple form when written in the Lagrangian variables (compare (4.32) with (2.58)).

From a practical point of view, that is from a viewpoint of an applied physicist, who must determine actual constitutive relations by performing appropriate experiments, the Helmholtz free energy is not necessarily the most convenient variable to match experiments with theory. These are rather the internal energy, the stress tensor, the polarization and magnetization per unit mass. In what follows, we shall demonstrate, firstly, how a consistent constitutive theory can be developed when starting from this end and, secondly, this approach will show that, ultimately one searches for the free energy also when using this more physical approach. The method will further demonstrate us that the constitutive relations for internal energy, entropy, free energy and stress can all be separated into two parts, one of which is of purely thermoelastic origin and can therefore be determined from thermoelastic experiments in the absense of electromagnetic fields. The second parts are then the respective effects due to the electromagnetic fields. We shall outline the procedure for the case that $\mathbb{P}_\alpha$ and $\mathbb{M}_\alpha$ are chosen as independent electromagnetic field variables. At the center of the following derivation lies the Gibbs relation (4.32) or the identities (4.31), which are the basis for the derivation of the latter.

Note that the relations (4.31) imply integrability conditions, which can easily be derived by cross differentiations. If these differentiations are performed the following chain of identities is obtained: (here $\mathbb{M}_\alpha \equiv \mathbb{M}^C_\alpha$)

$$\frac{1}{\theta} = \frac{-\dfrac{\partial \mathbb{E}_\alpha}{\partial \theta}}{\dfrac{\partial U}{\partial \mathbb{P}_\alpha/\rho_o} - \mathbb{E}_\alpha} = \frac{-\dfrac{\partial \mathbb{H}_\alpha}{\partial \theta}}{\dfrac{\partial U}{\partial \mu_o \mathbb{M}_\alpha/\rho_o} - \mathbb{H}_\alpha} = \frac{-\dfrac{\partial\, {}^{II}T^P_{\alpha\beta}}{\partial \theta}}{2\rho_o \dfrac{\partial U}{\partial C_{\alpha\beta}} - {}^{II}T^P_{\alpha\beta}} ,$$

(over)

$$(4.33)\qquad \frac{\partial \mathbb{E}_\alpha}{\partial \mu_o \mathbb{M}_\beta/\rho_o} = \frac{\partial \mathbb{H}_\beta}{\partial \mathbb{P}_\alpha/\rho_o}\,, \quad \frac{\partial \mathbb{E}_\alpha}{\partial \mathbb{P}_\beta/\rho_o} = \frac{\partial \mathbb{E}_\beta}{\partial \mathbb{P}_\alpha/\rho_o}\,, \quad \frac{\partial \mathbb{H}_\alpha}{\partial \mu_o \mathbb{M}_\beta/\rho_o} = \frac{\partial \mathbb{H}_\beta}{\partial \mu_o \mathbb{M}_\alpha/\rho_o}\,,$$

$$2\rho_o \frac{\partial \mathbb{E}_\alpha}{\partial C_{\beta\gamma}} = \frac{\partial II_T P_{\beta\gamma}}{\partial \mathbb{P}_\alpha/\rho_o}\,, \quad 2\rho_o \frac{\partial \mathbb{H}_\alpha}{\partial C_{\beta\gamma}} = \frac{\partial II_T P_{\beta\gamma}}{\partial \mu_o \mathbb{M}_\alpha/\rho_o}\,, \quad \frac{\partial II_T P_{\alpha\beta}}{\partial C_{\gamma\delta}} = \frac{\partial II_T P_{\gamma\delta}}{\partial C_{\alpha\beta}}\,.$$

It follows that the relations on the right of $(4.33)^{1,2,3}$ are all functions of the temperature alone. All the more, they are exactly $1/\theta$ and are equal. Such relations, of course, reduce the effective labor of the experimentalist considerably.

The next step consists in the integration of the identities (4.33). To this end, notice that $(4.33)^{1,2}$ can also be written in the form

$$(4.34)\qquad \frac{\partial U}{\partial \mathbb{P}_\alpha/\rho_o} = -\theta^2 \frac{\partial}{\partial\theta}\left(\frac{\mathbb{E}_\alpha}{\theta}\right), \qquad \frac{\partial U}{\partial \mu_o \mathbb{M}_\alpha/\rho_o} = -\theta^2 \frac{\partial}{\partial\theta}\left(\frac{\mathbb{H}_\alpha}{\theta}\right).$$

As we shall see in a moment, these equations allow us to decompose U into two parts of which one is due to the electromagnetic fields, whereas the other is the internal energy of the body in the absense of the fields. To see this, we introduce the 6-tuples

$$(4.35)\qquad x_A = (\mathbb{P}_\alpha/\rho_o, \mu_o \mathbb{M}_\alpha/\rho_o),\ f_A = (\mathbb{E}_\alpha, \mathbb{H}_\alpha),\ (A = 1,2,\ldots,6;\ \alpha = 1,2,3)\,.$$

The two equations (4.34) then combine to give

$$(4.36)\qquad \frac{\partial U}{\partial x_A} = -\theta^2 \frac{\partial}{\partial\theta}\left(\frac{f_A}{\theta}\right) = F_A, \quad A = 1,2,\ldots,6\,.$$

If the functions f_A (or F_A) are known functions of their arguments x_A, $C_{\alpha\beta}$ and θ, equation (4.36) is a set of six partial differential equations for U. Of course, in order that such a system is integrable the functions f_A must satisfy integrability conditions, namely

$$(4.37)\qquad \frac{\partial f_A}{\partial x_B} = \frac{\partial f_B}{\partial x_A}\,, \quad \text{or} \quad \frac{\partial F_A}{\partial x_B} = \frac{\partial F_B}{\partial x_A}\,.$$

In the above case these are identical with $(4.33)^{4,5,6}$ and consequently, (4.36) can indeed be integrated. To construct the solution of (4.36) we write it in vector form.

$$\nabla U = \underline{F}\,,$$

which is more suggestive, because it shows the resemblance with the relation between the potential energy U and a conservative force $\underline{F}$, the conservatism of which is assured by (4.37). The value of U in a point $\underline{x}$ can then be found from the following line integral

$$U(\underline{x}) = \int_{\underline{0}}^{\underline{x}} \underline{F}(\underline{x}').d\underline{x}' + \text{constant} ,$$

which is independent of the path transversed from $\underline{0}$ to $\underline{x}$. Choosing the straight line from $\underline{0}$ to $\underline{x}$, we thus can take

$$\underline{x}' = \underline{x}s, \quad 0 \leq s \leq 1 ,$$

and then obtain

$$U(\underline{x}) = \underline{x}.\int_0^1 \underline{F}(\underline{x}s)ds + \text{constant} .$$

Proceeding in an analogous way with (4.36), we find the following solution:

$$(4.38) \qquad U = U(x_A, C_{\alpha\beta}, \theta) = -\theta^2 \frac{\partial}{\partial\theta}\left(\frac{I}{\theta}\right) + U^o(C_{\alpha\beta}, \theta) ,$$

where $U^o(C_{\alpha\beta}, \theta)$ replaces the integration constant in the preceding analysis and I stands for

$$(4.39) \qquad I = I(x_A, C_{\alpha\beta}, \theta) = x_A \int_0^1 f_A(x_B s, C_{\alpha\beta}, \theta)ds .$$

From this relation it is obvious that I is zero for vanishing electromagnetic fields ($f_A = 0$). Furthermore, by differentiating (4.39) with respect to x_A it immediately follows that

$$(4.40) \qquad f_A = \frac{\partial I}{\partial x_A} .$$

With the representation (4.39) part of our goal is achieved, namely that the internal energy U is separated into two parts. U^o is the specific internal energy when the electromagnetic fields vanish. The term involving I is the correction due to the presence of the electromagnetic fields.

Not all the identities (4.33) have been exploited when constructing the solution (4.38), (4.39). For instance, we still must explore the conditions $(4.33)^3$. They can also be written in the form

$$(4.41) \qquad \frac{1}{\theta^2}\frac{\partial U^o}{\partial C_{\alpha\beta}} = \frac{\partial}{\partial\theta}\left\{\frac{1}{\theta}\frac{\partial I}{\partial C_{\alpha\beta}} - \frac{II_T P_{\alpha\beta}}{2\rho_o\theta}\right\} .$$

This is simplified if we introduce the purely thermoelastic part of the second Piola-Kirchhoff stress tensor ${}^{II}\overset{\circ}{T}{}^{P}_{\alpha\beta}$ by

$$(4.42) \qquad {}^{II}\overset{\circ}{T}{}^{P}_{\alpha\beta} = {}^{II}\overset{\circ}{T}{}^{P}_{\alpha\beta}(C_{\gamma\delta},\theta) = {}^{II}T^{P}_{\alpha\beta}(x_A = 0, C_{\gamma\delta},\theta) \ .$$

Then (4.41) with I = 0 delivers the relation

$$\frac{1}{\theta^2}\frac{\partial U^o}{\partial C_{\alpha\beta}} = -\frac{\partial}{\partial\theta}\left(\frac{\partial\, {}^{II}\overset{\circ}{T}{}^{P}_{\alpha\beta}}{2\rho_o\theta}\right)$$

whence follows

$$\frac{\partial}{\partial\theta}\{\frac{1}{\theta}[\frac{\partial I}{\partial C_{\alpha\beta}} + \frac{1}{2\rho_o}({}^{II}\overset{\circ}{T}{}^{P}_{\alpha\beta} - {}^{II}T^{P}_{\alpha\beta})]\} = 0 \ ,$$

or after integration

$$(4.43) \qquad {}^{II}T^{P}_{\alpha\beta} = {}^{II}\overset{\circ}{T}{}^{P}_{\alpha\beta} + 2\rho_o \frac{\partial I}{\partial C_{\alpha\beta}} + \theta\tau_{\alpha\beta}(x_A, C_{\gamma\delta}) \ .$$

With the aid of $(4.33)^{7,8}$ and (4.40), $\tau_{\alpha\beta}$ can be shown to be independent of x_A, so that

$$\tau_{\alpha\beta} = \tau_{\alpha\beta}(C_{\gamma\delta}) \ .$$

Taking this result into account in equation (4.43) and evaluating the latter at zero magnetic fields leads with the aid of (4.42) to the conclusion that $\tau_{\alpha\beta}$ must be zero. Hence,

$$(4.44) \qquad {}^{II}T^{P}_{\alpha\beta} = {}^{II}\overset{\circ}{T}{}^{P}_{\alpha\beta} + 2\rho_o \frac{\partial I}{\partial C_{\alpha\beta}} \ ,$$

an equation which separates the stress into a thermoelastic and a field part. (4.40) and (4.44) may now be used to write the Gibbs relation (4.32) in the form

$$d\{\eta + \frac{\partial I}{\partial\theta}\} = \frac{1}{\theta}\{dU^o - \frac{1}{2\rho_o}\, {}^{II}\overset{\circ}{T}{}^{P}_{\alpha\beta} dC_{\alpha\beta}\} \ ,$$

whence, since the right-hand side is independent of the electromagnetic fields, it follows that

$$(4.45) \qquad \eta = \eta^o(C_{\alpha\beta},\theta) - \frac{\partial I}{\partial\theta} \ .$$

Thus, it has also been possible to separate the field part from the thermoelastic part of the entropy.

After the introduction of

$$\psi^o = \Psi^o(C_{\alpha\beta},\theta) \ ,$$

as the free energy in case of vanishing electromagnetic fields, a nice interpretation of I follows from $(4.26)^1$ and (4.45). Accordingly,

$$(4.46) \qquad \frac{\partial}{\partial\theta}(\hat{\Psi} - I) = -\eta^o = \frac{\partial\Psi^o}{\partial\theta} \ ,$$

and, hence,

$$\hat{\Psi}(x_A,C_{\alpha\beta},\theta) = \Psi^o(C_{\alpha\beta},\theta) + I(x_A,C_{\alpha\beta},\theta) + g(x_A,C_{\alpha\beta})$$

Use of (4.40), $(4.26)^{2,3}$ and the definition of Ψ^o then shows that $g \equiv 0$, so that

$$(4.47) \qquad \hat{\Psi}(x_A,C_{\alpha\beta},\theta) = \Psi^o(C_{\alpha\beta},\theta) + I(x_A,C_{\alpha\beta},\theta) \ .$$

ψ^o denotes the free energy, when there are no electromagnetic fields present. I is therefore the field part of the Helmholtz free energy.
There is another possible separation of the internal energy, entropy, free energy and the stresses, which, from a practical point of view is as important as the above one. We mean the separation into contributions due to "rigid body processes" and the corrections due to deformations. This approach gives a natural separation of rigid body electrodynamics from that of deformable bodies. The idea is to commence the integration of $(4.33)^{1,2,3}$ with $(4.33)^3$, which with the definitions

$$(4.48) \qquad \begin{aligned} y_A &= (C_{11},C_{22},C_{33},C_{23},C_{31},C_{12}) \ , \\ g_A &= ({}^{II}T^P_{11},{}^{II}T^P_{22},{}^{II}T^P_{33},{}^{II}T^P_{23},{}^{II}T^P_{31},{}^{II}T^P_{12}) \ , \quad (A = 1,2,\ldots,6) \end{aligned}$$

may be written as

$$(4.49) \qquad \frac{\partial U}{\partial y_A} = -\theta^2 \frac{\partial}{\partial\theta}\left(\frac{g_A}{2\rho_o\theta}\right) \ ,$$

where

$$\frac{\partial g_A}{\partial y_B} = \frac{\partial g_B}{\partial y_A} \ ,$$

holds. Using the same approach as before the following equations can be derived:

$$(4.50) \qquad U(x_A,y_A,\theta) = U^R(x_A,\theta) - \theta^2 \frac{\partial}{\partial\theta}\left(\frac{J}{\theta}\right) \ ,$$

where

$$(4.51) \qquad J = J(x_A, y_A, \theta) = \frac{y_A}{2\rho_o} \int_0^1 g_A(x_B, y_B s, \theta) ds \ ,$$

and, moreover,

$$g_A = 2\rho_o \frac{\partial J}{\partial y_A} \ ,$$

$$(4.52) \qquad f_A = f_A^R(x_A, \theta) + \frac{\partial J}{\partial x_A} \ , \qquad \eta = \eta^R(x_A, \theta) - \frac{\partial J}{\partial \theta} \ ,$$

$$\hat{\Psi}(x_A, y_A, \theta) = \Psi^R(x_A, \theta) + J(x_A, y_A, \theta) \ ,$$

where we still note that for the derivation of (4.52) we have written the Gibbs relation (4.32) in the form

$$d(\eta + \frac{\partial J}{\partial \theta}) = \frac{1}{\theta}(dU^R - f_A^R dx_A) \ .$$

In these equations the quantities carrying a superscript R are the internal energy U^R, the electromagnetic fields f_A^R, the entropy η^R and the free energy Ψ^R for zero deformation. They constitute the rigid body contributions and form the constitutive relations of rigid body electrodynamics. All constitutive properties that can be traced back to deformation are contained in the function J.

This completes the constitutive theory. The integration procedure has shown that all difficulties rest on an appropriate determination of the functions I or J (or more generally of ψ).

Of course, the same procedure can also be taken when some of the dependent and independent variables are interchanged. Since the details of the pertinent calculations are the same as above we leave them to the reader.

Before we proceed we would like to mention, that there are still further possibilities of separating the various effects. These can easily be derived if either only $(4.33)^1$, or else $(4.33)^2$ is used in the integration process. If the results contained in all these combinations are put together, the free energy can, for instance, be written as a sum of four parts, one of which is the field-free energy. The second term involves, apart from the deformation, only the electric effects (polarization); in the third term then only magnetization appears. Only in the fourth term does there occur electromagnetic coupling. Hence we have shown that the separation of the interaction phenomena into the four physically clearly defined parts is indeed possible - and justified.

For later reference, we would like to list here also the constitutive equations for the case that, instead of $\mathbb{P}_\alpha/\rho_o$ and $\mu_o\mathbb{M}_\alpha/\rho_o$, $\mathbb{E}_\alpha$ and $\mathbb{H}_\alpha$ are the independent variables. In other words, we shall postulate constitutive equations of the form

$$(4.53) \qquad C = \bar{C}(C_{\alpha\beta}, \mathbb{E}_\alpha, \mathbb{H}_\alpha, \theta, \theta_{,\alpha}, \mathbb{Q}) ,$$

which can easily be obtained from the preceding ones by the Legendre transformation

$$(4.54) \qquad \bar{\psi} = U - \eta\theta - \frac{1}{\rho_o}\mathbb{E}_\alpha\mathbb{P}_\alpha - \frac{\mu_o}{\rho_o}\mathbb{H}_\alpha\mathbb{M}^C_\alpha = \bar{\varepsilon} - \eta\theta = \bar{\Psi}(C_{\alpha\beta}, \mathbb{E}_\alpha, \mathbb{H}_\alpha, \theta) ,$$

which leads to the relations

$$(4.55) \qquad \begin{aligned} \eta &= -\frac{\partial\bar{\Psi}}{\partial\theta} , \\ \mathbb{P}_\alpha &= -\rho_o \frac{\partial\bar{\Psi}}{\partial\mathbb{E}_\alpha} , \\ \mu_o\mathbb{M}^C_\alpha &= -\rho_o \frac{\partial\bar{\Psi}}{\partial\mathbb{H}_\alpha} , \\ II^P_{T_{\alpha\beta}} &= 2\rho_o \frac{\partial\bar{\Psi}}{\partial C_{\alpha\beta}} , \end{aligned}$$

and to the Gibbs relation

$$d\eta = \frac{1}{\theta}\{d\bar{\varepsilon} - \frac{1}{2\rho_o} II^P_{T_{\alpha\beta}} dC_{\alpha\beta} + \frac{\mathbb{P}_\alpha}{\rho_o} d\mathbb{E}_\alpha + \frac{\mu_o\mathbb{M}_\alpha}{\rho_o} d\mathbb{H}_\alpha\} .$$

We conclude this Section with a few complementary remarks. Firstly, in the above constitutive theory we have introduced the classical expression for the entropy flux, and we have done this already when treating this model in the Eulerian description. That this is correct was proved by Hutter [2], and indeed he obtains exactly the same Gibbs relation. Secondly, taking for ψ, or $\bar{\psi}$, in both models the same function of the form (4.25), or (4.54), and for the energy flux the same function of the form (4.23), or (4.53), we automatically guarantee that the two models are thermodynamically equivalent. Finally, we mention that it is particularly easy to linearize the equations in the Lagrangian formulation. We shall demonstrate this in Chapter V.

If we were to follow the order of presentation of the last Chapter, the Lagrangian description of the Maxwell-Minkowski formulation should now follow. For didactic reasons, we shall treat this formulation last.

4.3 THE MATERIAL DESCRIPTION OF THE STATISTICAL AND THE LORENTZ FORMULATION (MODELS IV AND V)

As we have seen in Chapter 2 already, the statistical formulation and the Lorentz formulation are very similar, and indeed in Chapter 3 we demonstrated how the two models could be brought into a one-to-one correspondence. We therefore expect the two models to show a close interrelation in the Lagrangian formulation as well. This is indeed the case and we shall give corroboration for this below.

To begin with, recall that the Maxwell equations in Chapter 1 were stated in terms of the variables $\mathcal{E}_i$, D_i, $\mathcal{H}_i$ and B_i and that these quantities were related to the statistical and Lorentzian variables in (2.104) and (2.119), respectively. Using also the relations

$$(4.56)\qquad P_i^S = P_i^L = P_i , \qquad \mathcal{M}_i = M_i^S + e_{ijk}\dot{x}_j P_k = M_i^L ,$$

it is then easy to show that the Maxwell equations (1.21), (1.22), (1.26) and (1.27) may be written in the form

$$\frac{d}{dt}\int_{S} B_i da_i + \int_{\partial S} \mathcal{E}_i d\ell_i = 0 ,$$

$$\int_{\partial V} B_i da_i = 0 ,$$

$$(4.57)\qquad -\frac{d}{dt}\int_{S} D_i^a da_i + \int_{\partial S} \mathcal{H}_i^a d\ell_i = \int_{S} \mathcal{J}_i da_i + \frac{d}{dt}\int_{S} P_i da_i + \int_{\partial S} M_i^L d\ell_i ,$$

$$\int_{\partial V} D_i^a da_i = \int_{V} \mathcal{Q}\, dv - \int_{\partial V} P_i da_i ,$$

where we have also introduced the auxiliary quantities

$$(4.58)\qquad D_i^a := \varepsilon_o E_i , \qquad \mathcal{H}_i^a := H_i^a - e_{ijk}\dot{x}_j D_k^a ,$$

with

$$(4.59)\qquad H_i^a := \frac{1}{\mu_o} B_i .$$

Note that even in a non-relativistic formulation

$$H_i^a \neq \mathcal{H}_i^a ,$$

but that in this approximation (recall the rules stated just after (2.70))

$$D_i^a = \varepsilon_o E_i, \qquad \mu_o H_i^a = \mu_o \mathcal{H}_i^a ,$$

(4.60)

$$\frac{1}{\mu_o} B_i B_i = H_i^a B_i = \mathcal{H}_i^a B_i .$$

We remind the reader that E_i and B_i are the electric field strength and magnetic flux density as they occur in the Minkowski, statistical and the Lorentz formulations. Furthermore, as it was convenient to introduce auxiliary fields for the two dipole models, so it is here, and the equations $(4.58)^1$ and (4.59) represent nothing but the familiar Maxwell-Lorentz aether relations.

Before we pass on to the presentation of the material description of the equations (4.57) a few words are in order regarding the interpretation of the various terms in (4.57). Firstly, there is a set of homogeneous equations and another one that is inhomogeneous through the presence of free charge and free current terms. As is seen from these, both polarization and magnetization manifest themselves as distributions of surface charges and surface currents, respectively. Secondly, we have expressed the integral laws (4.57) in terms of Lorentzian variables, but it is not difficult to write them in terms of the variables of the statistical description by simply invoking the transformations (4.56). The only term that changes in this formal substitution is

$$(4.61) \qquad \int_{\partial S} M_i^L d\ell_i = \int_{\partial S} M_i^S d\ell_i + \int_{\partial S} e_{ijk} \dot{x}_j P_k d\ell_i .$$

It is not convenient to absorb the second member on the right-hand side in the term

$$\int_{\partial S} \mathcal{H}_i^a da_i ,$$

as it occurs on the left-hand side of $(4.57)^3$, because in that case $\mathcal{H}_i^a$ would also be expressed in terms which describe the material behavior. Thus, we conclude that in both the statistical and the Lorentz formulations

$$\int_{\partial S} M_i^L d\ell_i$$

is the proper (magnetization) current, and this is one reason for us to regard the Lorentzian description as more advantageous than the statistical description, for it is the former which will directly lead to a material counterpart of the Eulerian variables. In the statistical formulation this current term must be attributed to magnetization as well as polarization.

To arrive at the Lagrangian counterpart of the equations (4.57) one need only transform the integrals back to the reference configuration. If this is done, one obtains

$$\frac{d}{dt}\int_{S_R} \mathbb{B}_\alpha dA_\alpha + \int_{\partial S_R} \mathbb{E}_\alpha dL_\alpha = 0 \ ,$$

$$\int_{\partial V_R} \mathbb{B}_\alpha dA_\alpha = 0 \ ,$$

(4.62)
$$-\frac{d}{dt}\int_{S_R} \mathbb{D}^a_\alpha dA_\alpha + \int_{\partial S_R} \mathbb{H}^a_\alpha dL_\alpha = \int_{S_R} \mathbb{J}_\alpha dA_\alpha + \frac{d}{dt}\int_{S_R} \mathbb{P}_\alpha dA_\alpha + \int_{\partial S_R} \mathbb{M}^L_\alpha dL_\alpha \ ,$$

$$\int_{\partial V_R} \mathbb{D}^a_\alpha dA_\alpha = \int_{V_R} \mathbb{Q}\, dV - \int_{\partial V_R} \mathbb{P}_\alpha dA_\alpha \ ,$$

and from these one may derive in the usual way the conservation law of charges in the form

$$(4.63) \qquad \frac{d}{dt}\int_{V_R} \mathbb{Q}\, dV + \int_{\partial V_R} \mathbb{J}_\alpha dA_\alpha = 0 \ .$$

All newly introduced variables are already defined in (4.6), except for $\mathbb{B}_\alpha$, $\mathbb{H}^a_\alpha$ and $\mathbb{M}^L_\alpha$, which are given by

$$\mathbb{B}_\alpha = JF^{-1}_{\alpha i}B_i, \qquad B_i = J^{-1}F_{i\alpha}\mathbb{B}_\alpha \ ,$$

(4.64)
$$\mathbb{H}^a_\alpha = F_{i\alpha}H^a_i \operatorname{sgn} J, \qquad H^a_i = F^{-1}_{\alpha i}\mathbb{H}^a_\alpha \operatorname{sgn} J \ ,$$

$$\mathbb{M}^L_\alpha = F_{i\alpha}M^L_i \operatorname{sgn} J, \qquad M^L_i = F^{-1}_{\alpha i}\mathbb{M}^L_\alpha \operatorname{sgn} J \ .$$

Specifically, we wish to point out the difference between $\mathbb{M}^L_\alpha$ and $\mathbb{M}^C_\alpha$, and in fact these two fields are related by

$$(4.65) \qquad \mathbb{M}^L_\alpha = \frac{1}{|J|} C_{\alpha\beta}\mathbb{M}^C_\beta \ .$$

For sufficiently smooth fields the equations (4.62) become

$$
(4.66)\quad
\begin{aligned}
&\mathbb{B}_{\alpha,\alpha} = 0 ,\\
&\dot{\mathbb{B}}_{\alpha} + e_{\alpha\beta\gamma}\mathbb{E}_{\gamma,\beta} = 0 ,\\
&\mathbb{D}^{a}_{\alpha,\alpha} = \mathbb{Q} - \mathbb{P}_{\alpha,\alpha} ,\\
&-\dot{\mathbb{D}}^{a}_{\alpha} + e_{\alpha\beta\gamma}\mathbb{H}^{a}_{\gamma,\beta} = \mathbb{J}_{\alpha} + \dot{\mathbb{P}}_{\alpha} + e_{\alpha\beta\gamma}\mathbb{M}^{L}_{\gamma,\beta} ,\\
&\dot{\mathbb{Q}} + \mathbb{J}_{\alpha,\alpha} = 0 .
\end{aligned}
$$

These equations are the Maxwell equations in the material description, as they naturally emerge from the Lorentz or the statistical formulations. As was the case in the Chu formulation, they contain two auxiliary variables which can be expressed in terms of $\mathbb{B}_{\alpha}$ and $\mathbb{E}_{\alpha}$. Indeed, using the relations (4.58) and the transformation rules for E_{i}, D^{a}_{i}, H^{a}_{i} and B_{i}, a straightforward calculation shows that in a nonrelativistic approximation

$$
(4.67)\quad
\begin{aligned}
&\mathbb{D}^{a}_{\alpha} = \varepsilon_{o}|J|C^{-1}_{\alpha\beta}\mathbb{E}_{\beta} ,\\
&\mathbb{H}^{a}_{\alpha} = \frac{1}{|J|}\{\frac{1}{\mu_{o}}\, C_{\alpha\beta}\mathbb{B}_{\beta} - \varepsilon_{o}e_{\mu\beta\gamma}C_{\alpha\beta}F_{j\gamma}\dot{x}_{j}\mathbb{E}_{\mu}\} .
\end{aligned}
$$

Recognize, as was already the case in the material description of the two dipole models that it is through the Maxwell-Lorentz aether relations that the formal linearity of the equations (4.66) is destroyed. Nevertheless, the equations (4.66) appear in a form which is identical to the one in the statistical description when spatial coordinates are used (see (2.105)). Variables are, however, different ones and so are the configurations.
One could, if one so desired, write (4.66) also as

$$
(4.68)\quad
\begin{aligned}
&\mathbb{B}_{\alpha,\alpha} = 0 ,\\
&\dot{\mathbb{B}}_{\alpha} + e_{\alpha\beta\gamma}\mathbb{E}_{\gamma,\beta} = 0 ,\\
&\mathbb{D}_{\alpha,\alpha} = \mathbb{Q} ,\\
&-\dot{\mathbb{D}}_{\alpha} + e_{\alpha\beta\gamma}\mathbb{H}_{\gamma,\beta} = \mathbb{J}_{\alpha} ,\\
&\dot{\mathbb{Q}} + \mathbb{J}_{\alpha,\alpha} = 0 ,
\end{aligned}
$$

where

$$
(4.69)\quad \mathbb{D}_{\alpha} := \mathbb{D}^{a}_{\alpha} + \mathbb{P}_{\alpha} \qquad \text{and} \qquad \mathbb{H}_{\alpha} := \mathbb{H}^{a}_{\alpha} - \mathbb{M}^{L}_{\alpha} ,
$$

and would in this way formally arrive at a material Minkowski formulation. We shall come back to this in the next Section. Note that (4.68) agrees with (1.47) and thus $\mathbb{D}_{\alpha}$ and $\mathbb{H}_{\alpha}$ also agree with $\mathbb{D}_{\alpha}$ and $\mathbb{H}_{\alpha}$ introduced there.

The balance laws (4.57) imply also jump conditions which can easily be derived from (1.60) by simply invoking (4.69). This yields

$$(4.70)\quad \begin{aligned} &[\![\mathbb{B}_\alpha]\!]N_\alpha = 0\,, \qquad e_{\alpha\beta\gamma}[\![\mathbb{E}_\beta]\!]N_\gamma + [\![\mathbb{B}_\alpha W_N]\!] = 0\,,\\ &[\![\mathbb{D}^a_\alpha + \mathbb{P}_\alpha]\!]N_\alpha = 0\,, \qquad e_{\alpha\beta\gamma}[\![\mathbb{H}^a_\beta - \mathbb{M}^L_\beta]\!]N_\gamma - [\![(\mathbb{D}^a_\alpha + \mathbb{P}_\alpha)W_N]\!] = 0\,,\\ &[\![\mathbb{J}_\alpha]\!]N_\alpha + [\![\mathbb{Q}W_N]\!] = 0\,. \end{aligned}$$

To complete the description, we must also derive the Lagrangian versions of the body force, body couple and energy supply expressions of electromagnetic origin. For this purpose we simply express these quantities, which in (2.106) and (2.123) are written in terms of the Eulerian variables, in the Lagrangian fields $\mathbb{E}_\alpha$, $\mathbb{B}_\alpha$, $\mathbb{P}_\alpha$, $\mathbb{M}^L_\alpha$ etc. Starting from (2.123), we obtain for the Lorentz formulation <u>(model V)</u>

$$(4.71)\quad \begin{aligned} &{}^V(\rho_o F^e_i) = F^{-1}_{\alpha i}\{(\mathbb{Q} - \mathbb{P}_{\beta,\beta})\mathbb{E}_\alpha + e_{\alpha\beta\gamma}(\mathbb{J}_\beta + \dot{\mathbb{P}}_\beta)\mathbb{B}_\gamma + (\mathbb{M}^L_{\alpha,\beta} - \mathbb{M}^L_{\beta,\alpha})\mathbb{B}_\beta\}\,,\\ &{}^V(\rho_o L^e_{ij}) = 0\,,\\ &{}^V(\rho_o r^e) = \mathbb{J}_\alpha\mathbb{E}_\alpha + \dot{\mathbb{P}}_\alpha\mathbb{E}_\alpha + e_{\alpha\beta\gamma}\mathbb{M}^L_{\gamma,\beta}\mathbb{E}_\alpha\,. \end{aligned}$$

For the derivation of the corresponding expressions of the statistical model it is convenient to make use of the relation (compare (3.31))

$$(4.72)\quad {}^{IV}(\rho F^e_i) = {}^V(\rho F^e_i) + (E_iP_j - M_iB_j + \delta_{ij}M_kB_k)_{,j}\,,$$

an expression which can be derived from (2.106) and (2.123). Alternatively the expressions for the body couple and energy supply follow from $(2.106)^2$ and $(2.106)^3$, respectively, the latter written in the form

$$(4.73)\quad {}^{IV}(\rho r^e) = J_iE_i + (\overset{*}{P}_i + P_j\dot{x}_{i,j})E_i - (\overset{*}{B}_i - B_i\dot{x}_{j,j} + B_j\dot{x}_{i,j})M_i\,.$$

Transforming the above expressions to referential coordinates we obtain for <u>model IV</u>

$$(4.74)\quad \begin{aligned} &{}^{IV}(\rho_o F^e_i) = F^{-1}_{\alpha i}\{(\mathbb{Q} - \mathbb{P}_{\beta,\beta})\mathbb{E}_\alpha + e_{\alpha\beta\gamma}(\mathbb{J}_\beta + \dot{\mathbb{P}}_\beta)\mathbb{B}_\gamma +\\ &\qquad + (\mathbb{M}^L_{\alpha,\beta} - \mathbb{M}^L_{\beta,\alpha})\mathbb{B}_\beta\} + \{F^{-1}_{\alpha i}(\mathbb{E}_\alpha\mathbb{P}_\beta - \mathbb{M}^L_\alpha\mathbb{B}_\beta + \delta_{\alpha\beta}\mathbb{M}^L_\gamma\mathbb{B}_\gamma)\}_{,\beta}\,,\\ &{}^{IV}(\rho_o L^e_{ij}) = F_{[i\alpha}F^{-1}_{\beta j]}(\mathbb{P}_\alpha\mathbb{E}_\beta - \mathbb{B}_\alpha\mathbb{M}^L_\beta)\,,\\ &{}^{IV}(\rho_o r^e) = \mathbb{J}_\alpha\mathbb{E}_\alpha + \dot{\mathbb{P}}_\alpha\mathbb{E}_\alpha - \mathbb{M}^L_\alpha\dot{\mathbb{B}}_\alpha + \dot{F}_{k\gamma}\{(\mathbb{E}_\beta\mathbb{P}_\gamma - \mathbb{M}^L_\beta\mathbb{B}_\gamma)F^{-1}_{\beta k} + \mathbb{M}^L_\beta\mathbb{B}_\beta F^{-1}_{\gamma k}\}\,. \end{aligned}$$

We would like to point out that the body force expression $(4.71)^1$ is seemingly different from the one given in [1]. However, when using the identity

$$(4.75) \qquad F^{-1}_{\nu i,\mu}\mathbb{B}_\mu \mathbb{M}^L_\nu - F_{k\delta}F^{-1}_{\mu i}F^{-1}_{\nu k,\mu}\mathbb{B}_\delta \mathbb{M}^L_\nu = F_{k\mu}(F^{-1}_{\nu i,k} - F^{-1}_{\nu k,i})\mathbb{B}_\mu \mathbb{M}^L_\nu = 0 ,$$

which was not observed in [1], the body force expressions turn out to be identical.

We see that the body force expressions in the two formulations differ by a term that is the divergence of a tensor. It corresponds to the divergence term occurring already in the Eulerian body force. Particularly interesting is a comparison of the electromagnetic energy supply terms listed in $(4.71)^3$ and $(4.74)^3$. Using $(4.66)^2$ it can be shown that ${}^{IV}(\rho_o r^e)$ may also be written as

$$(4.76) \qquad \begin{aligned} {}^{IV}(\rho_o r_e) = {} & \mathbb{J}_\alpha \mathbb{E}_\alpha + \dot{\mathbb{P}}_\alpha \mathbb{E}_\alpha + e_{\alpha\beta\gamma}\mathbb{M}^L_{\gamma,\beta}\mathbb{E}_\alpha + \\ & + (e_{\alpha\beta\gamma}\mathbb{E}_\beta \mathbb{M}^L_\gamma)_{,\alpha} + \dot{F}_{k\gamma}\{(\mathbb{E}_\beta \mathbb{P}_\gamma - \mathbb{M}^L_\beta \mathbb{B}_\gamma)F^{-1}_{\beta k} + \mathbb{M}^L_\beta \mathbb{B}_\beta F^{-1}_{\gamma k}\} . \end{aligned}$$

Recognizing that, as already said several times before, for a comparison of the energy supplies of two formulations we need not compare the term $\rho_o r^e$ alone but rather the combination $(\rho_o r^e + t_{ij}\dot{x}_{i,j})$. Thus when performing this comparison we see that the last term of expression (4.76) for ${}^{IV}(\rho_o r^e)$ originates from the difference in the stress tensors of the models IV and V. Indeed with the aid of (3.31) we easily show that

$$(4.77) \qquad {}^{IV}T_{i\alpha} = {}^{V}T_{i\alpha} + F^{-1}_{\beta i}(\mathbb{P}_\alpha \mathbb{E}_\beta - \mathbb{B}_\alpha \mathbb{M}^L_\beta) - \mathbb{M}^L_\gamma \mathbb{M}^L_\gamma F^{-1}_{\alpha i} .$$

But the term

$$(e_{\alpha\beta\gamma}\mathbb{E}_\beta \mathbb{M}^L_\gamma)_{,\alpha}$$

cannot be explained in this way. Consequently either the energy fluxes Q_α or the entropy fluxes Φ_α of the two formulations must differ. Indeed, we may choose as was already done in the Eulerian description

$$(4.78) \qquad Q^S_\alpha = Q^L_\alpha + e_{\alpha\beta\gamma}\mathbb{E}_\beta \mathbb{M}^L_\gamma ,$$

and then obtain

$$(4.79) \qquad \Phi^S_\alpha = \frac{Q^S_\alpha}{\theta} = \frac{Q^L_\alpha + e_{\alpha\beta\gamma}\mathbb{E}_\beta \mathbb{M}^L_\gamma}{\theta} = \Phi^L_\alpha .$$

This form of the entropy flux vector was proved to be correct by Hutter [1] in a theory of viscous isotropic thermoelastic, polarizable and magnetizable solids. Here we treat (4.78) as a postulate.

When we compare the expressions for the electromagnetic energy supply $(4.71)^3$ and $(4.74)^3$ with the corresponding Eulerian expressions, $(2.106)^3$ and $(2.123)^3$, we recognize that the Lorentzian expression ${}^{V}(\rho_o r^e)$ is formally the same as its spatial counterpart ${}^{V}(\rho r^e)$. This property is not shared by ${}^{IV}(\rho_o r^e)$ and ${}^{IV}(\rho r^e)$.

Finally, we note that the expressions for the electromagnetic body couple and body force are simpler in the Lorentz formulation than in the statistical formulation. All this, of course, are reasons which make the Lorentz formulation to be (formally) superior to the statistical one, although as we have shown, they are entirely equivalent.

We proceed with the jump conditions for momentum and energy of matter and fields as they are listed in their general form in (1.67). Therefore, it suffices to evaluate $T^M_{i\alpha}$, G_i, Π_α and Ω. These quantities are obtained if use is made of the transformations (4.16) for t^M_{ij}, g_i, π_i and ω, which in (2.111) and (2.126) are given for the statistical and the Lorentz formulation, respectively. When the indicated transformations are performed the following relations are obtained:

i) in the Lorentz formulation <u>(model V)</u>

$$
\begin{aligned}
{}^{V}T^M_{i\alpha} &= F^{-1}_{\beta i}(\mathbb{D}^a_\alpha \mathbb{E}_\beta + \mathbb{B}_\alpha \mathbb{H}^a_\beta) - \tfrac{1}{2} F^{-1}_{\alpha i}(\mathbb{D}^a_\beta \mathbb{E}_\beta + \mathbb{B}_\beta \mathbb{H}^a_\beta) , \\
{}^{V}G_i &= 0 , \\
{}^{V}\Pi_\alpha &= -e_{\alpha\beta\gamma}\mathbb{E}_\beta \mathbb{H}^a_\gamma + F^{-1}_{\beta i}(\mathbb{D}^a_\alpha \mathbb{E}_\beta + \mathbb{B}_\alpha \mathbb{H}^a_\beta)\dot{x}_i - \tfrac{1}{2} F^{-1}_{\alpha i}(\mathbb{D}^a_\beta \mathbb{E}_\beta + \mathbb{B}_\beta \mathbb{H}^a_\beta)\dot{x}_i = \\
&= -e_{\alpha\beta\gamma}\mathbb{E}_\beta \mathbb{H}^a_\gamma + {}^{V}T^M_{i\alpha}\dot{x}_i , \\
{}^{V}\Omega &= -\tfrac{1}{2}(\mathbb{D}^a_\alpha \mathbb{E}_\alpha + \mathbb{B}_\alpha \mathbb{H}^a_\alpha) ,
\end{aligned}
\tag{4.80}
$$

ii) in the statistical formulation <u>(model IV)</u>

$$
\begin{aligned}
{}^{IV}T^M_{i\alpha} &= {}^{V}T^M_{i\alpha} + F^{-1}_{\beta i}(\mathbb{P}_\alpha \mathbb{E}_\beta - \mathbb{B}_\alpha \mathbb{M}^L_\beta + \delta_{\alpha\beta}\mathbb{B}_\gamma \mathbb{M}^L_\gamma) , \\
{}^{IV}G_i &= 0 , \\
{}^{IV}\Pi_\alpha &= {}^{V}\Pi_\alpha + e_{\alpha\beta\gamma}\mathbb{E}_\beta \mathbb{M}^L_\gamma + F^{-1}_{\beta i}(\mathbb{P}_\alpha \mathbb{E}_\beta - \mathbb{B}_\alpha \mathbb{M}^L_\beta + \delta_{\alpha\beta}\mathbb{B}_\gamma \mathbb{M}^L_\gamma)\dot{x}_i = \\
&= -e_{\alpha\beta\gamma}\mathbb{E}_\beta(\mathbb{H}^a_\gamma - \mathbb{M}^L_\gamma) + {}^{IV}T^M_{i\alpha}\dot{x}_i ,
\end{aligned}
\tag{4.81}
$$

(over)

$$^{IV}V_{\Omega} = {}^{V}V_{\Omega} .$$

Substituting the above expressions into (1.67), what emerges are the jump conditions for momentum and energy of matter and fields.

It remains to formulate the constitutive theory of the two models. To this end we derive the reduced entropy inequality for model V by eliminating $\rho_o r^{ext}$ from the energy equation and the entropy inequality, taking thereby into account that $\rho_o r^e$ is given by $(4.71)^3$ and the entropy flux by (4.79), respectively. In this way we obtain

$$(4.82) \qquad -\rho_o\dot{\psi} - \rho_o\eta\dot{\theta} + {}^{V}T_{i\alpha}\dot{F}_{i\alpha} + \mathbb{E}_{\alpha}\dot{\mathbb{P}}_{\alpha} - \mathbb{M}^{L}_{\alpha}\dot{\mathbb{B}}_{\alpha} + \mathbb{E}_{\alpha}\mathbb{J}_{\alpha} - \frac{Q^{S}_{\alpha}\theta_{,\alpha}}{\theta} \geq 0 ,$$

where ${}^{V}T_{i\alpha}$ is the first Piola-Kirchhoff stress tensor in the Lorentz formulation. It should be noticed that in (4.82) the heat flux vector Q^{S}_{α} is used instead of Q^{L}_{α}.

Constitutive relations are now written as

$$(4.83) \qquad C = \overset{+}{C}(C_{\alpha\beta}, \mathbb{P}_{\alpha}/\rho_o, \mathbb{B}_{\alpha}, \theta, \theta_{,\alpha}, \mathbb{Q}) .$$

A short calculation then shows, as usual, that

$$(4.84) \qquad \psi = U - \eta\theta = \overset{+}{\Psi}(C_{\alpha\beta}, \mathbb{P}_{\alpha}/\rho_o, \mathbb{B}_{\alpha}, \theta) ,$$

and

$$(4.85) \qquad \begin{aligned} \eta &= -\frac{\partial\overset{+}{\Psi}}{\partial\theta} , \\ \mathbb{E}_{\alpha} &= \frac{\partial\overset{+}{\Psi}}{\partial(\mathbb{P}_{\alpha}/\rho_o)} , \\ \mathbb{M}^{L}_{\alpha} &= -\rho_o\frac{\partial\overset{+}{\Psi}}{\partial\mathbb{B}_{\alpha}} , \\ {}^{V}T^{P}_{\alpha\beta} &= 2\rho_o\frac{\partial\overset{+}{\Psi}}{\partial C_{\alpha\beta}} . \end{aligned}$$

Here, $T^{P}_{\alpha\beta}$ is the second Piola-Kirchhoff stress tensor which is defined in terms of $T_{i\alpha}$ in (1.40).

With these results the energy equation reduces to its ultimate form

$$(4.86) \qquad \rho_o\theta\dot{\eta} = \mathbb{E}_{\alpha}\mathbb{J}_{\alpha} - Q^{S}_{\alpha,\alpha} + \rho_o r^{ext} .$$

On the other hand, the reduced entropy inequality becomes

$$(4.87)\qquad \mathbb{E}_\alpha \mathbb{J}_\alpha - \frac{Q^S_\alpha \theta_{,\alpha}}{\theta} \geq 0 \ ,$$

and the Gibbs relation reads

$$(4.88)\qquad d\eta = \frac{1}{\theta}\{dU - \frac{1}{2\rho_o}\,{}^V T^P_{\alpha\beta} dC_{\alpha\beta} - \mathbb{E}_\alpha d(\frac{\mathbb{P}_\alpha}{\rho_o}) + \frac{\mathbb{M}^L_\alpha}{\rho_o}\, d\mathbb{B}_\alpha\} \ ,$$

an equation that was also derived in [1].
The results for model IV and V are identical, except for the stress tensors the relation of which is given in (4.77).
It follows from the latter and $(4.85)^4$ that

$$(4.89)\qquad {}^{IV}T^P_{\alpha\beta} = 2\rho_o \frac{\partial \overset{+}{\psi}}{\partial C_{\alpha\beta}} + (\mathbb{P}_\alpha \mathbb{E}_\gamma - \mathbb{B}_\alpha \mathbb{M}^L_\gamma) C^{-1}_{\beta\gamma} + \mathbb{M}^L_\gamma \mathbb{M}^L_\gamma C^{-1}_{\alpha\beta} \ .$$

With this constitutive relation for the second Piola-Kirchhoff stress tensor, the balance law of angular momentum for the statistical formulation is satisfied identically.

For later use, we wish to give constitutive relations also of the form

$$(4.90)\qquad C = \overset{\vee}{C}(C_{\alpha\beta}, \mathbb{E}_\alpha, \mathbb{B}_\alpha, \theta, \theta_{,\alpha}, \mathbb{Q}) \ .$$

With the Legendre transformation

$$(4.91)\qquad \overset{\vee}{\psi} = U - \eta\theta - \frac{1}{\rho_o} \mathbb{E}_\alpha \mathbb{P}_\alpha = \overset{\vee}{\varepsilon} - \eta\theta = \overset{\vee}{\psi}(C_{\alpha\beta}, \mathbb{E}_\alpha, \mathbb{B}_\alpha, \theta) \ ,$$

this immediately leads to

$$(4.92)\qquad \begin{aligned} \eta &= -\frac{\partial \overset{\vee}{\psi}}{\partial \theta} \ , \\ \mathbb{P}_\alpha &= -\rho_o \frac{\partial \overset{\vee}{\psi}}{\partial \mathbb{E}_\alpha} \ , \\ \mathbb{M}^L_\alpha &= -\rho_o \frac{\partial \overset{\vee}{\psi}}{\partial \mathbb{B}_\alpha} \ , \\ {}^V T^P_{\alpha\beta} &= 2\rho_o \frac{\partial \overset{\vee}{\psi}}{\partial C_{\alpha\beta}} \ , \\ d\eta &= \frac{1}{\theta}\{d\overset{\vee}{\varepsilon} - \frac{1}{2\rho_o}\,{}^V T^P_{\alpha\beta} dC_{\alpha\beta} + \frac{\mathbb{P}_\alpha}{\rho_o}\, d\mathbb{E}_\alpha + \frac{\mathbb{M}^L_\alpha}{\rho_o}\, d\mathbb{B}_\alpha\} \ . \end{aligned}$$

As was done in the Chu formulation we could now, if we so desired, also explore the consequences implied by the Gibbs relation and would then again be able to show that constitutive relations and free energy are separable into a thermoelastic and a field part or else, a rigid body part and a part due to deformation. The procedure is entirely analogous to the one shown before, and therefore we leave the pertinent details to the reader.

4.4 THE MATERIAL DESCRIPTION OF THE MAXWELL - MINKOWSKI FORMULATION

We saw in Section 2.3 that the spatial electromagnetic field variables in the Maxwell-Minkowski formulation are formally the same as those in the statistical description, and that the latter are closely related to the variables in the Lorentz formulation. We therefore adopt the same material electromagnetic variables as in the last Section, which are given by (4.64). There is only one difference that must be noted. Magnetization and polarization are regarded as auxiliary variables while the dielectric displacement and the magnetic field strength are considered basic. The Lagrangian counterparts of these fields can easily be read off from (4.69), and it is not difficult to show that $\mathbb{D}_\alpha$ and $\mathbb{H}_\alpha$ are related to H_i and D_i according to

$$\text{(4.93)}\quad \begin{aligned} \mathbb{D}_\alpha &= |J| F^{-1}_{\alpha i} D_i , & D_i &= \frac{1}{|J|} F_{i\alpha} \mathbb{D}_\alpha , \\ \mathbb{H}_\alpha &= F_{i\alpha} H_i \operatorname{sgn} J , & H_i &= F^{-1}_{\alpha i} \mathbb{H}_\alpha \operatorname{sgn} J . \end{aligned}$$

The Maxwell equations, expressed in $\mathbb{E}_\alpha$, $\mathbb{B}_\alpha$, $\mathbb{D}_\alpha$ and $\mathbb{H}_\alpha$, are already given in (4.68) and we refrain from repeating them here.

Before passing on to the presentation of electromagnetic body force, body couple and energy supply, one remark concerning the auxiliary fields must be made. In view of the properties of this formulation as just outlined one would expect $\mathbb{P}_\alpha$ and $\mathbb{M}^L_\alpha$ as auxiliary variables. However, since the description becomes formally much simpler when $\mathbb{M}^C_\alpha$ is used instead, we shall prefer to use the latter. In that case the auxiliary fields may be obtained from (4.69), (4.65) and (4.67); they are

$$\text{(4.94)}\quad \begin{aligned} \mathbb{P}_\alpha &= \mathbb{D}_\alpha - \varepsilon_o |J| C^{-1}_{\alpha\beta} \mathbb{E}_\beta , \\ \mu_o \mathbb{M}^C_\alpha &= \mathbb{B}_\alpha - \mu_o |J| C^{-1}_{\alpha\beta} \mathbb{H}_\beta . \end{aligned}$$

In Section 2.3 the expressions for the electromagnetic body force, body couple and energy supply are derived directly from the global energy balance law (2.77). When written in material form the latter reads

$$
\begin{aligned}
&\frac{d}{dt}\int_{V_R}\{\rho_o U + \frac{1}{2}\rho_o \dot{x}_i\dot{x}_i\}dV - \int_{V_R}\{\rho_o r^{ext} + \rho_o F_i^{ext}\dot{x}_i\}dV + \\
&- \int_{\partial V_R}\{T_{i\alpha}\dot{x}_i - Q_\alpha\}dA_\alpha = \frac{d}{dt}\int_{V_R}\{-\frac{1}{2}|J|C^{-1}_{\alpha\beta}(\varepsilon_o \mathbb{E}_\alpha\mathbb{E}_\beta + \mu_o \mathbb{H}_\alpha\mathbb{H}_\beta)\}dV + \\
&+ \int_{\partial V_R}\{-e_{\alpha\beta\gamma}\mathbb{E}_\beta\mathbb{H}_\gamma + F^{-1}_{\beta i}(\mathbb{D}_\alpha\mathbb{E}_\beta + \mathbb{B}_\alpha\mathbb{H}_\beta)\dot{x}_i + \\
&- \frac{1}{2}|J|F^{-1}_{\alpha i}C^{-1}_{\beta\gamma}(\varepsilon_o\mathbb{E}_\beta\mathbb{E}_\gamma + \mu_o\mathbb{H}_\beta\mathbb{H}_\gamma)\dot{x}_i\}dA_\alpha \ .
\end{aligned} \tag{4.95}
$$

From this equation the expressions for ${}^{III}\Omega$ and ${}^{III}\Pi_\alpha$ can be read off directly. They are equal to the integrands of the first and the second integral on the right-hand side of (4.95).

With the aid of the Maxwell equations (4.68) and the relations (4.94) this balance law can be transformed into the form

$$
\begin{aligned}
&\int_{V_R}\{[\rho_o\ddot{x}_i - \rho_o F_i^{ext} - T_{i\alpha,\alpha} - \{F^{-1}_{\beta i}(\mathbb{D}_\alpha\mathbb{E}_\beta + \mathbb{B}_\alpha\mathbb{H}_\beta) + \\
&- \frac{1}{2}|J|F^{-1}_{\alpha i}C^{-1}_{\beta\gamma}(\varepsilon_o\mathbb{E}_\beta\mathbb{E}_\gamma + \mu_o\mathbb{H}_\beta\mathbb{H}_\gamma)\}_{,\alpha}]\dot{x}_i + \rho_o\dot{U} - \rho_o r^{ext} + Q_{\alpha,\alpha} + \\
&-\mathbb{J}_\alpha\mathbb{E}_\alpha - \dot{\mathbb{P}}_\alpha\mathbb{E}_\alpha - \mu_o\dot{\mathbb{M}}^C_\alpha\mathbb{H}_\alpha - [T_{i\alpha} + F^{-1}_{\beta i}(\mathbb{P}_\alpha\mathbb{E}_\beta + \mu_o\mathbb{M}^C_\alpha\mathbb{H}_\beta)]\dot{F}_{i\alpha}\}dV \ .
\end{aligned} \tag{4.96}
$$

By applying invariance requirements as was done in Section 2.3, it is possible to derive from (4.96) local balance equations of linear and angular momentum and energy in a material formulation. Comparing these equations with those given in (1.39) then yields the material versions of the electromagnetic body force, body couple and energy supply. In the derivation of the angular momentum equation it must be observed that $\mathbb{P}_\alpha$ and $\mathbb{M}^C_\alpha$ are objective quantities under the Euclidean transformation group (in contrast to $\dot{P}_i$ and $\dot{M}_i$) this because $\mathbb{P}_\alpha$ and $\mathbb{M}^C_\alpha$ are objective scalars. In this way we obtain

$$
\begin{aligned}
{}^{III}(\rho_o F^e_i) &= \{F^{-1}_{\beta i}(\mathbb{D}_\alpha\mathbb{E}_\beta + \mathbb{B}_\alpha\mathbb{H}_\beta) - \frac{1}{2}|J|F^{-1}_{\alpha i}C^{-1}_{\beta\gamma}(\varepsilon_o\mathbb{E}_\beta\mathbb{E}_\gamma + \mu_o\mathbb{H}_\beta\mathbb{H}_\gamma)\}_{,\alpha} = \\
&= F^{-1}_{\alpha i}\{\mathbb{Q}\,\mathbb{E}_\alpha + e_{\alpha\beta\gamma}\mathbb{J}_\beta\mathbb{B}_\gamma + \mathbb{P}_\beta\mathbb{E}_{\beta,\alpha} + \mu_o\mathbb{M}^C_\beta\mathbb{H}_{\beta,\alpha} + \\
&+ e_{\alpha\beta\gamma}(\mathbb{D}_\beta\dot{\mathbb{B}}_\gamma + \dot{\mathbb{D}}_\beta\mathbb{B}_\gamma)\} + F^{-1}_{\beta i,j}F_{j\alpha}(\mathbb{P}_\alpha\mathbb{E}_\beta + \mu_o\mathbb{M}^C_\alpha\mathbb{H}_\beta) \ , \\
{}^{III}(\rho_o L^e_{ij}) &= F_{[i\alpha}F^{-1}_{\beta j]}(\mathbb{P}_\alpha\mathbb{E}_\beta + \mu_o\mathbb{M}^C_\alpha\mathbb{H}_\beta) \ , \\
{}^{III}(\rho_o r^e) &= \mathbb{J}_\alpha\mathbb{E}_\alpha + \dot{\mathbb{P}}_\alpha\mathbb{E}_\alpha + \mu_o\dot{\mathbb{M}}^C_\alpha\mathbb{H}_\alpha + F^{-1}_{\beta i}(\mathbb{P}_\alpha\mathbb{E}_\beta + \mu_o\mathbb{M}^C_\alpha\mathbb{H}_\beta)\dot{F}_{i\alpha} \ .
\end{aligned} \tag{4.97}
$$

Of course, these expressions can also be obtained from their spatial versions (2.80), (2.81) and (2.82), by transforming the latter into their Lagrangian counterparts.

Furthermore, from (4.97)[1] and (4.95) the electromagnetic momentum G_i, Maxwell stress $T^M_{i\alpha}$, electromagnetic energy flux Π_α and energy density Ω are obtained as

$$
\begin{aligned}
{}^{III}T^M_{i\alpha} &= F^{-1}_{\beta i}(\mathbb{D}_\alpha\mathbb{E}_\beta + \mathbb{B}_\alpha\mathbb{H}_\beta) - \tfrac{1}{2}|J|F^{-1}_{\alpha i}C^{-1}_{\beta\gamma}(\varepsilon_o\mathbb{E}_\beta\mathbb{E}_\gamma + \mu_o\mathbb{H}_\beta\mathbb{H}_\gamma)\ , \\
{}^{III}G_i &= 0\ , \\
{}^{III}\Pi_\alpha &= -e_{\alpha\beta\gamma}\mathbb{E}_\beta\mathbb{H}_\gamma + F^{-1}_{\beta i}(\mathbb{D}_\alpha\mathbb{E}_\beta + \mathbb{B}_\alpha\mathbb{H}_\beta)\dot{x}_i - \tfrac{1}{2}|J|F^{-1}_{\alpha i}C^{-1}_{\beta\gamma}(\varepsilon_o\mathbb{E}_\beta\mathbb{E}_\gamma + \mu_o\mathbb{H}_\beta\mathbb{H}_\gamma)\dot{x}_i = \\
&= -e_{\alpha\beta\gamma}\mathbb{E}_\beta\mathbb{H}_\gamma + {}^{III}T^M_{i\alpha}\dot{x}_i\ , \\
{}^{III}\Omega &= -\tfrac{1}{2}|J|C^{-1}_{\alpha\beta}(\varepsilon_o\mathbb{E}_\alpha\mathbb{E}_\beta + \mu_o\mathbb{H}_\alpha\mathbb{H}_\beta)\ .
\end{aligned}
\tag{4.98}
$$

Substitution of (4.98) into (1.67) then yields the jump conditions for momentum and energy of matter and fields.

It remains to formulate the constitutive theory. To this end, the reduced entropy inequality must be derived. Introducing

$$\psi = U - \eta\theta - \frac{1}{\rho_o}\mathbb{E}_\alpha\mathbb{P}_\alpha\ , \tag{4.99}$$

and

$$\Phi_\alpha = \frac{Q_\alpha}{\theta}\ , \tag{4.100}$$

we obtain, as usual,

$$
\begin{aligned}
&-\rho_o\dot{\psi} - \rho_o\eta\dot{\theta} - \mathbb{P}_\alpha\dot{\mathbb{E}}_\alpha + \mu_o\dot{\mathbb{M}}^C_\alpha\mathbb{H}_\alpha + \\
&+ [T_{i\alpha} + F^{-1}_{\beta i}(\mathbb{P}_\alpha\mathbb{E}_\beta + \mu_o\mathbb{M}^C_\alpha\mathbb{H}_\beta)]\dot{F}_{i\alpha} + \mathbb{J}_\alpha\mathbb{E}_\alpha - \frac{Q_\alpha\theta_{,\alpha}}{\theta} \geq 0\ .
\end{aligned}
\tag{4.101}
$$

Assuming constitutive relations of the form

$$C = \tilde{C}(C_{\alpha\beta},\mathbb{E}_\alpha,\frac{\mu_o\mathbb{M}^C_\alpha}{\rho_o},\theta,\theta_{,\alpha},\mathbb{Q})\ , \tag{4.102}$$

a short calculation shows that

$$\psi = \tilde{\Psi}(C_{\alpha\beta},\mathbb{E}_\alpha,\frac{\mu_o\mathbb{M}^C_\alpha}{\rho_o},\theta)\ , \tag{4.103}$$

and that

$$
(4.104)\quad
\begin{aligned}
\eta &= -\frac{\partial\tilde{\Psi}}{\partial\theta}\,,\\
\mathbb{P}_\alpha &= -\frac{\partial\tilde{\Psi}}{\partial\mathbb{E}_\alpha}\,,\\
\mathbb{H}_\alpha &= \frac{\partial\tilde{\Psi}}{\partial\mu_o\mathbb{M}^C_\alpha/\rho_o}\,,\\
{}^{III}T_{i\alpha} &= 2\rho_o\frac{\partial\tilde{\Psi}}{\partial C_{\alpha\beta}}F_{i\beta} - F^{-1}_{\beta i}(\mathbb{P}_\alpha\mathbb{E}_\beta + \mu_o\mathbb{M}^C_\alpha\mathbb{H}_\beta)\,.
\end{aligned}
$$

Moreover, it is easily shown that with the relation $(4.104)^4$ the balance law of moment of momentum is satisfied identically. Furthermore,

$$
(4.105)\quad {}^{III}T^P_{\alpha\beta} = 2\rho_o\frac{\partial\tilde{\Psi}}{\partial C_{\alpha\beta}} - C^{-1}_{\beta\gamma}(\mathbb{P}_\alpha\mathbb{E}_\gamma + \mu_o\mathbb{M}^C_\alpha\mathbb{H}_\gamma)
$$

When use is made of (4.104), (4.101) reduces to the residual inequality

$$
\mathbb{E}_\alpha\mathbb{J}_\alpha - \frac{Q_\alpha\theta_{,\alpha}}{\theta} \geq 0\,.
$$

On the other hand, the energy balance reduces in form to relation (4.30), and (4.99) and (4.104) imply

$$
(4.106)\quad
\begin{aligned}
\frac{\partial\eta}{\partial\theta} &= \frac{1}{\theta}\frac{\partial\tilde{\varepsilon}}{\partial\theta}\,,\\
\frac{\partial\eta}{\partial\mathbb{E}_\alpha} &= \frac{1}{\theta}\Big[\frac{\partial\tilde{\varepsilon}}{\partial\mathbb{E}_\alpha} + \mathbb{P}_\alpha\Big]\,,\\
\frac{\partial\eta}{\partial\mu_o\mathbb{M}^C_\alpha/\rho_o} &= \frac{1}{\theta}\Big[\frac{\partial\tilde{\varepsilon}}{\partial\mu_o\mathbb{M}^C_\alpha/\rho_o} - \mathbb{H}_\alpha\Big]\,,\\
\frac{\partial\eta}{\partial C_{\alpha\beta}} &= \frac{1}{\theta}\Big[\frac{\partial\tilde{\varepsilon}}{\partial C_{\alpha\beta}} - \frac{1}{2\rho_o}\{{}^{III}T^P_{\alpha\beta} + C^{-1}_{\beta\gamma}(\mathbb{P}_\alpha\mathbb{E}_\gamma + \mu_o\mathbb{M}^C_\alpha\mathbb{H}_\gamma)\}\Big]\,,
\end{aligned}
$$

where

$$
(4.107)\quad \tilde{\varepsilon} := U - \frac{1}{\rho_o}\mathbb{E}_\alpha\mathbb{P}_\alpha = \tilde{\Psi} - \eta\theta\,,
$$

from which the Gibbs relation

$$
(4.108)\quad
\begin{aligned}
d\eta = \frac{1}{\theta}\Big[&d\tilde{\varepsilon} + \mathbb{P}_\alpha d\mathbb{E}_\alpha - \mathbb{H}_\alpha d\Big(\frac{\mu_o\mathbb{M}^C_\alpha}{\rho_o}\Big) +\\
&- \frac{1}{2\rho_o}\{{}^{III}T^P_{\alpha\beta} + C^{-1}_{\beta\gamma}(\mathbb{P}_\alpha\mathbb{E}_\gamma + \mu_o\mathbb{M}^C_\alpha\mathbb{H}_\gamma\}dC_{\alpha\beta}\Big]
\end{aligned}
$$

may be derived.

When the functional ψ as defined by (4.99) is replaced by the Helmholtz free energy

$$(4.109) \qquad \psi = U - \eta\theta = \hat{\Psi}(C_{\alpha\beta}, \frac{\mathbb{P}_\alpha}{\rho_o}, \frac{\mu_o \mathbb{M}_\alpha^C}{\rho_o}, \theta) \ ,$$

the constitutive equations (4.104) only change in that $\tilde{\Psi}$ becomes $\hat{\Psi}$, and that the second equation must be replaced by

$$(4.110) \qquad \mathbb{E}_\alpha = \frac{\partial\hat{\Psi}}{\partial\mathbb{P}_\alpha/\rho_o} \ .$$

A comparison of the results of this Section with those of Section 4.2 then immediately shows that the Lagrangian formulations of the models I and III are completely equivalent. As we have already proved the thermodynamical equivalence of the models I and II in Section 4.2 and of the models IV and V in Section 4.3, we need for a comparison of the Lagrangian formulations of the various models only consider one model out of each of the following groups:

i) models I, II and III,

ii) models IV and V.

This comparison between the models II and V will be made in Section 4.6.

4.5 THERMOSTATIC EQUILIBRIUM - CONSTITUTIVE RELATIONS FOR ENERGY FLUX AND ELECTRIC CURRENT

In the above, we derived constitutive equations for entropy, stress and two electromagnetic field vectors, all of which turned out to be derivable from a free energy. We did not present constitutive relations for the free current $\mathbb{J}_\alpha$ and the energy flux vector Q_α. These must be given separately and they are restricted by the residual inequality

$$(4.111) \qquad \gamma := \mathbb{E}_\alpha \mathbb{J}_\alpha - \frac{Q_\alpha \theta_{,\alpha}}{\theta} \geq 0 \ .$$

Here, Q_α denotes the heat flux in all but the Lorentz formulation, where it must be replaced by the right-hand side of (4.78).

As was the case in the Eulerian description, the exploitation of (4.111) depends on whether we are dealing with an electrical conductor or insulator. Hence, we shall discuss the two cases separately.

a) For an electrical insulator ($\mathbb{J}_\alpha = 0$) the residual inequality (4.111) reduces to

$$\gamma = -\frac{Q_\alpha \theta_{,\alpha}}{\theta} \geq 0 \; . \tag{4.112}$$

In thermostatic equilibrium, which will again be characterized by an index $|_E$, that is for time-independent processes with uniform temperature,

$$\gamma\Big|_E = 0 \; .$$

Since γ is non-negative in general, it thus assumes its minimum for thermostatic equilibrium. Of necessity then

$$\left.\frac{\partial\gamma}{\partial\theta_{,\alpha}}\right|_E = 0 \; , \qquad \left.\frac{\partial^2\gamma}{\partial\theta_{,\alpha}\partial\theta_{,\beta}}\right|_E \text{ is positive-semi definite} \; , \tag{4.113}$$

or, when expressed in terms of Q_α

$$Q_\alpha\Big|_E = 0 \; , \qquad \left.\frac{\partial Q_{(\alpha}}{\partial\theta_{,\beta)}}\right|_E \text{ is negative-semi definite} \; . \tag{4.114}$$

To see what consequences these relations impose on Q_α we consider the case in which constitutive relations are prescribed in the form

$$C = \check{C}(C_{\alpha\beta}, \mathbb{E}_\alpha, \mathbb{B}_\alpha, \theta, \theta_{,\alpha}, \mathbb{Q}) \; . \tag{4.115}$$

A necessary and sufficient condition for the energy flux vector Q_α to vanish in thermostatic equilibrium is to write

$$Q_\alpha = -\check{\kappa}_{\alpha\beta}(C_{\gamma\delta}, \mathbb{E}_\gamma, \mathbb{B}_\gamma, \theta, \theta_{,\gamma}, \mathbb{Q})\theta_{,\beta} \; , \tag{4.116}$$

and this implies that $\check{\kappa}_{(\alpha\beta)}\big|_E$ must be positive-semi definite. Nothing can be said about the skewsymmetric part $\check{\kappa}_{[\alpha\beta]}\big|_E$, but when one restricts oneself to small deformations and small deviations from thermostatic equilibrium it follows from the Onsager relations that

$$\check{\kappa}_{[\alpha\beta]}(\delta_{\gamma\delta}, \mathbb{E}_\gamma, \mathbb{B}_\gamma, \theta, 0, \mathbb{Q}) = 0 \; . \tag{4.117}$$

b) In an electrical conductor we may write

$$Q_\alpha = \check{Q}_\alpha(C_{\beta\gamma}, \mathbb{E}_\beta, \mathbb{B}_\beta, \theta, \theta_{,\beta}, \mathbb{Q}) \; , \qquad \mathbb{J}_\alpha = \check{\mathbb{J}}_\alpha(C_{\beta\gamma}, \mathbb{E}_\beta, \mathbb{B}_\beta, \theta, \theta_{,\beta}, \mathbb{Q}) \; . \tag{4.118}$$

Thermostatic equilibrium is defined here as a time-independent process with uniform temperature and vanishing electric field strength $\mathbb{E}_\alpha$. Hence,

$$\gamma\Big|_E = 0 ,$$

must hold here too, and thus the following conditions emerge

$$\frac{\partial\gamma}{\partial\mathbb{E}_\alpha}\Big|_E = 0, \quad \text{and} \quad \frac{\partial\gamma}{\partial\theta_{,\alpha}}\Big|_E = 0 ,$$

(4.119)

$$\begin{pmatrix} \dfrac{\partial^2\gamma}{\partial\mathbb{E}_\alpha\partial\mathbb{E}_\beta} & \dfrac{\partial^2\gamma}{\partial\mathbb{E}_\alpha\partial\theta_{,\beta}} \\ \dfrac{\partial^2\gamma}{\partial\mathbb{E}_\alpha\partial\theta_{,\beta}} & \dfrac{\partial^2\gamma}{\partial\theta_{,\alpha}\partial\theta_{,\beta}} \end{pmatrix}\Bigg|_E \quad \text{is positive-semi definite .}$$

Of necessity then

(4.120) $$\mathbb{J}_\alpha\Big|_E = 0, \quad \text{and} \quad Q_\alpha\Big|_E = 0 ,$$

as well as

(4.121) $$\begin{pmatrix} \dfrac{\partial\check{\mathbb{J}}_{(\alpha}}{\partial\mathbb{E}_{\beta)}} & \dfrac{\partial\check{\mathbb{J}}_\alpha}{\partial\theta_{,\beta}} - \dfrac{1}{\theta}\dfrac{\partial\check{Q}_\beta}{\partial\mathbb{E}_\alpha} \\ \dfrac{\partial\check{\mathbb{J}}_\alpha}{\partial\theta_{,\beta}} - \dfrac{1}{\theta}\dfrac{\partial\check{Q}_\beta}{\partial\mathbb{E}_\alpha} & -\dfrac{1}{\theta}\dfrac{\partial\check{Q}_{(\alpha}}{\partial\theta_{,\beta)}} \end{pmatrix}\Bigg|_E \quad \text{is positive-semi definite .}$$

These conditions are satisfied provided that

(4.122)

$$Q_\alpha = -\kappa_{\alpha\beta}\theta_{,\beta} + \beta^{(Q)}_{\alpha\beta}\mathbb{E}_\beta ,$$

$$\mathbb{J}_\alpha = \beta^{(\mathbb{J})}_{\alpha\beta}\theta_{,\beta} + \sigma_{\alpha\beta}\mathbb{E}_\beta ,$$

and that

(4.123) $$\begin{pmatrix} \sigma_{(\alpha\beta)} & \beta^{(\mathbb{J})}_{\alpha\beta} - \dfrac{1}{\theta}\beta^{(Q)}_{\beta\alpha} \\ \beta^{(\mathbb{J})}_{\alpha\beta} - \dfrac{1}{\theta}\beta^{(Q)}_{\beta\alpha} & \dfrac{1}{\theta}\kappa_{(\alpha\beta)} \end{pmatrix}\Bigg|_E \quad \text{is positive-semi definite .}$$

Necessary conditions for (4.123) to be satisfied are that $\sigma_{(\alpha\beta)}$ and $\kappa_{(\alpha\beta)}$ are positive definite, but these conditions are by no means sufficient. Conditions of sufficiency include for instance also

(4.124) $$\sigma_{[\alpha\beta]}\Big|_E = 0, \ \kappa_{[\alpha\beta]}\Big|_E = 0, \ \text{and} \ \beta^{(\mathbb{J})}_{\alpha\beta}\Big|_E = \frac{1}{\theta}\beta^{(Q)}_{\beta\alpha}\Big|_E .$$

These represent the well-known Onsager relations.

4.6 RECAPITULATION AND COMPARISON

In the preceding Sections we presented the material versions of the two dipole models, of the statistical and Lorentz formulations and of the Maxwell-Minkowski formulation. The basic principle in this derivation consisted in transforming known equations into the Lagrangian form by introducing new variables which are particularly convenient in this material description. Since this Chapter is heavily loaded with partly complicated formulas, it might be advantageous when we recollect the basic ideas here.

Key idea behind the Lagrangian description is to derive the equations of electromechanical interactions in a form which can directly be used in the theory of solid bodies. All equations should therefore be referred to the reference configuration. While such a formulation is well-known in continuum mechanics, it is hardly used in electrodynamics. This is the reason why most theories of solids of electromechanical interactions are treated in the spatial description.

Whereas the advantages of the material description will be described extensively in Chapter 5, we would like to draw the reader's attention here on formal differences and similarities of the various formulations only. To this end, consider the Maxwell equations first. They are listed in (4.8) and (4.66) for the two-dipole and the Lorentz or statistical models, respectively. For clarity of presentation they will be repeated here. In the Chu formulation they read

$$\begin{aligned}
&\mathbb{B}^a_{\alpha,\alpha} = -\mu_o \mathbb{M}^C_{\alpha,\alpha} \,, \\
&\dot{\mathbb{B}}^a_\alpha + e_{\alpha\beta\gamma}\mathbb{E}_{\gamma,\beta} = -\mu_o \dot{\mathbb{M}}^C_\alpha \,, \\
(4.125)\quad &\mathbb{D}^a_{\alpha,\alpha} = \mathbb{Q} - \mathbb{P}^C_{\alpha,\alpha} \,, \\
&-\dot{\mathbb{D}}^a_\alpha + e_{\alpha\beta\gamma}\mathbb{H}_{\gamma,\beta} = \mathbb{J}_\alpha + \dot{\mathbb{P}}^C_\alpha \,, \\
&\dot{\mathbb{Q}} + \mathbb{J}_{\alpha,\alpha} = 0 \,,
\end{aligned}$$

whereas in the statistical and Lorentz formulation they are

$$\begin{aligned}
&\mathbb{B}_{\alpha,\alpha} = 0 \,, \\
&\dot{\mathbb{B}}_\alpha + e_{\alpha\beta\gamma}\mathbb{E}_{\gamma,\beta} = 0 \,, \\
(4.126)\quad &\mathbb{D}^a_{\alpha,\alpha} = \mathbb{Q} - \mathbb{P}^L_{\alpha,\alpha} \,,
\end{aligned}$$

(over)

$$-\dot{\mathbb{D}}^a_\alpha + e_{\alpha\beta\gamma}\mathbb{H}^a_{\gamma,\beta} = \mathbb{J}_\alpha + \mathbb{P}^L_\alpha - e_{\alpha\beta\gamma}\mathbb{M}^L_{\gamma,\beta} ,$$
$$\dot{\mathbb{Q}} + \mathbb{J}_{\alpha,\alpha} = 0 .$$

Both sets of equations allow the presentation of the Lagrangian version of the Minkowski formulation by introducing either

(4.127) $\mathbb{B}_\alpha = \mathbb{B}^a_\alpha + \mu_o \mathbb{M}^C_\alpha , \qquad \mathbb{D}_\alpha = \mathbb{D}^a_\alpha + \mathbb{P}^C_\alpha$, or

(4.128) $\mathbb{H}_\alpha = \mathbb{H}^a_\alpha - \mathbb{M}^L_\alpha , \qquad \mathbb{D}_\alpha = \mathbb{D}^a_\alpha + \mathbb{P}^L_\alpha$.

This then yields

(4.129) $\mathbb{P}^C_\alpha = \mathbb{P}^L_\alpha = \mathbb{P}_\alpha ,$ and $\mathbb{M}^C_\alpha = |J| C^{-1}_{\alpha\beta}\mathbb{M}^L_\beta$

so that the Lagrangian form of the Maxwell equations in the Minkowski formulation becomes

$$\mathbb{B}_{\alpha,\alpha} = 0 ,$$
$$\dot{\mathbb{B}}_\alpha + e_{\alpha\beta\gamma}\mathbb{E}_{\gamma,\beta} = 0 ,$$
(4.130) $$\mathbb{D}_{\alpha,\alpha} = \mathbb{Q} ,$$
$$-\dot{\mathbb{D}}_\alpha + e_{\alpha\beta\gamma}\mathbb{H}_{\gamma,\beta} = \mathbb{J}_\alpha ,$$
$$\dot{\mathbb{Q}} + \mathbb{J}_{\alpha,\alpha} = 0 .$$

The equations (4.125), (4.126) and (4.130) are the only ones that emerge in the Lagrangian description from all presently known electromagnetic formulations. We have presented in Chapter 2 four different formulations, but there are others (for instance the Ampère formulation of Penfield and Haus [7]) and all these reduce to (4.125), (4.126) or (4.130). The Lagrangian description has therefore reduced this number to at most three, and the differences among these models are particularly transparent in the Lagrangian description. The Chu formulation differs from the statistical and Lorentz formulation only in the choice of magnetization. This resulted in the selection of different auxiliary fields, which are related to the basic fields via the Maxwell-Lorentz aether relations (4.9) and (4.67). As a result, $\mathbb{H}_\alpha$ and $\mathbb{E}_\alpha$ are the two basic electromagnetic field vectors in the Chu formulation, while $\mathbb{B}_\alpha$ and $\mathbb{E}_\alpha$ are those of the statistical or Lorentz formulation. In this connection it is interesting to note that the Lagrangian electric field $\mathbb{E}_\alpha$ is the same in all formulations. This was not so in the spatial description where E^C_i and $E^L_i = E^S_i$ are different

variables. The same is also true for the Lagrangian magnetic field strength $\mathbb{H}_\alpha$; it is the same variable in both, the Chu and the Minkowski formulation. Another noteworthy point is that the Maxwell equations are formally the same as those in the spatial version of the statistical or Minkowski formulation. Because the transformations (4.127) and (4.128) relating the systems (4.125) and (4.126) to (4.130) are so simple, it is now also plausible why the Maxwell-Minkowski formulation served as connecting piece between the two-dipole and the statistical and the Lorentz formulations.
In performing actual calculations in electromagnetism it is advantageous to introduce electromagnetic potentials by satisfying those Maxwell equations identically that do not involve the charge $\mathbb{Q}$ and the current $\mathbb{J}_\alpha$. Mathematically this is achieved by introducing two potentials A_α and Φ, such that

$$(4.131) \qquad \mathbb{B}_\alpha = e_{\alpha\beta\gamma} A_{\gamma,\beta}, \qquad \text{and} \qquad \mathbb{E}_\alpha = \Phi_{,\alpha} - \dot{A}_\alpha \; .$$

Since $\mathbb{B}_\alpha$ is a basic field in all but the Chu formulation, the replacement of the Maxwell equations by the corresponding equations for the potentials A_α and Φ and for the charge $\mathbb{Q}$ is easier in the Lorentz, the statistical and the Maxwell-Minkowski formulation than it is in the Chu formulation. This may be regarded as a disadvantage of the Chu formulation.

The above discussion is only concerned with the Maxwell equations and leaves all mechanical balance laws aside. Yet, the basic advantages of the different formulations are drawn from the expressions of electromagnetic body force, body couple and energy supply. These are listed in (4.11) and (4.12) (two-dipole models), (4.71) (Lorentz formulation), (4.74) (statistical formulation) and (4.78) (Maxwell-Minkowski formulation). Of all these formulations the two dipole model with symmetric stress tensor and the Lorentz formulation led to electromagnetic body force and energy supply expressions which are particularly simple and easy to interpret (see the corresponding expressions (4.12) and (4.71)). This is not so for all other formulations, although it is known that their interaction terms have a clear physical meaning and are based on a clear method of derivation. It is also interesting to note that the two most simple formulations are those with no electromagnetic body couple and with symmetric Cauchy stress. Hence, since all formulations are non-relativistically equivalent anyhow, future calculations should be performed with either one of these simplest formulations. The argument on electromagnetic potentials given above favors the Lorentz formulation. We shall come back to this point in Chapter 5.

As was done for the Maxwell equations, we also wish to recapitulate the mechanical balance equations and to comment on the equivalence properties in the five presented models. Starting with the Chu-models, we showed in Section 4.2 already that these two models are completely equivalent and, therefore, we only list the balance equations for model II:

$$\rho_o \ddot{x}_i = {}^{II}T_{i\alpha,\alpha} + F^{-1}_{\alpha i}\{(\mathbb{Q} - \mathbb{P}_{\beta,\beta})\mathbb{E}_\alpha + e_{\alpha\beta\gamma}(\mathbb{J}_\beta + \dot{\mathbb{P}}_\beta)\mathbb{B}^a_\gamma - \mu_o \mathbb{M}^C_{\beta,\beta}\mathbb{H}_\alpha\} + + \rho_o F^{ext}_i \ .$$

$$(4.132) \qquad {}^{II}T_{[i\alpha}F_{j]\alpha} = 0 \ ,$$

$$\rho_o \theta\dot{\eta} = \mathbb{J}_\alpha \mathbb{E}_\alpha - Q_{\alpha,\alpha} + \rho_o r^{ext}$$

Here the first and the second equation follow by substituting $(4.12)^{1,2}$ into $(1.39)^1$ and $(1.39)^2$.

The corresponding mechanical balance equations for model I are most easily obtained if in (4.132) the stress tensor ${}^{II}T_{i\alpha}$ is replaced by ${}^{I}T_{i\alpha}$; this is achieved through use of (4.22).

Next we pass on to model V, or the Lorentz model. Its mechanical balance laws emerge if (4.71) and (4.85) are substituted into (1.39). When this is done care must be observed that the heat flux vector Q_α in (4.85) is correctly selected; it is Q^S_α of the statistical model, rather than Q^L_α as originally introduced in the Lorentz model. We thus obtained

$$\rho_o \ddot{x}_i = {}^{V}T_{i\alpha,\alpha} + F^{-1}_{\alpha i}\{(\mathbb{Q} - \mathbb{P}_{\beta,\beta})\mathbb{E}_\alpha + e_{\alpha\beta\gamma}(\mathbb{J}_\beta + \dot{\mathbb{P}}_\beta)\mathbb{B}_\gamma + + (\mathbb{M}^L_{\alpha,\beta} - \mathbb{M}^L_{\beta,\alpha})\mathbb{B}_\beta\} + \rho_o F^{ext}_i \ ,$$

$$(4.133) \qquad {}^{V}T_{[i\alpha}F_{j]\alpha} = 0 \ ,$$

$$\rho_o \theta\dot{\eta} = \mathbb{J}_\alpha \mathbb{E}_\alpha - Q_{\alpha,\alpha} + \rho_o r^{ext} \ .$$

The next model is model IV, but again there is no need to list the results for this model explicitly, because the only difference between the models IV and V lies in the stress tensors and consequently the balance equations for model IV may be obtained by merely introducing into (4.133) ${}^{IV}T_{i\alpha}$ as given by (4.88). Needless to state that the two models have been shown to be equivalent already in Section 4.3.

There remains the presentation of model III. Because of its complexity in relation to other models it is presented last. Its mechanical balance equations emerge when (4.97) is substituted in (1.39). This yields

$$\rho_o \ddot{x}_i = {}^{III}T_{i\alpha,\alpha} + F^{-1}_{\alpha i}\{\mathbb{Q}\,\mathbb{E}_\alpha + e_{\alpha\beta\gamma}\mathbb{J}_\beta\mathbb{B}_\gamma + \mathbb{P}_\beta\mathbb{E}_{\beta,\alpha} + $$

$$+ \mu_o\overset{C}{\mathbb{M}}_\beta\mathbb{H}_{\beta,\alpha} + e_{\alpha\beta\gamma}(\mathbb{D}_\beta\dot{\mathbb{B}}_\gamma + \dot{\mathbb{D}}_\beta\mathbb{B}_\gamma)\} +$$

(4.134)
$$+ F^{-1}_{\beta i,j}F_{j\alpha}(\mathbb{P}_\alpha\mathbb{E}_\beta + \mu_o\overset{C}{\mathbb{M}}_\alpha\mathbb{H}_\beta) + \rho_o F^{ext}_i ,$$

$${}^{III}T_{[i\alpha}F_{j]\alpha} = F_{[i\alpha}F^{-1}_{\beta j]}(\mathbb{P}_\alpha\mathbb{E}_\beta + \mu_o\overset{C}{\mathbb{M}}_\alpha\mathbb{H}_\beta) ,$$

$$\rho_o\theta\dot{\eta} = \mathbb{J}_\alpha\mathbb{E}_\alpha - Q_{\alpha,\alpha} + \rho_o r^{ext} .$$

As already stated in Section 4.4, model III is identical to model I, and, hence, the only difference in the formulations of the models III and II stems from the difference in the definition of the stress tensors, which are related to each other according to (see (4.105) or (4.22))

(4.135) $${}^{III}T_{i\alpha} = {}^{II}T_{i\alpha} - F^{-1}_{\beta i}(\mathbb{P}_\alpha\mathbb{E}_\beta + \mu_o\overset{C}{\mathbb{M}}_\alpha\mathbb{H}_\beta)$$

From the above considerations we conclude that the only necessary step for a completion of the comparison of the five models is a comparison of the models II and V. To perform this comparison, notice that in view of (4.132) and (4.133) the balance laws of moment of momentum and energy are identical in the two formulations. Consequently at most the balance laws of linear momentum can differ; and the difference can at most be a difference in stress. This difference must, furthermore, be such that the balance of moment of momentum is met; in other words the difference between the second Piola-Kirchhoff tensors ${}^{II}T^P_{\alpha\beta}$ and ${}^{V}T^P_{\alpha\beta}$ must be a symmetric tensor.
In order to find this difference the balance equations $(4.132)^1$ and $(4.133)^2$ must be used. Yet, this step becomes much easier if for the electromagnetic body force the respective representation in terms of the Maxwell stress tensor is used, see $(4.17)^1$ and $(4.74)^1$. We then obtain

$${}^{V}T_{i\alpha} - {}^{II}T_{i\alpha} = {}^{II}T^M_{i\alpha} - {}^{V}T^M_{i\alpha} =$$

(4.136)

$$= F^{-1}_{\beta i}(\mathbb{B}^a_\alpha\mathbb{H}_\beta - \mathbb{B}_\alpha\mathbb{H}^a_\beta) - \frac{1}{2}F^{-1}_{\alpha i}(\mathbb{B}^a_\beta\mathbb{H}_\beta - \mathbb{B}_\beta\mathbb{H}^a_\beta) .$$

To eliminate from this expression the auxiliary fields use must be made of (4.9), (4.67) and (4.94); we may then derive the expressions

$$\mathbb{B}^a_\alpha\mathbb{H}_\beta = \mu_o|J|C^{-1}_{\alpha\gamma}\mathbb{H}_\gamma\mathbb{H}_\beta ,$$

(4.137) $$\mathbb{B}_\alpha\mathbb{H}^a_\beta = \frac{1}{\mu_o}|J^{-1}|C_{\beta\gamma}\mathbb{B}_\gamma\mathbb{B}_\alpha = \mu_o|J|C^{-1}_{\alpha\gamma}\mathbb{H}_\gamma\mathbb{H}_\beta + \mu_o\overset{C}{\mathbb{M}}_\alpha\mathbb{H}_\beta +$$

$$+ \mu_o C^{-1}_{\alpha\gamma}C_{\beta\delta}\mathbb{H}_\gamma\overset{C}{\mathbb{M}}_\delta + \mu_o|J^{-1}|C_{\beta\gamma}\overset{C}{\mathbb{M}}_\alpha\overset{C}{\mathbb{M}}_\gamma .$$

Substitution into (4.136) yields

$$
\begin{aligned}
{}^{V}T_{i\alpha} &= {}^{II}T_{i\alpha} - \mu_o F^{-1}_{\beta i}\,{}^{C}\mathbb{M}_\alpha \mathbb{H}_\beta - \mu_o C^{-1}_{\alpha\gamma} F_{i\beta}\,\mathbb{H}_\gamma\,{}^{C}\mathbb{M}_\beta + \\
&\quad -\mu_o |J^{-1}| F_{i\beta}\,{}^{C}\mathbb{M}_\alpha\,{}^{C}\mathbb{M}_\beta + \mu_o F^{-1}_{\alpha i}\,{}^{C}\mathbb{M}_\beta \mathbb{H}_\beta + \frac{1}{2}\mu_o |J^{-1}| F^{-1}_{\alpha i} C_{\beta\gamma}\,{}^{C}\mathbb{M}_\beta\,{}^{C}\mathbb{M}_\gamma ,
\end{aligned}
\tag{4.138}
$$

and after a multiplication with $F^{-1}_{\beta i}$

$$
\begin{aligned}
{}^{V}T^{P}_{\alpha\beta} &= {}^{II}T^{P}_{\alpha\beta} - \mu_o C^{-1}_{\beta\gamma}\,{}^{C}\mathbb{M}_\alpha \mathbb{H}_\gamma - \mu_o C^{-1}_{\alpha\gamma}\,{}^{C}\mathbb{M}_\beta \mathbb{H}_\gamma + \mu_o C^{-1}_{\alpha\beta}\,{}^{C}\mathbb{M}_\gamma \mathbb{H}_\gamma + \\
&\quad -\mu_o |J^{-1}|\,{}^{C}\mathbb{M}_\alpha\,{}^{C}\mathbb{M}_\beta + \frac{1}{2}\mu_o |J^{-1}| C^{-1}_{\alpha\beta} C_{\gamma\delta}\,{}^{C}\mathbb{M}_\gamma\,{}^{C}\mathbb{M}_\delta ,
\end{aligned}
\tag{4.139}
$$

from which it is easily seen that the difference between ${}^{V}T^{P}_{\alpha\beta}$ and ${}^{II}T^{P}_{\alpha\beta}$ forms indeed a symmetric tensor. Needless to say that this relation corresponds to an equation that was established in Chapter 3 and relates the Cauchy stresses ${}^{V}t_{ij}$ and ${}^{II}t_{ij}$ (see the Table of Chapter 3).
In (4.139) the Lorentzian stress ${}^{V}T^{P}_{\alpha\beta}$ is expressed in terms of ${}^{II}T^{P}_{\alpha\beta}$ and the Chu-variables. The reverse relation expresses ${}^{II}T^{P}_{\alpha\beta}$ in terms of ${}^{V}T^{P}_{\alpha\beta}$ and the Lorentz-variables. This latter expression is obtained from (4.65), $(4.69)^2$ and $(4.67)^2$ and reads

$$
\begin{aligned}
{}^{II}T^{P}_{\alpha\beta} &= {}^{V}T^{P}_{\alpha\beta} + C^{-1}_{\alpha\gamma}\,{}^{L}\mathbb{M}_\gamma\,{}^{L}\mathbb{B}_\beta + C^{-1}_{\beta\gamma}\,{}^{L}\mathbb{M}_\gamma\,{}^{L}\mathbb{B}_\alpha - C^{-1}_{\alpha\beta}\,{}^{L}\mathbb{M}_\gamma\,{}^{L}\mathbb{B}_\gamma + \\
&\quad -\mu_o |J| C^{-1}_{\alpha\gamma} C^{-1}_{\beta\delta}\,{}^{L}\mathbb{M}_\gamma\,{}^{L}\mathbb{M}_\delta + \frac{1}{2}\mu_o |J| C^{-1}_{\alpha\beta} C^{-1}_{\gamma\delta}\,{}^{L}\mathbb{M}_\gamma\,{}^{L}\mathbb{M}_\delta .
\end{aligned}
\tag{4.140}
$$

The above equations and similar relations of the previous models are <u>necessary conditions</u> which must be satisfied in order that the models be equivalent. If these relations hold equivalence goes as far as the local balance laws of linear and angular momentum, energy and the corresponding jump conditions are concerned. The conditions are not sufficient, however, because the constitutive relations and therefore also the thermodynamic requirements impose further conditions. In the following these thermodynamic requirements will be discussed.
It follows from the foregoing considerations that in the constitutive theories of the models I, II and III the internal energies must be identical, i.e.

$$
{}^{I}(\rho_o U) = {}^{II}(\rho_o U) = {}^{III}(\rho_o U) , \tag{4.141}
$$

if the models are to be equivalent. The same holds for the models IV and V:

$$
{}^{IV}(\rho_o U) = {}^{V}(\rho_o U) . \tag{4.142}
$$

However, and as could already be expected from the corresponding results of Chapter 3, the relation (4.139) can only be satisfied simultaneously with the constitutive equations (4.26) and (4.84) provided that the internal energies for the models II and V differ. To derive the corresponding relation note that according to Chapter 3

$$^{V}(\rho U) = {}^{II}(\rho U) - \mu_o H_i M_i - \frac{1}{2}\mu_o M_i M_i ,$$

which in Lagrangian notation reads

$$(4.143) \qquad ^{V}(\rho_o U) = {}^{II}(\rho_o U) - \mu_o \mathbb{H}_\alpha \mathbb{M}^C_\alpha - \frac{1}{2}\mu_o |J^{-1}| C_{\alpha\beta} \mathbb{M}^C_\alpha \mathbb{M}^C_\beta .$$

If for the models II and V we use now constitutive relations of the form (4.53) and (4.90) and if we introduce the Helmholtz free energies (4.54) and (4.91), then (4.143) is equivalent to

$$(4.144) \qquad ^{V}(\rho_o \overset{V}{\Psi}) = {}^{II}(\rho_o \bar{\Psi}) - \frac{1}{2}\mu_o |J^{-1}| C_{\alpha\beta} \mathbb{M}^C_\alpha \mathbb{M}^C_\beta .$$

This equation may now be used in (4.55) and (4.92) to obtain expressions for the entropy, polarization, magnetization and the stress in the respective formulations. These must then satisfy the identities derived above, and if they do, the models are equivalent. The proof on the basis of (4.144), (4.55) and (4.92) is straightforward and thus we leave it to the reader. Yet, we shall come back to the consequences of this statement when defining material coefficients in the various formulations.

Finally, we would like to point out that a major goal of the derivation of the material description was in the presentation of a correct approach in linearizing field equations. The Lagrangian formulation is useful from just this practical point of view. Indeed, once the equations are known in their material description the transformations

$$\frac{\partial}{\partial x_i} \rightarrow \frac{\partial}{\partial X_\alpha} \quad \text{and} \quad \frac{\partial}{\partial t} \rightarrow \frac{d}{dt}$$

need no longer be performed, because all variables are per se already functions of X_α and t. This means that any perturbation approximation is much easier to be carried out when the equations are written in material rather than spatial coordinates. All linearization procedures performed so far were in the spatial description (see e.g. Toupin [4], Hutter and Pao [7] and van de Ven [8]). The treatments of Toupin and Hutter and Pao are approximate, however, insofar as they contain ad hoc assumptions that cannot be justified on the basis of non-relativistic arguments. This was pointed out by van de Ven, who also presents the correct solution. All the difficulties Toupin and Hutter and Pao were faced with disappear in the material description and no ad hoc assumptions must be introduced here. Corroboration of this will be given in the next Chapter.

4.7 APPROACH TO A UNIFIED CONSTITUTIVE THEORY

In the preceding Section we demonstrated that all theories of deformable bodies in the electromagnetic fields which have the complexity of thermoelastic polarizable andmagnetizable solids are non-relativistically equivalent, provided that the respective constitutive functions for the internal energy or the free energy satisfy certain relationships. Similar equivalence statements were established also in the Eulerian description so that a new proof was not a necessity except, perhaps, that it led to equivalence conditions expressed in the Lagrangian variables; these are very useful relationships. To recapitulate them briefly, recall that in order to enforce equivalence of the formulations I, II and III the free energies must in all these formulations be the same functions of the same variables. A similar statement also holds for the models IV and V, but to achieve equivalence between the groups (I,II,III) and (IV,V) the corresponding free energies must be related by (4.144), viz.

$$\rho_o \check{\Psi} = \rho_o \bar{\Psi} - \frac{1}{2}\mu_o |J^{-1}| C_{\alpha\beta} \mathbb{M}^C_\alpha \mathbb{M}^C_\beta \tag{4.145}$$

Here, $\bar{\Psi}$ is the free energy function in any one of the formulations I, II or III, and, correspondingly $\check{\Psi}$ is that of the models IV and V; they are defined in (4.54) and (4.91), respectively. Accordingly, and apart from a dependency on $C_{\alpha\beta}$ and θ, $\check{\Psi}$ is a function of $(\mathbb{E}_\alpha, \mathbb{B}_\alpha)$, whereas $\bar{\Psi}$ depends on $(\mathbb{E}_\alpha, \mathbb{H}_\alpha)$ as does $\mathbb{M}^C_\alpha$, because (see (4.55)[3])

$$\mathbb{M}^C_\alpha = - \frac{\rho_o}{\mu_o} \frac{\partial \bar{\Psi}}{\partial \mathbb{H}_\alpha} .$$

Equation (4.145) can therefore also be written as

$$\begin{aligned} \check{\Psi}(C_{\alpha\beta}, \mathbb{E}_\alpha, \mathbb{B}_\alpha, \theta) &= \bar{\Psi}(C_{\alpha\beta}, \mathbb{E}_\alpha, \mathbb{H}_\alpha, \theta) + \\ &- \frac{1}{2}\frac{\rho_o}{\mu_o} |J^{-1}| C_{\alpha\beta} \left(\frac{\partial \bar{\Psi}}{\partial \mathbb{H}_\alpha} \frac{\partial \bar{\Psi}}{\partial \mathbb{H}_\beta}\right)(C_{\alpha\beta}, \mathbb{E}_\alpha, \mathbb{B}_\alpha, \theta) \end{aligned} \tag{4.146}$$

In this identity the left-hand side is a function of $\mathbb{B}_\alpha$, in contrast to the expression on the right-hand side, which is a function of $\mathbb{H}_\alpha$. In view of

(4.94)[2] and (4.55)[3] we may, however, express $\mathbb{B}_\alpha$ as

$$\mathbb{B}_\alpha = -\rho_o \frac{\partial\bar{\Psi}}{\partial\mathbb{H}_\alpha} + \mu_o|J|C^{-1}_{\alpha\beta}\mathbb{H}_\beta \ , \tag{4.147}$$

so that

$$\begin{aligned}&\check{\Psi}(C_{\alpha\beta},\mathbb{E}_\alpha,\mu_o|J|C^{-1}_{\alpha\beta}\mathbb{H}_\beta - \rho_o \frac{\partial\bar{\Psi}}{\partial\mathbb{H}_\alpha},\theta) = \\ &= \bar{\Psi}(C_{\alpha\beta},\mathbb{E}_\alpha,\mathbb{H}_\alpha,\theta) - \frac{1}{2}\frac{\rho_o}{\mu_o}|J^{-1}|C_{\alpha\beta}\frac{\partial\bar{\Psi}}{\partial\mathbb{H}_\alpha}\frac{\partial\bar{\Psi}}{\partial\mathbb{H}_\beta} \ ,\end{aligned} \tag{4.148}$$

in which the arguments in $\partial\bar{\Psi}/\partial\mathbb{H}_\alpha$ are, of course, the same as in $\bar{\Psi}$. For given functions $\bar{\Psi}$ and $\check{\Psi}$, equation (4.148) must be satisfied identically if the formulations (I,II,III) and (IV,V) are to be equivalent. If, on the other hand only one of the functions $\check{\Psi}$ or $\bar{\Psi}$ is given, then (4.148) is a functional differential equation to determine the other. The solution to this will, in general, be very complex. We shall not try to solve it for a given function $\bar{\Psi}$, say. We shall rather exploit (4.148) as common forms of the free energies $\bar{\Psi}$ and $\check{\Psi}$. A very popular procedure is to write these functions as polynomial expressions of their independent variables. These polynomials must be regarded as truncated Taylor series expansions about a state of zero electromagnetic fields and zero deformation of a general functional relationship for the free energy. Now, the equivalence of the various models was guaranted in the above for the general theory using a representation for the constitutive equations of the free energy not restricted by any means. Otherwise stated, if one attempts to establish equivalence statements between two given theories for which the free energies are polynomials truncated at the quadratic terms, the two theories might very well be non-equivalent simply because equivalence would in one formulation require the inclusion of cubic, quartic or even higher order terms.

The point just raised is important, because it illustrates that, strictly, fully equivalent theories might become non-equivalent, because in both formulations one insists in too restrictive energy expressions. Nontheless, we can keep the usual polynomial representations and still claim equivalence, but if we do so this is only in the following restricted sense: We formally interpret the polynomials as truncated Taylor series expansions. With the use of the latter equivalence of two formulations can be established exactly; it amounts to a comparison of the polynomial coefficients. If in these Taylor series expansions we restrict ourselves to terms of a certain order, then equivalence can be established approximately.

To illustrate the above point more clearly, consider the following somewhat academic example: Let

$$f(x,y) = \sum_{\mu,\nu=0}^{\infty} a_{\mu\nu} x^{\mu} y^{\nu}, \qquad g(x,z) = \sum_{\mu,\nu=0}^{\infty} b_{\mu\nu} x^{\mu} z^{\nu},$$

and assume that for some reason $z = x + y$ and $f = g$. Then by the binomial theorem we may set

$$z^{\nu} = (x + y)^{\nu} = \sum_{k=0}^{\nu} \binom{\nu}{k} x^{k} y^{\nu-k} ,$$

and therefore

$$\sum_{\mu,\nu=0}^{\infty} a_{\mu\nu} x^{\mu} y^{\nu} = \sum_{\mu,\nu=0}^{\infty} b_{\mu\nu} \sum_{k=0}^{\nu} \binom{\nu}{k} x^{k+\mu} y^{\nu-k} .$$

This last equation can also be written as

$$\sum_{\mu,\nu=0}^{\infty} \{ \sum_{k=0}^{\mu} \binom{\nu+k}{k} b_{(\mu-k)(\nu+k)} - a_{\mu\nu} \} x^{\mu} y^{\nu} = 0 ,$$

so that the coefficient functions must satisfy the relations

$$a_{\mu\nu} = \sum_{k=0}^{\mu} \binom{\nu+k}{k} b_{(\mu-k)(\nu+k)} .$$

These identities must hold for all positive integers μ and ν. If we truncate the above polynomial representations at $\nu = N$ and $\mu = M$ they can still be satisfied, but then f_{MN} and g_{MN}, which denote the truncated expressions

$$f_{MN}(x,y) = \sum_{\mu=0}^{M} \sum_{\nu=0}^{N} a_{\mu\nu} x^{\mu} y^{\nu}, \qquad g_{MN}(x,z) = \sum_{\mu=0}^{M} \sum_{\nu=0}^{N} b_{\mu\nu} x^{\mu} z^{\nu},$$

are no longer exactly equal, but only in the sense that

$$f_{MN} = g_{MN} + O(x^{M+1}) + O(y^{N+1}) ,$$

holds. For instance, if

$$f_{12}(x,y) = a_{00} + a_{01}y + a_{02}y^2 + a_{10}x + a_{11}xy + a_{12}xy^2 ,$$

and

$$g_{12}(x,z) = b_{00} + b_{01}z + b_{02}z^2 + b_{10}x + b_{11}xz + b_{12}xz^2 ,$$

and g_{12} is expressed in terms of x and y, then g_{12} should contain quadratic and cubic terms in x, whence follows that f_{12} and g_{12} cannot be identical except in the above mentioned approximate sense. In fact, one obtains

$$g_{12} = f_{12} + (b_{11} + b_{02} + 2b_{12}y)x^2 + b_{12}x^3 .$$

The above considerations may look somewhat artificial to the novel reader, yet they are important, and we would like to illustrate them using the most simple example that accounts for magnetoelastic interactions. For that purpose we restrict ourselves to conditions of isotropy and to free energies which are at most of quartic order in the electric and magnetic field quantities and of quadratic order in the temperature difference

$$\theta = \theta - \theta_o .$$

Here, θ_o denotes a reference temperature. Moreover, we assume small deformations, so that it suffices to write the free energy as a quadratic function of the Lagrangian deformation tensor

(4.149) $$E_{\alpha\beta} = \frac{1}{2}(C_{\alpha\beta} - \delta_{\alpha\beta}) .$$

With these limitations we may choose the following representations for the free energy functions: for model II

(4.150)
$$\begin{aligned}\bar{\Psi} = {} & \frac{1}{2\rho_o}\,{}_2\bar{\chi}^{(m)}\mathbb{H}_\alpha\mathbb{H}_\alpha + \frac{1}{4\rho_o}\,{}_4\bar{\chi}^{(m)}(\mathbb{H}_\alpha\mathbb{H}_\alpha)^2 + \frac{1}{2\rho_o}\,{}_2\bar{\chi}^{(e)}\mathbb{E}_\alpha\mathbb{E}_\alpha + \\ & + \frac{1}{4\rho_o}\,{}_4\bar{\chi}^{(e)}(\mathbb{E}_\alpha\mathbb{E}_\alpha)^2 - \frac{1}{2}\bar{c}\theta^2 + \\ & + \frac{1}{2\rho_o}\bar{L}^{(m)}\mathbb{H}_\alpha\mathbb{H}_\alpha\theta + \frac{1}{2\rho_o}\bar{L}^{(e)}\mathbb{E}_\alpha\mathbb{E}_\alpha\theta + \\ & + [\frac{1}{2\rho_o}\bar{b}^{(m)}_{\alpha\beta\gamma\delta}\mathbb{H}_\alpha\mathbb{H}_\beta + \frac{1}{2\rho_o}\bar{b}^{(e)}_{\alpha\beta\gamma\delta}\mathbb{E}_\alpha\mathbb{E}_\beta - \bar{\nu}\delta_{\gamma\delta}\theta]E_{\gamma\delta} + \\ & + \frac{1}{2\rho_o}\bar{c}_{\alpha\beta\gamma\delta}E_{\alpha\beta}E_{\gamma\delta} ,\end{aligned}$$

for model V

(4.151)
$$\begin{aligned}\check{\Psi} = {} & + \frac{1}{2\rho_o}\,{}_2\check{\chi}^{(m)}\mathbb{B}_\alpha\mathbb{B}_\alpha + \frac{1}{4\rho_o}\,{}_4\check{\chi}^{(m)}(\mathbb{B}_\alpha\mathbb{B}_\alpha)^2 + \frac{1}{2\rho_o}\,{}_2\check{\chi}^{(e)}\mathbb{E}_\alpha\mathbb{E}_\alpha + \\ & + \frac{1}{4\rho_o}\,{}_4\check{\chi}^{(e)}(\mathbb{E}_\alpha\mathbb{E}_\alpha)^2 - \frac{1}{2}\check{c}\theta^2 + \\ & + \frac{1}{2\rho_o}\check{L}^{(m)}\mathbb{B}_\alpha\mathbb{B}_\alpha\theta + \frac{1}{2\rho_o}\check{L}^{(e)}\mathbb{E}_\alpha\mathbb{E}_\alpha\theta +\end{aligned}$$

(over)

$$+ \left[\frac{1}{2\rho_o} \check{b}^{(m)}_{\alpha\beta\gamma\delta}\mathbb{B}_\alpha\mathbb{B}_\beta + \frac{1}{2\rho_o} \check{b}^{(e)}_{\alpha\beta\gamma\delta}\mathbb{E}_\alpha\mathbb{E}_\beta - \check{\nu}\delta_{\gamma\delta}\theta\right]E_{\gamma\delta} +$$

$$+ \frac{1}{2\rho_o} \check{c}_{\alpha\beta\gamma\delta}E_{\alpha\beta}E_{\gamma\delta} \; .$$

Before we proceed it is worthwhile to look at these expressions more closely. The polynomial representations for the free energy start with quadratic terms. A constant term is left out, because it is immaterial, and linear terms are discarded, because at zero deformation, zero temperature difference and zero electromagnetic fields no stress and no polarization and magnetization should be present. In other words we assume the body to possess a natural unstrained state. Cubic terms are also present in (4.150) and (4.151), but in this regard the polynomials are not complete, since products like $E_{\alpha\beta}E_{\gamma\delta}\mathbb{H}_\varepsilon$, $E_{\alpha\beta}E_{\gamma\delta}\mathbb{E}_\varepsilon$ and $E_{\alpha\beta}E_{\gamma\delta}\theta$ are missing. We believe that these omissions are justified, because usually the field free elasticities are much more important than their change by the electromagnetic fields. We have further neglected electromagnetic coupling terms.

In what follows our aim is to investigate in what sense the above representations would allow an identical satisfaction of (4.145). In view of the above general remarks we expect that a full equivalence cannot be established, and that differences will show up in the higher order terms of (4.150) and (4.151). To find the equivalence conditions we substitute (4.150) into the first term on the right-hand side of (4.145) and rewrite the latter by expressing the second term as a function of the independent variables of $\bar{\Psi}$. To this end, note that

$$(4.152) \qquad |J^{-1}|C_{\alpha\beta} = \delta_{\alpha\beta}(1 - E_{\gamma\gamma}) + 2E_{\alpha\beta} + O(E^2) = \delta_{\alpha\beta} + n_{\alpha\beta\gamma\delta}E_{\gamma\delta} + O(E^2) \; ,$$

where

$$(4.153) \qquad n_{\alpha\beta\gamma\delta} = -\delta_{\alpha\beta}\delta_{\gamma\delta} + \delta_{\alpha\gamma}\delta_{\beta\delta} + \delta_{\alpha\delta}\delta_{\beta\gamma}$$

In view of $(4.55)^3$ and (4.150) we may also write

$$(4.154) \qquad \mu_o\mathbb{M}^C_\alpha = -\rho_o \frac{\partial\bar{\Psi}}{\partial\mathbb{H}_\alpha} = -{}_2\bar{\chi}^{(m)}\mathbb{H}_\alpha - {}_4\bar{\chi}^{(m)}\mathbb{H}_\beta\mathbb{H}_\beta\mathbb{H}_\alpha - \bar{L}^{(m)}\mathbb{H}_\alpha\theta - \bar{b}^{(m)}_{\alpha\beta\gamma\delta}\mathbb{H}_\beta E_{\gamma\delta} \; .$$

Using these relations in (4.145) the right-hand side of the latter may be written as

$$\check{\Psi} = \frac{1}{2\rho_o}\left(1 - \frac{{}_2\bar{\chi}^{(m)}}{\mu_o}\right){}_2\bar{\chi}^{(m)}\mathbb{H}_\alpha\mathbb{H}_\alpha + \frac{1}{4\rho_o}\,{}_4\bar{\chi}^{(m)}\left(1 - 4\,\frac{{}_2\bar{\chi}^{(m)}}{\mu_o}\right)(\mathbb{H}_\alpha\mathbb{H}_\alpha)^2 +$$

(over)

$$
\begin{aligned}
&+ \frac{1}{2\rho_o}\, {}_2\bar{\chi}^{(e)} \mathbb{E}_\alpha \mathbb{E}_\alpha + \frac{1}{2\rho_o}\, {}_4\bar{\chi}^{(e)} (\mathbb{E}_\alpha \mathbb{E}_\alpha)^2 - \frac{1}{2}\, \bar{c}\theta^2 + \frac{1}{2\rho_o}\Big(1 - 2\, \frac{{}_2\bar{\chi}^{(m)}}{\mu_o}\Big)\bar{L}^{(m)} \mathbb{H}_\alpha \mathbb{H}_\alpha \theta + \\
&+ \frac{1}{2\rho_o}\, \bar{L}^{(e)} \mathbb{E}_\alpha \mathbb{E}_\alpha \theta + \frac{1}{2\rho_o}\Big[\Big(1 - 2\, \frac{{}_2\bar{\chi}^{(m)}}{\mu_o}\Big)\bar{b}^{(m)}_{\alpha\beta\gamma\delta} - \frac{{}_2\bar{\chi}^{(m)2}}{\mu_o}\, n_{\alpha\beta\gamma\delta}\Big] \mathbb{H}_\alpha \mathbb{H}_\beta E_{\gamma\delta} + \\
(4.155)\quad &+ \frac{1}{2\rho_o}\, \bar{b}^{(e)}_{\alpha\beta\gamma\delta} \mathbb{E}_\alpha \mathbb{E}_\beta E_{\gamma\delta} - \bar{\nu}\theta E_{\gamma\gamma} + \frac{1}{2\rho_o}\, \bar{c}_{\alpha\beta\gamma\delta} E_{\alpha\beta} E_{\gamma\delta} + \\
&- \frac{\bar{L}^{(m)2}}{2\rho_o\mu_o} \mathbb{H}_\alpha \mathbb{H}_\alpha \theta^2 + \frac{\bar{L}^{(m)}}{\rho_o\mu_o}\big(-{}_2\bar{\chi}^{(m)} n_{\alpha\beta\gamma\delta} - \bar{b}^{(m)}_{\alpha\beta\gamma\delta}\big) \mathbb{H}_\alpha \mathbb{H}_\beta E_{\gamma\delta}\theta + \\
&- \frac{\bar{L}^{(m)2}}{2\rho_o\mu_o}\, n_{\alpha\beta\gamma\delta} \mathbb{H}_\alpha \mathbb{H}_\beta E_{\gamma\delta}\theta^2 + \text{higher order terms} .
\end{aligned}
$$

Here and henceforth all terms of order

$$\mathbb{H}_\varepsilon E_{\alpha\beta} E_{\gamma\delta},\ \mathbb{E}_\varepsilon E_{\alpha\beta} E_{\gamma\delta},\ \theta E_{\alpha\beta} E_{\gamma\delta} ,$$

and higher are neglected. This is consistent with our basic assumption that third order terms involving the square of the deformation tensor are discarded. In contrast to (4.150) or (4.151) the above representation for $\check{\Psi}$ contains also mixed fourth order terms (the last three expressions on the right-hand side). For the reasons described above, these should be neglected as well. The occurrence of these higher order terms is corroboration for our earlier statement that full equivalence might simply be impossible, because the free energies in the respective formulations are too restrictive. Yet it is apparent from the above calculation that a full equivalence could be achieved if the entire Taylor series expansion would be kept.
The functional relationship for $\check{\Psi}$ is still not in the appropriate form for a comparison with (4.151). For that purpose $\mathbb{B}_\alpha$ in (4.151) must be replaced by $\mathbb{H}_\alpha$. This replacement is accomplished by using (4.94) and (4.154), and it yields

$$
\begin{aligned}
\mathbb{B}_\alpha &= \mu_o(1 - {}_2\bar{\chi}^{(m)})\mathbb{H}_\alpha - {}_4\bar{\chi}^{(m)} \mathbb{H}_\beta \mathbb{H}_\beta \mathbb{H}_\alpha - \bar{L}^{(m)} \mathbb{H}_\alpha \theta - (\bar{b}^{(m)}_{\alpha\beta\gamma\delta} + \mu_o n_{\alpha\beta\gamma\delta}) \mathbb{H}_\beta E_{\gamma\delta} = \\
(4.156)\quad &= \mu \mathbb{H}_\alpha - {}_4\bar{\chi}^{(m)} \mathbb{H}_\beta \mathbb{H}_\beta \mathbb{H}_\alpha - \bar{L}^{(m)} \mathbb{H}_\alpha \theta - (\bar{b}^{(m)}_{\alpha\beta\gamma\delta} + \mu_o n_{\alpha\beta\gamma\delta}) \mathbb{H}_\beta E_{\gamma\delta} ,
\end{aligned}
$$

where we have set

$$(4.157)\quad \mu_o - {}_2\bar{\chi}^{(m)} = \mu, \qquad \text{or} \qquad {}_2\bar{\chi}^{(m)} = \mu_o - \mu .$$

If (4.156) is substituted into (4.151) and the resulting expression is rearranged (thereby neglecting third and fourth order terms as was done above) we obtain

$$\begin{aligned}\check{\Psi} = {} & \frac{1}{2\rho_o}\,\mu^2\,{}_2\check{\chi}^{(m)}\mathbb{H}_\alpha\mathbb{H}_\alpha + \frac{1}{4\rho_o}(\mu^4\,{}_4\check{\chi}^{(m)} - 4\mu\,{}_2\check{\chi}^{(m)}\,{}_4\bar{\chi}^{(m)})(\mathbb{H}_\alpha\mathbb{H}_\alpha)^2 + \\ & + \frac{1}{2\rho_o}\,{}_2\check{\chi}^{(e)}\mathbb{E}_\alpha\mathbb{E}_\alpha + \frac{1}{4\rho_o}\,{}_4\check{\chi}^{(e)}(\mathbb{E}_\alpha\mathbb{E}_\alpha)^2 - \frac{1}{2}\check{c}\theta^2 + \\ & + \frac{1}{2\rho_o}(\mu^2\check{L}^{(m)} - 2\mu\bar{L}^{(m)}\,{}_2\check{\chi}^{(m)})\mathbb{H}_\alpha\mathbb{H}_\alpha\theta + \frac{1}{2\rho_o}\check{L}^{(e)}\mathbb{E}_\alpha\mathbb{E}_\alpha\theta + \\ & + \frac{1}{2\rho_o}[\mu^2\check{b}^{(m)}_{\alpha\beta\gamma\delta} - 2\mu(\bar{b}^{(m)}_{\alpha\beta\gamma\delta} + \mu_o n_{\alpha\beta\gamma\delta})\,{}_2\check{\chi}^{(m)}]\mathbb{H}_\alpha\mathbb{H}_\beta E_{\gamma\delta} + \\ & + \frac{1}{2\rho_o}\check{b}^{(e)}_{\alpha\beta\gamma\delta}\mathbb{E}_\alpha\mathbb{E}_\beta E_{\gamma\delta} - \check{\nu}\theta E_{\gamma\gamma} + \frac{1}{2\rho_o}\check{c}_{\alpha\beta\gamma\delta}E_{\alpha\beta}E_{\gamma\delta}\,,\end{aligned} \tag{4.158}$$

which is expressed in terms of the coefficients $(\check{\ })$ as well as $(\bar{\ })$. Identifying (4.155) with (4.158) now yields

$$\begin{aligned} & \mu_o\mu\,{}_2\check{\chi}^{(m)} = {}_2\bar{\chi}^{(m)} = \mu_o - \mu =: -\mu_o\,{}_2\chi^{(m)}, \quad {}_2\check{\chi}^{(e)} = {}_2\bar{\chi}^{(e)} =: -{}_2\chi^{(e)}, \\ & {}_4\check{\chi}^{(m)} = \frac{1}{\mu^4}\,{}_4\bar{\chi}^{(m)}, \quad {}_4\bar{\chi}^{(e)} = {}_4\check{\chi}^{(e)} =: -{}_4\chi^{(e)}\,, \\ & \check{c} = \bar{c} =: \frac{c_W}{\theta_o}\,, \quad \mu^2\check{L}^{(m)} = \bar{L}^{(m)}, \quad \check{L}^{(e)} = \bar{L}^{(e)} =: L^{(e)}\,, \\ & \check{b}^{(m)}_{\alpha\beta\gamma\delta} = \frac{1}{\mu^2}\bar{b}^{(m)}_{\alpha\beta\gamma\delta} - \frac{\mu_o\,{}_2\chi^{(m)}}{\mu^2}(2 + {}_2\chi^{(m)})n_{\alpha\beta\gamma\delta}, \quad \check{b}^{(e)}_{\alpha\beta\gamma\delta} = \bar{b}^{(e)}_{\alpha\beta\gamma\delta} = b^{(e)}_{\alpha\beta\gamma\delta}\,, \\ & \check{\nu} = \bar{\nu} = \nu, \quad \check{c}_{\alpha\beta\gamma\delta} = \bar{c}_{\alpha\beta\gamma\delta} = c_{\alpha\beta\gamma\delta}\,.\end{aligned} \tag{4.159}$$

Several of the coefficients occurring in the above equations can, without confusion, be given specific names, as for instance

${}_2\chi^{(m)}$	magnetic susceptibility,
μ	magnetic permeability,
${}_2\chi^{(e)}, {}_4\chi^{(e)}$	second and fourth order electric susceptibility,
c_W	specific heat,
$L^{(e)}$	thermoelectric constant,
$b^{(e)}_{\alpha\beta\gamma\delta}$	electrostrictive constants,
ν	thermoelastic constants,
$c_{\alpha\beta\gamma\delta}$	elastic constants.

More difficulties arise, however, for $L^{(m)}$, because it is not clear whether $\bar{L}^{(m)}$ or $\check{L}^{(m)}$ should be called thermomagnetic constant. A similar statement also holds for the anisotropy coefficients ${}_4\bar{\chi}^{(m)}$ and ${}_4\check{\chi}^{(m)}$. Since the difference between $\bar{b}^{(m)}_{\alpha\beta\gamma\delta}$ and $\check{b}^{(m)}_{\alpha\beta\gamma\delta}$ is essential the situation is even more drastic for $b^{(m)}_{\alpha\beta\gamma\delta}$, the coefficients usually attributed with the notion of magnetostriction. In the Chu formulation one is inclined to call $\bar{b}^{(m)}_{\alpha\beta\gamma\delta}$ a magnetostrictive constant; in the Lorentz formulation this is the case for $\check{b}^{(m)}_{\alpha\beta\gamma\delta}$ instead.

For isotropic materials the relations (4.159)^8 can still somewhat be simplified, since for this special group of materials

$$(4.160) \qquad b^{(m)}_{\alpha\beta\gamma\delta} = b^{(m)}_1 \delta_{\alpha\beta}\delta_{\gamma\delta} + b^{(m)}_2 (\delta_{\alpha\gamma}\delta_{\beta\delta} + \delta_{\alpha\delta}\delta_{\beta\gamma})$$

must hold. Substituting this into (4.159)^8 we obtain

$$(4.161) \qquad \begin{aligned} \check{b}^{(m)}_1 &= \frac{1}{\mu^2}[\bar{b}^{(m)}_1 + \mu_o\, {}_2\chi^{(m)}({}_2\chi^{(m)} + 2)] \,, \\ \check{b}^{(m)}_2 &= \frac{1}{\mu^2}[\bar{b}^{(m)}_1 - \mu_o\, {}_2\chi^{(m)}({}_2\chi^{(m)} + 2)] \,. \end{aligned}$$

From these expressions it is now evident that a unique definition of magnetostrictive constants is not possible by merely establishing the respective polynomial representation of the free energy function. What would be needed, is a simple experiment in which magnetostrictive effects could uniquely be defined and in which the magnetostrictive constants could be measured, which then must be independent of the model chosen.

At this point we must warn the reader to take the above conclusions as the ultimate truth. The results are special insofar as they hold for isotropic bodies and no electromagnetic coupling. For instance the fact that all electric coefficients in the energy expressions are identical is a consequence of the fact that electromagnetic coupling terms have been omitted (these are terms like $\bar{\chi}^{(em)}\mathbb{E}_\alpha\mathbb{H}_\alpha$ and $\bar{b}^{(em)}_{\alpha\beta\gamma\delta}\mathbb{E}_\alpha\mathbb{H}_\beta E_{\gamma\delta}$. Any generalization to these more complicated polynomial expressions is straightforward, however, and will be left to the reader.

There still remains the evaluation and comparison of the constitutive relations (4.55) and (4.92). To begin with, let us look at the entropy more closely. According to (4.55)^1 and (4.150) we obtain

$$(4.162) \qquad {}^{II}\eta = -\frac{\partial\bar{\Psi}}{\partial\theta} = c_W\frac{\theta}{\theta_o} - \frac{1}{2\rho_o}\bar{L}^{(m)}\mathbb{H}_\alpha\mathbb{H}_\alpha - \frac{L^{(e)}}{2\rho_o}\mathbb{E}_\alpha\mathbb{E}_\alpha + \nu E_{\gamma\gamma} \,.$$

On the other hand, straightforward evaluation of (4.92)^1 on the basis of (4.151) gives

$$(4.163) \qquad {}^{V}\eta = -\frac{\partial\check{\Psi}}{\partial\theta} = c_W\frac{\theta}{\theta_o} - \frac{1}{2\rho_o}\check{L}^{(m)}\mathbb{B}_\alpha\mathbb{B}_\alpha - \frac{L^{(e)}}{2\rho_o}\mathbb{E}_\alpha\mathbb{E}_\alpha + \nu E_{\gamma\gamma} \,,$$

which with the aid of (4.156) and (4.159) becomes

$$(4.164)\qquad {}^{V}\eta = c_W \frac{\theta}{\theta_o} - \frac{1}{2\rho_o} \bar{L}^{(m)} \{\delta_{\alpha\beta}(1 - \frac{2}{\mu}\, {}_4\bar{\chi}^{(m)} \mathbb{H}_\gamma \mathbb{H}_\gamma - \frac{2}{\mu} \bar{L}^{(m)} \theta) + - \frac{2}{\mu}(\bar{b}^{(m)}_{\alpha\beta\gamma\delta} + \mu_o n_{\alpha\beta\gamma\delta}) E_{\gamma\delta}\} \mathbb{H}_\alpha \mathbb{H}_\beta - \frac{L^{(e)}}{2\rho_o} \mathbb{E}_\alpha \mathbb{E}_\alpha + \nu E_{\gamma\gamma} + \text{higher order terms.}$$

Mere comparison of (4.162) and (4.164) shows that ${}^{II}\eta$ differs from ${}^{V}\eta$. Hence, use of (4.159) has not led to identical expressions for the entropy. The difference arises, because of the terms $(\mathbb{H}_\alpha \mathbb{H}_\alpha)^2$, $\mathbb{H}_\alpha \mathbb{H}_\alpha \theta$, $\mathbb{H}_\alpha \mathbb{H}_\beta E_{\gamma\delta}$, and still higher order terms, which in the energy expression can be traced back to the fourth order terms $(\mathbb{H}_\alpha \mathbb{H}_\alpha)^2 \theta$, $\mathbb{H}_\alpha \mathbb{H}_\alpha \theta^2$ and $\mathbb{H}_\alpha \mathbb{H}_\beta E_{\alpha\beta} \theta$. In the process of the transformation of the energy expressions these fourth order terms were omitted (and they must be, if the polynomials are interpreted as truncated Taylor series expansions).

Similar discrepancies also occur when the other constitutive quantities derivable from the free energy are determined. These are given by

$$(4.165)\qquad \mathbb{P}_\alpha = {}^{II}\mathbb{P}_\alpha = {}^{V}\mathbb{P}_\alpha = {}_2\chi^{(e)} \mathbb{E}_\alpha + {}_4\chi^{(e)} \mathbb{E}_\beta \mathbb{E}_\beta \mathbb{E}_\alpha - L^{(e)} \mathbb{E}_\alpha \theta - b^{(e)}_{\alpha\beta\gamma\delta} \mathbb{E}_\beta E_{\gamma\delta} ,$$

$$(4.166)\qquad \mu_o \mathbb{M}^C_\alpha = \mu_o\, {}_2\chi^{(m)} \mathbb{H}_\alpha - {}_4\bar{\chi}^{(m)} \mathbb{H}_\beta \mathbb{H}_\beta \mathbb{H}_\alpha - \bar{L}^{(m)} \mathbb{H}_\alpha \theta - \bar{b}^{(m)}_{\alpha\beta\gamma\delta} \mathbb{H}_\beta E_{\gamma\delta} ,$$

$$(4.167)\qquad \mathbb{M}^L_\alpha = \frac{{}_2\chi^{(m)}}{\mu} \mathbb{B}_\alpha - {}_4\check{\chi}^{(m)} \mathbb{B}_\beta \mathbb{B}_\beta \mathbb{B}_\alpha - \check{L}^{(m)} \mathbb{B}_\alpha \theta - \check{b}^{(m)}_{\alpha\beta\gamma\delta} \mathbb{B}_\beta E_{\gamma\delta} ,$$

$$(4.168)\qquad {}^{II}T^P_{\alpha\beta} = \bar{b}^{(m)}_{\gamma\delta\alpha\beta} \mathbb{H}_\gamma \mathbb{H}_\delta + b^{(e)}_{\gamma\delta\alpha\beta} \mathbb{E}_\gamma \mathbb{E}_\delta - \nu \delta_{\alpha\beta} \theta + c_{\alpha\beta\gamma\delta} E_{\gamma\delta} ,$$

$$(4.169)\qquad {}^{V}T^P_{\alpha\beta} = \check{b}^{(m)}_{\gamma\delta\alpha\beta} \mathbb{B}_\gamma \mathbb{B}_\delta + b^{(e)}_{\gamma\delta\alpha\beta} \mathbb{E}_\gamma \mathbb{E}_\delta - \nu \delta_{\alpha\beta} \theta + c_{\alpha\beta\gamma\delta} E_{\gamma\delta} .$$

They can be transformed into each other by neglecting all inconsistent terms as was done above for the entropy.

This completes our transformation of the theories of group (I,II,III) into those of group (IV,V). The calculations show that full equivalence is often destroyed by a too special choice of the energy functions. But the calculations have simultaneously demonstrated how the material coefficients of one theory can be related to those of another. These questions are of immense practical importance and will be reconsidered in Chapter 5.

REFERENCES

[1] Hutter, K., *On Thermodynamics and Thermostatics of Viscous Thermoelastic Solids in the Electromagnetic Fields. A Lagrangian Formulation*, Arch. Rat. Mech. Anal. 58 (1975), 339-368.

[2] Hutter, K., *A Thermodynamic Theory of Fluids and Solids in the Electromagnetic Fields*, Arch. Rat. Mech. Anal. 64 (1977), 269-298.

[3] Alblas, J.B., *General Theory of Electro- and Magneto-Elasticity*, in Electromagnetic Interactions in Elastic Solids, ed. by H. Parkus, Springer, Wien, 1978.

[4] Toupin, R.A., *A Dynamical Theory of Elastic Dielectrics*, Int. J. Eng. Sc. 1 (1963), 101-126.

[5] Prechtl, A., *On the Electrodynamics of Deformable Media. Part II*, Acta Mechanica 28 (1977), 273-294.

[6] Penfield, P. and H.A. Haus, *Electrodynamics of Moving Media*, The M.I.T. Press, Cambridge, Massachusetts, 1967.

[7] Hutter, K. and Y.H. Pao, *A Dynamic Theory for Magnetizable Elastic Solids with Thermal and Electrical Conduction*, J. of Elasticity 4 (1974), 89-114.

[8] Ven, A.A.F. van de, *Interaction of Electromagnetic and Elastic Fields in Solids*, Dr. of Science Thesis, University of Technology Eindhoven, the Netherlands, 1975.

5. LINEARIZATION

5.1 STATEMENT OF THE PROBLEM

The governing dynamical equations of field-matter interaction in thermoelastic materials as outlined in the previous Chapters are highly nonlinear. Generally, it is hardly possible to find exact solutions even if the most simple problems that are still of some physical relevance are attacked. As stated in the last Chapter already, one of the major disadvantages, namely that the equations are given in the spatial description, while boundary conditions for a solid body are usually prescribed in the reference configuration, has been removed by the introduction of a consistent material description for both the field equations and the jump and boundary conditions. Nevertheless, the resulting equations are still highly nonlinear, and this implies that some approximation scheme must be found.

To render the equations amenable to direct analysis, they will be linearized with respect to some intermediate state. We suppose that in the intermediate state the position of a material point, initially at $\underline{X}$, is given by $\underline{\xi}$ (ξ_α, $\alpha = 1,2,3$). The total motion of the particle from its initial position $\underline{X}$ to its final position $\underline{x}$ is then decomposed into the motion from the $\underline{X}$-state to the intermediate state, characterized by the displacement vector $\underline{\bar{U}}$,

$$(5.1) \qquad \underline{\bar{U}} = \underline{\xi} - \underline{X}$$

and the motion from the intermediate state to the present state with displacement $\underline{u}$

$$(5.2) \qquad \underline{u} = \underline{x} - \underline{\xi} ,$$

(i.e. from now on $\underline{u}$ is not the total displacement, but the displacement from $\underline{\xi}$ to $\underline{x}$ only).

In the above, $\underline{\xi}$ was not specified. In principle, any continuous map $\underline{X} \to \underline{\xi}$ can be considered suitable for this intermediate configuration of the body. Yet it will be assumed that the problem at hand suggests a natural definition of this state. For the time being we assume that it is known or at least determinable. We further assume that the position of a material point in the intermediate state $\underline{\xi}$ is close to its final position $\underline{x}$. Based on this closeness the motion and more generally, all fields may be decomposed into two parts. One of these parts represents the fields when $\underline{u} = \underline{0}$, the other one is due to the perturbations from the $\underline{\xi}$-state to the $\underline{x}$-state. Because these perturbations are assumed to be small, the governing equations can be linearized in these perturbations.

This linearization procedure can be performed in a completely consistent way. For instance, if a body is initially subjected to large biasing electromagnetic fields and wave propagation or vibration properties of this body are investigated, then our linearization procedure will be applicable (e.g. [1], [2] and [3,4]). Of the same nature are magnetoelastic stability problems ([5], [6]). To describe the linearization procedure into more detail, all field variables will be decomposed into two parts. Those in the intermediate state are labeled with an overhead bar, whereas the perturbations on this state are indicated by lower case letters, e.g.

(5.3) $\quad \mathbb{B}_\alpha = \bar{\mathbb{B}}_\alpha + b_\alpha, \qquad \mathbb{E}_\alpha = \bar{\mathbb{E}}_\alpha + e_\alpha ,$

where

(5.4) $\quad \dfrac{\| \underline{b} \|}{\| \underline{\mathbb{B}} \|} = O(\varepsilon), \qquad \dfrac{\| \underline{e} \|}{\| \underline{\mathbb{E}} \|} = O(\varepsilon)$, etc.

In (5.4) ε denotes a small positive quantity ($0 < \varepsilon << 1$). The norms in (5.4) may conveniently be defined as

(5.5) $\quad \| a \|^2 := \limsup_{0 \le \tau \le t} (a^2(\tau)) ,$

where $\tau = 0$ is the time at which the process started and t is the current time. Moreover, the displacements from $\underline{\xi}$ to $\underline{x}$ and their material time derivatives are assumed to be small in the sense that

(5.6) $\quad \| \dfrac{\partial u}{\partial \underline{\xi}} \| = O(\varepsilon) \qquad \text{and} \qquad \dfrac{\| \dot{\underline{u}} \|}{v_0} = O(\varepsilon) ,$

where v_0 is some characteristic wave speed.
In the Lagrangian description any differentiation with respect to $\underline{X}$ or t falls directly onto the respective variables as decomposed in (5.3). For instance

$$\mathbb{B}_{\alpha,\alpha} = \bar{\mathbb{B}}_{\alpha,\alpha} + b_{\alpha,\alpha} .$$

This is particularly easy and convenient, and to see this let us first briefly investigate the Eulerian formulation. In this case all equations must be traced back to the intermediate configuration. In particular, all derivatives with respect to the present coordinates must be expressed in terms of the intermediate coordinates by means of the transformation rules, which we shall now briefly outline, (although lateron they will not be used). To this end, let f be any physical quantity. It may be regarded as a function of the variables $(\underline{X},t)$, $(\underline{\xi},t)$ or $(\underline{x},t)$; thus

(5.7) $\quad f = \hat{f}(\underline{X},t) = \tilde{f}(\underline{\xi},t) = \check{f}(\underline{x},t) .$

Dependent on which representation we choose, we thus have

$$\frac{\partial \check{f}}{\partial x_i} = \frac{\partial \tilde{f}}{\partial \xi_\alpha}\frac{\partial \check{\xi}_\alpha}{\partial x_i} = (\delta_{i\alpha} - \frac{\partial \check{u}_\alpha}{\partial x_i})\frac{\partial \tilde{f}}{\partial \xi_\alpha} .$$

In particular with $f \equiv u_\alpha$ we obtain

$$\frac{\partial \check{u}_\alpha}{\partial x_i} = (\delta_{i\beta} - \frac{\partial \check{u}_\beta}{\partial x_i})\frac{\partial \tilde{u}_\alpha}{\partial \xi_\beta} \cong \delta_{i\beta}\frac{\partial \tilde{u}_\alpha}{\partial \xi_\beta} ,$$

whence follows

$$\frac{\partial \check{f}}{\partial x_i} \cong \delta_{i\beta}(\delta_{\alpha\beta} - \frac{\partial \tilde{u}_\alpha}{\partial \xi_\beta})\frac{\partial \tilde{f}}{\partial \xi_\alpha} . \tag{5.8}$$

Similarly, for the material time derivative we obtain

$$\frac{df}{dt} = \frac{\partial \hat{f}}{\partial t} = \frac{\partial \tilde{f}}{\partial t} + \frac{\partial \tilde{f}}{\partial \xi_\alpha}\frac{\partial \hat{\xi}_\alpha}{\partial t} = \frac{\partial \check{f}}{\partial t} + \frac{\partial \check{f}}{\partial x_i}\frac{\partial \hat{x}_i}{\partial t} ,$$

from which one easily deduces that

$$\frac{\partial \check{f}}{\partial t} = \frac{\partial \tilde{f}}{\partial t} + \frac{\partial \tilde{f}}{\partial \xi_\alpha}\frac{\partial \hat{\xi}_\alpha}{\partial t} - \frac{\partial \tilde{f}}{\partial \xi_\alpha}\frac{\partial \check{\xi}_\alpha}{\partial x_i}\frac{\partial \hat{x}_i}{\partial t} = \frac{\partial \tilde{f}}{\partial t} - \frac{\partial \tilde{f}}{\partial \xi_\alpha}(\frac{\partial \check{\xi}_\alpha}{\partial x_i}\frac{\partial \hat{x}_i}{\partial t} - \frac{\partial \hat{\xi}_\alpha}{\partial t}) .$$

Here, in each of the occurring functions we have indicated the functional dependences. With the obvious definitions

$$\frac{\partial \hat{\xi}_\alpha}{\partial t} := \dot{\xi}_\alpha \qquad \text{and} \qquad \frac{\partial \hat{x}_i}{\partial t} := \dot{x}_i = \dot{u}_i + \dot{\xi}_\alpha \delta_{i\alpha}$$

this now becomes

$$\begin{aligned} \frac{\partial \check{f}}{\partial t} &= \frac{\partial \tilde{f}}{\partial t} - [\delta_{i\beta}(\delta_{\alpha\beta} - \frac{\partial \tilde{u}_\alpha}{\partial \xi_\beta})(\dot{u}_i + \delta_{i\gamma}\dot{\xi}_\gamma) - \dot{\xi}_\alpha]\frac{\partial \tilde{f}}{\partial \xi_\alpha} = \\ &\cong \frac{\partial \tilde{f}}{\partial t} - [\dot{u}_\alpha - \dot{\xi}_\beta \frac{\partial \tilde{u}_\alpha}{\partial \xi_\beta}]\frac{\partial \tilde{f}}{\partial \xi_\alpha} . \end{aligned} \tag{5.9}$$

The above formulas (5.8) and (5.9) must be applied in all field equations whenever space and time derivatives of physical quantities occur. That this is very tedious can be seen from the fact that for a dynamical theory it has been tried by several authors in the past as e.g. Toupin [7], Hutter and Pao [8] and van de Ven [9]. Yet, except for the procedure of van de Ven none of these is completely correct. Toupin, [7], for instance, seems to replace (5.8) and (5.9) at certain places by the approximations

$$\frac{\partial}{\partial \underline{X}} \cong \frac{\partial}{\partial \underline{x}} \qquad \text{and} \qquad \frac{\partial}{\partial t} \cong \frac{d}{dt} .$$

Hutter and Pao, on the other hand, make full use of (5.8), but disregard (5.9) alltogether. This led to inconsistencies, which resulted in other ad hoc assumptions (as, for instance equation (5.7) in [8]).
In the Eulerian formulation the linearization of jump conditions is equally tedious. For, in this case they are described at the deformed surface in the $\underline{x}$-state, which first must be traced back to the surface in the intermediate state. Corroboration for the fact that this procedure is rather cumbersome can be found e.g. in [8] or [9]. As we shall see in Section 5.2.4, in a Lagrangian description the linearization of jump conditions is straightforward. Several other authors have also published linearization procedures, which are all less general than the one we shall present here, because these authors restrict themselves to (quasi-) magneto or electrostatic processes, to static intermediate states, or they simply delete jump conditions, etc. We confine ourselves to mentioning Alblas [10], Pao and Yeh [11], Tiersten [12], Baumhauer and Tiersten [13] and Jordan and Eringen [14].

The advantage of the Lagrangian formulation is that the transformations (5.8) and (5.9) need not be applied, because in this formulation all operations are already referred to the reference or initial state. Similarly, the jump conditions hold on the undeformed surface. Once this is realized, the linearization procedure turns out to be straightforward.
A second, but less direct advantage of the Lagrangian or material formulation is that it allows an immediate introduction of electromagnetic potentials for the perturbed electromagnetic fields, which is, although also possible, more elaborate in the Eulerian description. The reader may find corroboration for this by noting (as we shall show lateron) that in the Lagrangian description the homogeneous Maxwell-equations (i.e. (1.23)) remain homogeneous in the perturbed state, whereas they become inhomogeneous in the Eulerian description (cf. [8], p. 81).
This will then be our procedure. The nonlinear equations will be developed by consistently expanding all variables about the intermediate state and neglecting terms of order $O(\varepsilon^2)$. Two systems of equations emerge thereby, one for the equations in the intermediate state and a second one for the perturbed quantities, in which the quantities of the intermediate state appear as coefficients. From a practical point of view the construction of the solution of the perturbed equations is often more important than that of the intermediate state. Often the intermediate state serves only as a (static) biasing state deviations

from which can be calculated by solving the perturbed equations. For buckling problems, for instance, the prebuckled state is taken as intermediate state and its stability follows from the perturbed equations. Since, moreover, the perturbed equations are relatively insensitive to an exact determination of the intermediate fields, it would be advantageous to find an approximation scheme by which the coefficients could be determined to a sufficient degree of accuracy without making use of the exact intermediate equations. This is indeed often possible; we will demonstrate it below by introducing the so-called rigid body state. The purpose of its use is to have a quick access to the perturbed equations without having to solve the equations in the intermediate state exactly. To explain it we consider the special situation where deformations caused by the electromagnetic fields are small and where changes in temperature are small as well. If at the same time coupling terms are small, which means that small deformations or small temperature changes result in small changes in the electromagnetic fields, then the intermediate state is close to an undeformed state. This imaginary state will be called rigid body state. In this state the motion of the body, clearly, consists of a pure translation and rotation. The electromagnetic fields and the temperature distributions in this state are determinable from rigid body electrodynamics, mechanics and thermodynamics. The fields in this state will be denoted by a superscript $(\ )^{o}$, e.g. $\mathbb{B}^{o}_{\alpha}, \mathbb{E}^{o}_{\alpha}$, whereas subscripts $(\ )_{o}$ will indicate initial values prior to any motion. By mere definition we then have

$$\rho^{o} = \rho_{o} \ , \tag{5.10}$$

but generally neither $\theta^{o} = \theta_{o}$ nor $\underline{U}^{o} = \underline{0}$. If $\underline{U}^{o} = \underline{0}$, then rigid body motions are excluded. In most applications this will be the case.
Since deformations are assumed to be small, we may define a measure E, $0 < E \ll 1$, by

$$E := \| \bar{U}_{\alpha,\beta} \| \ , \tag{5.11}$$

with the aid of which we may conclude that

$$\frac{|\bar{\theta} - \theta^{o}|}{\theta^{o}} = O(E), \qquad \frac{\| \bar{\underline{B}} - \underline{B}^{o} \|}{\| \underline{B}^{o} \|} = O(E), \text{ etc.} \tag{5.12}$$

Returning to the perturbed equations, we notice that these equations are linear in the perturbed fields $\underline{u}$, $\underline{b}$, etc., with coefficients which depend on the values of the fields in the intermediate state. When the intermediate fields in these coefficients are replaced by the rigid body fields, errors of the order $O(E)$ are introduced. Because the perturbed equations are of the order $O(\varepsilon)$ themselves, the neglect of these terms ultimately means that the resulting equations are correct except for terms of order $O(E\varepsilon)$. In short: The linearized perturbation equations, in which the coefficients are referred to the rigid body state, represent an approximation bound to an error not larger than $O(E\varepsilon)$. It should further be emphasized that it is not justified, in general, to evaluate the intermediate fields themselves by applying the rigid body approximations. This would result in errors of order $O(E)$, which is larger than $O(E\varepsilon)$.

The approach just described to approximate the equations is more consistent than the usual small strain approximations. In particular there are two immediate advantages:

i) the consistency of the linearization is a proven property and not an a priori assumption (the neglects can be made explicit) and,
ii) The stresses (and all other fields) in the intermediate state can still be calculated to within any desired degree of exactness.

This latter point is of importance in particular for a consistent derivation of the equations governing magnetoelastic buckling of beams and plates (see [9], Ch. IX).

As said above, we are still free to solve the equations in the intermediate state as accurately as we please. For most practical purposes, however, a small-strain approximation will suffice. We shall return to this point in due course with the developments in this Chapter.

5.2 LINEARIZATION OF THE LORENTZ MODEL

1. Motivation for this Choice – Governing Equations

In this Section we shall linearize the field equations, constitutive relations and jump conditions of one particular interaction model, but before we present the details a justification for our choice of the model is in order. We saw that all models are equivalent; as a consequence, only practical considerations and reasons of convenience can guide us to prefer one particular model over any other one.

There are several reasons for the choice of model V. Firstly, the calculations in Chapter 4 showed that the structure of many formulas is preserved when they are transformed from the spatial description into their material counterpart. Secondly, several thermodynamic relations, such as the relation between stress and free energy or the Gibbs relation, are much more consize for formulations with a symmetric stress tensor than those with an unsymmetric one. This would leave us with model II and model V, but model V is again computationally advantageous, because the Gauss-Faraday law and the Gauss law are homogeneous equations. This makes the introduction of electromagnetic potentials much easier than it would be otherwise. We shall deviate in one respect from the original Lorentz-model, however, in that we shall use Q_α^S as energy-flux instead of Q_α^L. This essentially amounts to a different choice of the entropy flux (see Sections 2.5 and 4.2). Moreover, we shall exclude external sources, so that

$$\rho_o F_i^{ext} = \rho_o r^{ext} = 0 \ . \tag{5.13}$$

In the Lagrangian formulation and for the Lorentz model the governing equations read as follows

Maxwell-equations: (4.66)

$$\begin{aligned} &\mathbb{B}_{\alpha,\alpha} = 0 \ , \\ &\dot{\mathbb{B}}_\alpha + e_{\alpha\beta\gamma}\mathbb{E}_{\gamma,\beta} = 0 \ , \\ &\mathbb{D}^a_{\alpha,\alpha} = \mathbb{Q} - \mathbb{P}_{\alpha,\alpha} \ , \\ &-\dot{\mathbb{D}}^a_\alpha + e_{\alpha\beta\gamma}\mathbb{H}^a_{\gamma,\beta} = \mathbb{J}_\alpha + \dot{\mathbb{P}}_\alpha + e_{\alpha\beta\gamma}\mathbb{M}^L_{\gamma,\beta} \ , \\ &\dot{\mathbb{Q}} + \mathbb{J}_{\alpha,\alpha} = 0 \ , \end{aligned} \tag{5.14}$$

where

$$\begin{aligned} &\mathbb{D}^a_\alpha = \varepsilon_o |J| C^{-1}_{\alpha\beta}\mathbb{E}_\beta \ , \\ &\mathbb{H}^a_\alpha = \frac{1}{|J|}\{\frac{1}{\mu_o} C_{\alpha\beta}\mathbb{B}_\beta - \varepsilon_o e_{\mu\beta\gamma} C_{\alpha\beta} F_{j\gamma}\dot{x}_j\mathbb{E}_\mu\} \ . \end{aligned} \tag{5.15}$$

Balance of momentum: ((4.133)[1], with $F_i^{ext} = 0$)

$$\begin{aligned} \rho_o\ddot{x}_i - T_{i\alpha,\alpha} = \rho_o F_i^e = F^{-1}_{\alpha i}\{&(\mathbb{Q} - \mathbb{P}_{\beta,\beta})\mathbb{E}_\alpha + e_{\alpha\beta\gamma}(\mathbb{J}_\beta + \dot{\mathbb{P}}_\beta)\mathbb{B}_\gamma + \\ &+ (\mathbb{M}^L_{\alpha,\beta} - \mathbb{M}^L_{\beta,\alpha})\mathbb{B}_\beta\} \ . \end{aligned} \tag{5.16}$$

Balance of energy: ((4.133)[3], with $Q_\alpha = Q_\alpha^S$ and $r^{ext} = 0$)

$$\rho_o\theta\dot{\eta} = \mathbb{J}_\alpha\mathbb{E}_\alpha - Q_{\alpha,\alpha} \tag{5.17}$$

In the above, all variables are referred to the time-independent reference state characterized by the coordinates X_α. Superimposed dots thus represent time derivatives at fixed particles.

Constitutive relations:

In accordance with (4.90) we choose as independent variables

$$C_{\alpha\beta},\ \mathbb{E}_\alpha,\ \mathbb{B}_\alpha,\ \theta,\ \theta_{,\alpha}\ \text{and}\ \mathbb{Q}\ .$$

Following (4.91) and (4.92) we then have

$$\text{(5.18)}\qquad \overset{\vee}{\psi} = U - \theta\eta + \frac{1}{\rho_o}\mathbb{E}_\alpha\mathbb{P}_\alpha = \overset{\vee}{\psi}(C_{\alpha\beta},\mathbb{E}_\alpha,\mathbb{B}_\alpha,\theta)\ ,$$

and

$$\text{(5.19)}\qquad \begin{aligned} \eta &= -\frac{\partial\overset{\vee}{\psi}}{\partial\theta}\ , \\ \mathbb{P}_\alpha &= -\rho_o\frac{\partial\overset{\vee}{\psi}}{\partial\mathbb{E}_\alpha}\ , \\ \mathbb{M}^L_\alpha &= -\rho_o\frac{\partial\overset{\vee}{\psi}}{\partial\mathbb{B}_\alpha}\ , \\ T^P_{\alpha\beta} &= 2\rho_o\frac{\partial\overset{\vee}{\psi}}{\partial C_{\alpha\beta}}\ . \end{aligned}$$

Moreover, for an electrical conductor the constitutive relations for the electric current and the energy flux are of the form (see (4.122) with $(4.124)^3$)

$$\text{(5.20)}\qquad \begin{aligned} \mathbb{J}_\alpha &= \sigma_{\alpha\beta}\mathbb{E}_\beta + \beta_{\beta\alpha}\frac{\theta_{,\beta}}{\theta}\ , \\ Q_\alpha &= -\kappa_{\alpha\beta}\theta_{,\beta} + \beta_{\alpha\beta}\mathbb{E}_\beta\ , \end{aligned}$$

where, in general, the coefficient matrices are still functions of $C_{\alpha\beta}$, $\mathbb{E}_\alpha$, $\mathbb{B}_\alpha$, θ, $\theta_{,\alpha}$ and $\mathbb{Q}$. However, in accordance with the linearization as described in the derivation of (2.172) we shall assume these coefficients to be independent of $\mathbb{E}_\alpha$ and $\theta_{,\alpha}$. Moreover, we shall exclude an explicit occurrence of $\mathbb{Q}$. The equations (5.20) automatically guarantee that the current $\mathbb{J}_\alpha$ and heat flux Q_α vanish in thermostatic equilibrium; in short, if $\mathbb{E}_\alpha = 0$ and $\theta_{,\alpha} = 0$, then

$$\mathbb{J}_{\alpha|E} = 0 \quad\text{and}\quad Q_{\alpha|E} = 0\ .$$

Further conditions of thermostatic equilibrium are that $\sigma_{(\alpha\beta)}$ and $-\kappa_{(\alpha\beta)}$ must be positive-semi definite matrices and, if the Onsager relations are adopted, that the skew symmetric parts of $\sigma_{\alpha\beta}$ and $\kappa_{\alpha\beta}$ must vanish.

<u>Jump conditions</u>: ((4.70), (1.67), (4.80))

$$[\![\mathbb{B}_\alpha]\!]N_\alpha = 0\ ,\qquad e_{\alpha\beta\gamma}[\![\mathbb{E}_\beta]\!]N_\gamma + [\![\mathbb{B}_\alpha W_N]\!] = 0\ ,$$

$$[\![\mathbb{D}^a_\alpha + \mathbb{P}_\alpha]\!]N_\alpha = 0\ ,\qquad e_{\alpha\beta\gamma}[\![\mathbb{H}^a_\beta - \mathbb{M}^L_\beta]\!]N_\gamma - [\![(\mathbb{D}^a_\alpha + \mathbb{P}_\alpha)W_N]\!] = 0\ ,$$

$$[\![\mathbb{J}_\alpha]\!]N_\alpha - [\![QW_N]\!] = 0\ ,$$

(5.21) (over)

$$[\![\rho_o W_N]\!] = 0\ ,$$

$$[\![\rho_o \dot{x}_i W_N]\!] + [\![T_{i\alpha} + \overset{M}{T}_{i\alpha}]\!] N_\alpha = 0\ ,$$

$$[\![(\tfrac{1}{2}\rho_o \dot{x}_i \dot{x}_i + \rho_o U - \Omega) W_N]\!] + [\![(T_{i\alpha} + \overset{M}{T}_{i\alpha})\dot{x}_i - Q_\alpha - e_{\alpha\beta\gamma}\mathbb{E}_\beta(\mathbb{H}^a_\gamma - \overset{L}{\mathbb{M}}_\gamma)]\!] N_\alpha = 0.$$

In these relations N_α is the unit normal vector on the singular surface and W_N is the speed of propagation. $\overset{M}{T}_{i\alpha}$ and Ω are given in (4.80); for ease of reference they will be repeated here:

$$\overset{M}{T}_{i\alpha} = F^{-1}_{\beta i}\{\mathbb{D}^a_\alpha \mathbb{E}_\beta + \mathbb{B}_\alpha \mathbb{H}^a_\beta - \tfrac{1}{2}\delta_{\alpha\beta}(\mathbb{D}^a_\gamma \mathbb{E}_\gamma + \mathbb{B}_\gamma \mathbb{H}^a_\gamma)\}\ ,$$

(5.22)

$$\Omega = -\tfrac{1}{2}(\mathbb{D}^a_\alpha \mathbb{E}_\alpha + \mathbb{B}_\alpha \mathbb{H}^a_\alpha)\ .$$

In this Chapter we shall from now on restrict ourselves to singular surfaces of second order (see e.g. [17], Section 2.8). On such a surface, the deformation gradients and the velocity will be continuous, as is the density ρ_o, i.e.

$$[\![F_{i\alpha}]\!] = [\![\dot{x}_i]\!] = [\![\rho_o]\!] = 0\ .$$

In view of $(5.21)^6$ continuity of ρ_o also implies that the speed of propagation W_N is continuous,

$$[\![W_N]\!] = 0\ .$$

For the case that the singular surface is a real propagating surface, i.e. $W_N \neq 0$, the associated waves are called acceleration waves. Singular surfaces of second order include as a special case material surfaces for which $W_N = 0$, and also embrace the boundary of a solid body in a vacuum, if, as is usually the case, the vacuum is considered as a medium with zero density, which admits continuity of the velocities at the boundary. Surfaces on which tangential velocities may jump are, however, excluded.

Under the above restrictions the last two jump conditions can be simplified. In view of the continuity of ρ_o, $\dot{x}_i$ and W_N $(5.21)^7$ becomes

$$(5.23)\qquad [\![T_{i\alpha}]\!] N_\alpha = -[\![\overset{M}{T}_{i\alpha}]\!] N_\alpha\ ,$$

the right-hand side of which can be written in the following form

$$(5.24)\qquad [\![T_{i\alpha}]\!] N_\alpha = F^{-1}_{\gamma i}\{\langle\!\langle \mathbb{E}_\gamma \rangle\!\rangle [\![\mathbb{P}_\beta]\!] N_\beta + \langle\!\langle \mathbb{B}_\beta \rangle\!\rangle ([\![\overset{L}{\mathbb{M}}_\beta]\!] N_\gamma - [\![\overset{L}{\mathbb{M}}_\gamma]\!] N_\beta) + e_{\beta\gamma\delta}\langle\!\langle \mathbb{B}_\beta \rangle\!\rangle [\![\mathbb{P}_\delta]\!] W_N\}\ .$$

Here, the symbol $\langle\!\langle\ \rangle\!\rangle$ stands for the arithmetic mean of a quantity over the singular surface, i.e.

(5.25) $\langle\mathbb{E}_\alpha\rangle = \frac{1}{2}(\mathbb{E}^+_\alpha + \mathbb{E}^-_\alpha)$.

To prove (5.24), notice that in view of the relations (5.15) and of the continuity conditions of second order singular surfaces

(5.26) $\langle\mathbb{D}^a_\alpha\rangle[\![\mathbb{E}_\alpha]\!] = \langle\mathbb{E}_\alpha\rangle[\![\mathbb{D}^a_\alpha]\!]$, $\quad\langle\mathbb{H}^a_\alpha\rangle[\![\mathbb{B}_\alpha]\!]= \langle\mathbb{B}_\alpha\rangle[\![\mathbb{H}^a_\alpha]\!]$.

Moreover, the jump of the first term in the curly brackets of (5.22) may be written as

$$[\![\mathbb{D}^a_\alpha\mathbb{E}_\beta]\!]N_\alpha = (\langle\mathbb{D}^a_\alpha\rangle[\![\mathbb{E}_\beta]\!] + [\![\mathbb{D}^a_\alpha]\!]\langle\mathbb{E}_\beta\rangle)N_\alpha =$$
$$= \langle\mathbb{D}^a_\alpha\rangle N_\alpha[\![\mathbb{E}_\beta]\!] - \langle\mathbb{E}_\beta\rangle[\![\mathbb{P}_\alpha]\!]N_\alpha ,$$

in which use has been made of $(5.21)^3$. Using $(5.26)^1$ and $(5.21)^4$ it is then straightforward to show that in a non-relativistic approximation

$$[\![\mathbb{D}^a_\alpha\mathbb{E}_\beta]\!]N_\alpha - \frac{1}{2}[\![\mathbb{D}^a_\gamma\mathbb{E}_\gamma]\!]\delta_{\alpha\beta}N_\alpha = -\langle\mathbb{E}_\beta\rangle[\![\mathbb{P}_\alpha]\!]N_\alpha + e_{\alpha\beta\gamma}\langle\mathbb{D}^a_\alpha\rangle[\![\mathbb{B}_\gamma]\!]W_N =$$
$$= -\langle\mathbb{E}_\beta\rangle[\![\mathbb{P}_\alpha]\!]N_\alpha .$$

In a completely analogous way the remaining terms on the right-hand side of (5.22), (5.23) can be handled, so that, finally (5.24) emerges.
On the other hand, the first and fourth term in $(5.21)^8$ vanish for acceleration waves, and, furthermore, it can be shown with the use of $(5.21)^{2,4}$ that

(5.27) $[\![e_{\alpha\beta\gamma}\mathbb{E}_\beta(\mathbb{H}^a_\gamma - \mathbb{M}^L_\gamma)]\!]N_\alpha = \{[\![\mathbb{B}_\alpha]\!]\langle\mathbb{H}^a_\alpha - \mathbb{M}^L_\alpha\rangle + \langle\mathbb{E}_\alpha\rangle[\![\mathbb{D}^a_\alpha + \mathbb{P}_\alpha]\!]\}W_N$.

Finally, from $(5.22)^2$ and with the application of the relations (5.26) we find

(5.28) $[\![\Omega]\!] = -\langle\mathbb{E}_\alpha\rangle[\![\mathbb{D}^a_\alpha]\!] - \langle\mathbb{H}^a_\alpha\rangle[\![\mathbb{B}_\alpha]\!]$.

With these results the jump condition for the energy flux assumes the simple and elegant form

(5.29) $[\![Q_\alpha]\!]N_\alpha = \{\rho_o[\![U]\!] + \langle\mathbb{M}^L_\alpha\rangle[\![\mathbb{B}_\alpha]\!] - \langle\mathbb{E}_\alpha\rangle[\![\mathbb{P}_\alpha]\!]\}W_N$.

It should be noted here that the right-hand side vanishes when the surface is material.
Before we close, consider a surface separating a body from the vacuum. According to our interpretation of the vacuum as a medium with vanishing small mass density, and because we are looking at surfaces of second order the deformation tensor is continuous across such a surface so that on the vacuum side (+) one has

$$\mathbb{D}^{a+}_{\alpha} = (\varepsilon_{\circ} J C^{-1}_{\alpha\beta})^{+} \mathbb{E}^{+}_{\beta} = (\varepsilon_{\circ} J C^{-1}_{\alpha\beta})^{-} \mathbb{E}^{+}_{\beta} \ .$$

Similar statements also hold for $\mathbb{H}^{a+}_{\alpha}$. Note also that

$$\mathbb{E}^{+}_{\alpha} \neq E^{+}_{i}\delta_{i\alpha} \ ,$$

but rather

$$\mathbb{E}^{+}_{\alpha} = F^{-}_{i\alpha}(E^{+}_{i} + e_{ijk}\dot{x}^{-}_{j}B^{+}_{k}) \ ,$$

and that

$$\mathbb{B}^{+}_{\alpha} = (JF^{-1}_{\alpha i})^{-} B^{+}_{i} \ .$$

These facts should be born in mind, for otherwise incorrect results emerge. This completes the collection of the basic governing equations. Their linearized versions will be derived below.

2. Decomposition of the Balance Laws

Having presented the governing equations we now proceed with the decomposition of the balance laws. As was said several times before already, this decomposition is particularly easy in the material description. To corroborate this statement, let us consider the first of the Maxwell equations (5.14). Introducing (5.3) as decomposition for the magnetic induction $\mathbb{B}_{\alpha}$, equation $(5.14)^{1}$ may be written as

$$(\bar{\mathbb{B}}_{\alpha} + b_{\alpha})_{,\alpha} = \bar{\mathbb{B}}_{\alpha,\alpha} + b_{\alpha,\alpha} = 0$$

In view of the basic separation assumption, according to which the governing equations must be fulfilled for the intermediate state itself, we thus have

$$\bar{\mathbb{B}}_{\alpha,\alpha} = 0 \ ,$$

and, consequently,

$$b_{\alpha,\alpha} = 0 \ .$$

In these equations, both $\bar{\mathbb{B}}_{\alpha}$ and b_{α} are regarded as functions of X_{α} and t. Proceeding with all Maxwell equations as was explained above with the Gauss law it is easily shown that the equations, valid in the intermediate state, are given by

$$\bar{\mathbb{B}}_{\alpha,\alpha} = 0 \ ,$$
$$\dot{\bar{\mathbb{B}}}_{\alpha} + e_{\alpha\beta\gamma}\bar{\mathbb{E}}_{\gamma,\beta} = 0 \ ,$$

(over)

$$(5.30)\qquad \begin{aligned} &\bar{\mathbb{D}}^a_{\alpha,\alpha} = \bar{\mathbb{Q}} - \bar{\mathbb{P}}_{\alpha,\alpha} \ , \\ &-\dot{\bar{\mathbb{D}}}^a_\alpha + e_{\alpha\beta\gamma}\bar{\mathbb{H}}^a_{\gamma,\beta} = \bar{\mathbb{J}}_\alpha + \dot{\bar{\mathbb{P}}}_\alpha + e_{\alpha\beta\gamma}\bar{\mathbb{M}}^L_{\gamma,\beta} \ , \\ &\dot{\bar{\mathbb{Q}}} + \bar{\mathbb{J}}_{\alpha,\alpha} = 0 \ . \end{aligned}$$

These equations are obtained from the original Maxwell equations (5.14) by merely replacing in the latter all quantities by those carrying an overhead bar. On the other hand, the perturbed Maxwell equations are

$$(5.31)\qquad \begin{aligned} &b_{\alpha,\alpha} = 0 \ , \\ &\dot{b}_\alpha + e_{\alpha\beta\gamma}e_{\gamma,\beta} = 0 \ , \\ &d^a_{\alpha,\alpha} = q - p_{\alpha,\alpha} \ , \\ &-\dot{d}^a_\alpha + e_{\alpha\beta\gamma}h^a_{\gamma,\beta} = j_\alpha + \dot{p}_\alpha + e_{\alpha\beta\gamma}m_{\gamma,\beta} \ , \\ &\dot{q} + j_{\alpha,\alpha} = 0 \ , \end{aligned}$$

in which all lower case letters denote perturbed Lagrangian electromagnetic field quantities, the definitions being analogous to that for b_α. For instance,

$$e_\alpha = \mathbb{E}_\alpha - \bar{\mathbb{E}}_\alpha, \text{ etc.}$$

For convenience and since no confusion is possible here, we have used m_α for $(\mathbb{M}^L_\alpha - \bar{\mathbb{M}}^L_\alpha)$ instead of m^L_α. Whenever, in the sequel confusion with the Chu-magnetization becomes possible we shall use m^L_α, however.

Equations (5.31) are formally the same as the original Maxwell equations. This is no surprise, because the equations (5.14)[1-5] are written such that they appear in a linear form. Yet, they are nevertheless nonlinear; the nonlinearity is only covered by the use of the auxiliary fields $\mathbb{D}^a_\alpha$ and $\mathbb{H}^a_\alpha$. In the decomposition process of these quantities use must be made of the following relations, the proof of which is straightforward

$$(5.32)\qquad \begin{aligned} &x_i = \delta_{i\alpha}(\xi_\alpha + u_\alpha) \ , \\ &F_{i\alpha} = \frac{\partial x_i}{\partial X_\alpha} = \delta_{i\beta}\left(\frac{\partial \xi_\beta}{\partial X_\alpha} + \frac{\partial u_\beta}{\partial X_\alpha}\right) = \delta_{i\beta}(\bar{F}_{\beta\alpha} + u_{\beta,\alpha}) \ , \\ &F^{-1}_{\alpha i} = \frac{\partial X_\alpha}{\partial x_i} = \frac{\partial X_\alpha}{\partial \xi_\beta}\frac{\partial \xi_\beta}{\partial x_i} \simeq \bar{F}^{-1}_{\alpha\beta}(\delta_{i\beta} - \delta_{i\delta}\bar{F}^{-1}_{\gamma\delta}u_{\beta,\gamma}) \ , \\ &C_{\alpha\beta} = F_{i\alpha}F_{i\beta} \cong \bar{C}_{\alpha\beta} + \bar{F}_{\gamma\beta}u_{\gamma,\alpha} + \bar{F}_{\gamma\alpha}u_{\gamma,\beta} \ , \end{aligned}$$

(over)

$$C^{-1}_{\alpha\beta} = F^{-1}_{\alpha i}F^{-1}_{\beta i} \cong \bar{C}^{-1}_{\alpha\beta} - (\bar{C}^{-1}_{\beta\gamma}\bar{F}^{-1}_{\alpha\delta} + \bar{C}^{-1}_{\alpha\gamma}\bar{F}^{-1}_{\beta\delta})u_{\delta,\gamma} ,$$

$$J = \det(F_{i\alpha}) = \det[\bar{F}_{\beta\alpha}(\delta_{i\beta} + \delta_{i\delta}\bar{F}^{-1}_{\beta\gamma}u_{\delta,\gamma})] \cong \bar{J}(1 + \bar{F}^{-1}_{\gamma\delta}u_{\delta,\gamma}) .$$

Here,

$$\bar{F}_{\alpha\beta} = \frac{\partial \xi_\alpha}{\partial X_\beta} , \qquad \bar{F}^{-1}_{\alpha\beta} = \frac{\partial X_\alpha}{\partial \xi_\beta} , \qquad \bar{C}_{\alpha\beta} = \bar{F}_{i\alpha}\bar{F}_{i\beta} ,$$

(5.33)

$$\bar{C}^{-1}_{\alpha\beta} = \bar{F}^{-1}_{\alpha i}\bar{F}^{-1}_{\beta i}, \qquad \bar{J} = \det(\bar{F}_{i\alpha}) .$$

With these preliminary calculations the decomposition of the auxiliary fields $\mathbb{D}^a_\alpha$ and $\mathbb{H}^a_\alpha$ can now be performed. The results are:

$$\bar{\mathbb{D}}^a_\alpha = \varepsilon_o \bar{J}\bar{C}^{-1}_{\alpha\beta}\bar{\mathbb{E}}_\beta ,$$

(5.34)

$$\bar{\mathbb{H}}^a_\alpha = \frac{1}{\mu_o \bar{J}} \bar{C}_{\alpha\beta}\bar{\mathbb{B}}_\beta - \frac{\varepsilon_o}{\bar{J}} e_{\beta\gamma\delta}\bar{C}_{\alpha\beta}\bar{F}_{\varepsilon\gamma}\dot{\xi}_\varepsilon\bar{\mathbb{E}}_\delta ,$$

and

$$d^a_\alpha = \varepsilon_o \bar{J}\{\bar{C}^{-1}_{\alpha\beta}e_\beta + (\bar{C}^{-1}_{\alpha\beta}\bar{F}^{-1}_{\gamma\delta} - \bar{C}^{-1}_{\beta\gamma}\bar{F}^{-1}_{\alpha\delta} - \bar{C}^{-1}_{\alpha\gamma}\bar{F}^{-1}_{\beta\delta})\bar{\mathbb{E}}_\beta u_{\delta,\gamma}\} ,$$

$$h^a_\alpha = \frac{1}{\bar{J}}\{\frac{1}{\mu_o} \bar{C}_{\alpha\beta}b_\beta - \varepsilon_o e_{\beta\gamma\delta}\bar{F}_{\varepsilon\gamma}\dot{\xi}_\varepsilon\bar{C}_{\alpha\beta}e_\delta +$$

(5.35)

$$+ [(\frac{1}{\mu_o}\bar{\mathbb{B}}_\beta - e_{\beta\nu\mu}\bar{F}_{\varepsilon\nu}\dot{\xi}_\varepsilon\bar{\mathbb{E}}_\mu)(\bar{F}_{\delta\alpha}\delta_{\beta\gamma} + \bar{F}_{\delta\beta}\delta_{\alpha\gamma} - \bar{C}_{\alpha\beta}\bar{F}_{\gamma\delta}) +$$

$$- \varepsilon_o e_{\beta\gamma\mu}\bar{\mathbb{E}}_\mu\bar{C}_{\alpha\beta}\dot{\xi}_\delta]u_{\delta,\gamma} - \varepsilon_o e_{\beta\gamma\delta}\bar{\mathbb{E}}_\delta\bar{C}_{\alpha\beta}\bar{F}_{\varepsilon\gamma}\dot{u}_\varepsilon\} .$$

Here and henceforth we restrict ourselves to positive values of the Jacobian J and could therefore replace $|J|$ by J whenever it occurred. We further would like to emphasize that the perturbed fields d^a_α and h^a_α are expressed here in terms of the deformation and the electromagnetic fields in the intermediate state, all of which are functions of X_α and t, in general.
The deformation in the intermediate state may, in our approximation, be neglected and the intermediate fields be replaced by the fields in the rigid body state. If this is the case we may choose

$$(5.36) \qquad \xi^o_\alpha(\underline{X},t) = \Xi_\alpha(t) + R_{\alpha\beta}(t)X_\beta ,$$

where $\Xi_\alpha(t)$ can be identified with the coordinates of the center of mass of the body in its rigid body state (the position of the center of mass in the reference state, t = 0, is here taken as the origin of our coordinate system), and where $R_{\alpha\beta}(t)$ is a time-dependent proper orthogonal matrix, i.e.

$$R^{-1} = R^T \qquad \text{and} \qquad RR^T = I .$$

Differentiating both sides of (5.36) with respect to time gives

(5.37) $$\dot{\xi}^o_\alpha = \dot{\Xi}_\alpha + \dot{R}_{\alpha\beta}X_\beta = \dot{\Xi}_\alpha + e_{\alpha\beta\gamma}\Omega_\beta R_{\gamma\delta}X_\delta \ ,$$

where $\underline{\Omega}(t)$ is the angular velocity of the body which is given by

(5.38) $$\Omega_\alpha = -\frac{1}{2}\, e_{\alpha\beta\gamma}\dot{R}_{\beta\delta}R_{\gamma\delta} \ .$$

The approximate versions of the equations (5.35) can now be obtained by replacing in these equations all variables carrying an overhead bar, $(\bar{\cdot})$, by the variables in the rigid body state, $(\cdot)^o$. In other words the following replacements must be made

(5.39) $$\begin{aligned} &\bar{\mathbb{E}}_\alpha \to \mathbb{E}^o_\alpha \ , \qquad \bar{\mathbb{B}}_\alpha \to \mathbb{B}^o_\alpha \ , \qquad \bar{F}_{\alpha\beta} \to R_{\alpha\beta} , \qquad \bar{F}^{-1}_{\alpha\beta} \to R_{\beta\alpha} \ , \\ &\bar{J} \to 1, \qquad \bar{C}_{\alpha\beta} \to \delta_{\alpha\beta}, \qquad \bar{C}^{-1}_{\alpha\beta} \to \delta_{\alpha\beta} \ . \end{aligned}$$

With these we obtain for (5.35)

(5.40) $$\begin{aligned} d^a_\alpha &= \varepsilon_o\{e_\alpha + (\delta_{\alpha\beta}R_{\delta\gamma} - \delta_{\beta\gamma}R_{\delta\alpha} - \delta_{\alpha\gamma}R_{\delta\beta})\mathbb{E}^o_\beta u_{\delta,\gamma}\} \ , \\ h^a_\alpha &= \frac{1}{\mu_o} b_\alpha - \varepsilon_o e_{\alpha\beta\gamma}R_{\delta\beta}\dot{\xi}^o_\delta e_\gamma \ + \\ &+ [(\frac{1}{\mu_o}\mathbb{B}^o_\beta - e_{\beta\nu\mu}R_{\varepsilon\nu}\dot{\xi}^o_\varepsilon\mathbb{E}^o_\mu)(R_{\delta\alpha}\delta_{\beta\gamma} + R_{\delta\beta}\delta_{\alpha\gamma} - R_{\gamma\delta}\delta_{\alpha\beta}) \ + \\ &- \varepsilon_o e_{\alpha\gamma\mu}\mathbb{E}^o_\mu\dot{\xi}^o_\delta]u_{\delta,\gamma} - \varepsilon_o e_{\alpha\beta\gamma}\mathbb{E}^o_\gamma R_{\delta\beta}\dot{u}_\delta \end{aligned}$$

Still further simplifications can be found, if the intermediate state is not only close to a rigid body state, but deviates from the initial state by a small amount only (i.e. for a motionless rigid body state). In that case the equations (5.40) are further simplified by setting

$$R_{\alpha\beta} = \delta_{\alpha\beta} \qquad \text{and} \qquad \dot{\xi}^o_\alpha = 0 \ .$$

The next step in the simplification of the balance laws is the decomposition of the momentum and energy equations. To this end (5.16) and (5.17) will be written as

(5.41) $$\begin{aligned} &(\rho_o\ddot{\xi}_\alpha + \rho_o\ddot{u}_\alpha)\delta_{i\alpha} = \bar{T}_{i\alpha,\alpha} + \delta_{i\alpha}t_{\alpha\beta,\beta} + \rho_o\bar{F}^e_i + \rho_o f^e_\alpha\delta_{i\alpha} \ , \\ &\rho_o(\bar{\theta} + \theta)(\dot{\bar{\eta}} + \dot{s}) = (\mathcal{J}_\alpha + j_\alpha)(\bar{\mathbb{E}}_\alpha + e_\alpha) - (\bar{Q}_\alpha - q_\alpha)_{,\alpha} \end{aligned}$$

Apart from the perturbation variables already introduced before we have defined here

$$(5.42)\qquad \begin{aligned} &f^e_\alpha = \delta_{i\alpha}(F^e_i - \bar{F}^e_i), \qquad t_{\alpha\beta} = \delta_{i\alpha}(T_{i\beta} - \bar{T}_{i\beta}) ,\\ &s = \eta - \bar{\eta}, \qquad \theta = \Theta - \bar{\Theta}, \qquad q_\alpha = Q_\alpha - \bar{Q}_\alpha . \end{aligned}$$

Because we assume that the equations (5.41) must hold for the intermediate state, we have

$$(5.43)\qquad \begin{aligned} &\rho_o \ddot{\xi}_\alpha = \delta_{i\alpha}(\bar{T}_{i\beta,\beta} + \rho_o \bar{F}^e_i) ,\\ &\rho_o \overline{\Theta\dot{\eta}} = \bar{\mathbb{J}}_\alpha \bar{\mathbb{E}}_\alpha - \bar{Q}_{\alpha,\alpha} , \end{aligned}$$

as momentum and energy equations in this state and

$$(5.44)\qquad \begin{aligned} &\rho_o \ddot{u}_\alpha = t_{\alpha\beta,\beta} + \rho_o f^e_\alpha ,\\ &\rho_o(\theta\dot{\bar{\eta}} + \bar{\Theta}\dot{s}) = j_\alpha \bar{\mathbb{E}}_\alpha + \bar{\mathbb{J}}_\alpha e_\alpha - q_{\alpha,\alpha} , \end{aligned}$$

as the corresponding equations in the perturbed state.
An explicit representation of the body force expression is obtained from (5.16) and (5.32)[3], namely

$$(5.45)\qquad \rho_o \bar{F}^e_i = \delta_{i\delta}\bar{F}^{-1}_{\alpha\delta}\{(\bar{\mathbb{Q}} - \bar{\mathbb{P}}_{\beta,\beta})\bar{\mathbb{E}}_\alpha + e_{\alpha\beta\gamma}(\bar{\mathbb{J}}_\beta + \dot{\bar{\mathbb{P}}}_\beta)\bar{\mathbb{B}}_\gamma + (\bar{\mathbb{M}}^L_{\alpha,\beta} - \bar{\mathbb{M}}^L_{\beta,\alpha})\bar{\mathbb{B}}_\beta\} ,$$

as the expression for the body force in the intermediate state, and

$$(5.46)\qquad \begin{aligned} \rho_o f^e_\alpha = \bar{F}^{-1}_{\delta\alpha}\{&(\bar{\mathbb{Q}} - \bar{\mathbb{P}}_{\beta,\beta})e_\delta + (q - p_{\beta,\beta})\bar{\mathbb{E}}_\varepsilon +\\ &+ e_{\beta\gamma\delta}[(\bar{\mathbb{J}}_\beta + \dot{\bar{\mathbb{P}}}_\beta)b_\gamma + (j_\beta + \dot{p}_\beta)\bar{\mathbb{B}}_\gamma] +\\ &+ \bar{\mathbb{B}}_\beta(m_{\delta,\beta} - m_{\beta,\delta}) + b_\beta(\mathbb{M}^L_{\delta,\beta} - \mathbb{M}^L_{\beta,\delta})\} +\\ &- \bar{F}^{-1}_{\beta\gamma}\bar{F}^{-1}_{\delta\alpha}\{(\bar{\mathbb{Q}} - \bar{\mathbb{P}}_{\varepsilon,\varepsilon})\bar{\mathbb{E}}_\beta + e_{\beta\mu\nu}(\bar{\mathbb{J}}_\mu + \dot{\bar{\mathbb{P}}}_\mu)\bar{\mathbb{B}}_\nu + \bar{\mathbb{B}}_\varepsilon(\bar{\mathbb{M}}^L_{\beta,\varepsilon} - \bar{\mathbb{M}}^L_{\varepsilon,\beta})\}u_{\gamma,\delta} , \end{aligned}$$

as the corresponding expression in the perturbed state. When deformations in the intermediate state are neglected, but rigid motions are allowed, then $\bar{F}_{\alpha\beta}$ in (5.46) must simply be replaced by $R_{\alpha\beta}$ and, furthermore, all variables carrying an overhead bar must be replaced by the variables in the rigid body state. Moreover, for a motionless rigid body state $R_{\alpha\beta} = \delta_{\alpha\beta}$ and $\dot{\xi}^o_\alpha = 0$.

The balance laws of mechanics and electrodynamics are thus decomposed into equations valid in the intermediate state and those in the perturbed state. When in the latter the rigid body approximation is applied, the corresponding equations in this state are also needed. The governing equations consist of the Maxwell equations

$$\begin{aligned}
&\mathbb{B}^{o}_{\alpha,\alpha} = 0 \,, \\
&\dot{\mathbb{B}}^{o}_{\alpha} + e_{\alpha\beta\gamma}\mathbb{E}^{o}_{\gamma,\beta} = 0 \,, \\
(5.47)\qquad &\mathbb{D}^{ao}_{\alpha,\alpha} = \mathbb{Q}^{o} - \mathbb{P}^{o}_{\alpha,\alpha} \,, \\
&-\dot{\mathbb{D}}^{ao}_{\alpha} + e_{\alpha\beta\gamma}\mathbb{H}^{ao}_{\gamma,\beta} = \mathbb{J}^{o}_{\alpha} + \dot{\mathbb{P}}^{o}_{\alpha} + e_{\alpha\beta\gamma}\mathbb{M}^{Lo}_{\gamma,\beta} \,, \\
&\dot{\mathbb{Q}}^{o} + \mathbb{J}^{o}_{\alpha,\alpha} = 0 \,,
\end{aligned}$$

where

$$\begin{aligned}
&\mathbb{D}^{ao}_{\alpha} = \varepsilon_{o}\mathbb{E}^{o}_{\alpha} \,, \\
(5.48)\qquad &\mathbb{H}^{ao}_{\alpha} = \frac{1}{\mu_{o}}\mathbb{B}^{o}_{\alpha} - \varepsilon_{o}e_{\alpha\beta\gamma}R_{\delta\beta}\dot{\xi}^{o}_{\delta}\mathbb{E}^{o}_{\gamma} \,,
\end{aligned}$$

and of the energy equation

$$(5.49)\qquad \rho_{o}\theta^{o}\dot{\eta}^{o} = \mathbb{J}^{o}_{\alpha}\mathbb{E}^{o}_{\alpha} - Q^{o}_{\alpha,\alpha} \,,$$

suplemented by the equations of motion for the rigid body (which will be represented below), by the pertinent constitutive relations and by the jump conditions.

The equations (5.47)-(5.49), together with constitutive relations and jump conditions, suffice to determine the rigid body fields, provided that the rigid body state is motionless ($\dot{\xi}^{o}_{\alpha} = 0$) or has a prescribed motion. If this is not the case, the motion (i.e. $\Xi_{\alpha}(t)$ and $R_{\alpha\beta}(t)$ or $\Omega_{\alpha}(t)$) must be determined from momentum equations which can best be handled, if the balance laws of linear and angular momentum are recast into global form. We restrict ourselves to a rigid body placed in a vacuum, in which case these relations are

$$(5.50)\qquad \frac{d}{dt}\int_{\Omega} \rho_{o}\dot{\xi}^{o}_{\alpha}dV = \int_{\partial\Omega^{-}} \delta_{i\alpha}(T_{i\beta} + T^{M}_{i\beta})^{o}N_{\beta}dA + \int_{\Omega} \rho_{o}F^{ext}_{\alpha}dV \,,$$

(over)

$$\frac{d}{dt}\int_{\Omega} \rho_o(\xi^o_{[\alpha} - \Xi_{[\alpha})\dot{\xi}^o_{\beta]}dV = \int_{\partial\Omega^-} (\xi^o_{[\alpha} - \Xi_{[\alpha})(T_{i\gamma} + T^M_{i\gamma})^o\delta_{i\beta]}N_\gamma dA +$$

$$+ \int_{\Omega} \rho_o\{L^{ext}_{\alpha\beta} + (\xi^o_{[\alpha} - \Xi_{[\alpha})F^{ext}_{\beta]}\}dV .$$

Here, $T^M_{i\alpha}$ is the Maxwell stress tensor as defined in (5.22)[1] and F^{ext}_α and $L^{ext}_{\alpha\beta}$ are externally applied body forces and body couples with respect to the center of mass, respectively. Ω is the body manifold and $\partial\Omega^-$ the surface of Ω just inside the boundary of the body. According to the jump condition (5.23), which is applied for the boundary of the body separating the vacuum, it is possible to simplify the surface terms in (5.50), for

$$((T_{i\beta} + T^M_{i\beta})^o N_\beta)_{\partial\Omega^-} = (t^{(N)}_i + (T^M_{i\beta})^o N_\beta)_{\partial\Omega^+} ,$$

where $t^{(N)}_i$ are the surface tractions of other than electromagnetic origin, $(T^M_{i\beta})_{\partial\Omega^+}$ is the Maxwell stress tensor evaluated just outside the boundary of the body and where $\partial\Omega^+$ represents a surface just outside the boundary of the body.

With the aid of (5.37) we may then transform (5.50) into

(5.51)
$$M\ddot{\Xi}_\alpha = F^e_\alpha + F^{ext}_\alpha ,$$
$$\frac{d}{dt}(I_{\alpha\beta}\Omega_\beta) = \mathcal{L}^e_\alpha + \mathcal{L}^{ext}_\alpha ,$$

where

(5.52)
$$M = \int_{\Omega} \rho_o dV, \qquad I_{\alpha\beta} = \int_{\Omega} \rho_o\{\delta_{\alpha\beta}X_\gamma X_\gamma - R_{\alpha\gamma}R_{\beta\delta}X_\gamma X_\delta\}dV ,$$

are the total mass and the instantaneous central moments of inertia of the body,

(5.53)
$$F^e_\alpha = \int_{\partial\Omega^+} \delta_{i\alpha}(T^M_{i\beta})^o N_\beta dA ,$$
$$\mathcal{L}^e_\alpha = e_{\alpha\beta\gamma}\delta_{i\gamma}\int_{\partial\Omega^+} (\xi^o_\beta - \Xi_\beta)(T^M_{i\delta})^o N_\delta dA ,$$

are the resultant force and moment with respect to the center of mass of electromagnetic origin, and where

(5.54)
$$F^{ext}_\alpha = \int_{\partial\Omega^+} \delta_{i\alpha}t^{(N)}_i dA + \int_{\Omega} \rho_o F^{ext}_\alpha dV ,$$

(over)

$$\mathcal{L}_\alpha^{ext} = e_{\alpha\beta\gamma} \int_{\partial\Omega^+} (\xi_\beta^o - \Xi_\beta) t_i^{(N)} \delta_{i\gamma} dA + e_{\alpha\beta\gamma} \int_\Omega \rho_o \{L_{\beta\gamma}^{ext} + (\xi_\beta^o - \Xi_\beta) F_\gamma^{ext}\} dV ,$$

are the resultant force and moment with respect to the center of mass of the externally applied forces.

To summarize, the velocity of the center of mass $\dot{\Xi}_\alpha$ and the angular velocity Ω_α of a rigid body moving in an electromagnetic field are governed by equations (5.51), in which the electromagnetic surface forces and surface couples acting on the body are determined from the Maxwell stress tensor in the surrounding vacuum. Clearly, equations (5.51) must be solved along with the Maxwell equations and the energy equation in the rigid body state, (5.47)-(5.49). In these equations the angular velocity Ω_α and the matrix $R_{\alpha\beta}$ arise; these themselves are connected by equation (5.38).

This completes the decomposition of the electromagnetic and mechanical balance laws. In the following Sections the decomposed versions of the constitutive relations and jump conditions will be given. With these the zeroth order solution can be solved, and once this is done the first order perturbed problem can be attacked.

3. Decomposition of the Constitutive Equations

The constitutive equations for entropy, polarization, magnetization and stress are known, once the functional $\check{\Psi}$ as defined in (5.18) is specified. In order to make the theory complete constitutive equations for Q_α and $\mathcal{J}_\alpha$ must also be established.

Assuming $\check{\Psi}$ to be differentiable, we may expand it in terms of Taylor series about the intermediate state (eventually approximated by the rigid body state). The series is truncated and substituted into (5.19) and what emerges are expressions for entropy, polarization, magnetization and the Piola-Kirchhoff stress tensor, which all are linear in the perturbed quantities e_α, b_α, u_α etc. The expansion of the constitutive equations into Taylor series is straightforward and could be performed formally without specifying the energy functional $\check{\Psi}$. Such expressions are of little use, however, because ultimately one must specify the free energy anyhow. A common procedure, applicable for many practical purposes, is to assume $\check{\Psi}$ to be a polynomial in its independent variables, which can be truncated at a certain order.

A reasonable expression for $\check{\Psi}$, which contains all interaction effects at least to within first order terms is

$$
\begin{aligned}
\Psi = {} & \frac{1}{2\rho_o}\,{}_2\chi^{(m)}_{\alpha\beta}\mathbb{B}_\alpha\mathbb{B}_\beta + \frac{1}{4\rho_o}\,{}_4\chi^{(m)}_{\alpha\beta\gamma\delta}\mathbb{B}_\alpha\mathbb{B}_\beta\mathbb{B}_\gamma\mathbb{B}_\delta + \frac{1}{2\rho_o}\,{}_2\chi^{(e)}_{\alpha\beta}\mathbb{E}_\alpha\mathbb{E}_\beta + \\
& + \frac{1}{4\rho_o}\,{}_4\chi^{(e)}_{\alpha\beta\gamma\delta}\mathbb{E}_\alpha\mathbb{E}_\beta\mathbb{E}_\gamma\mathbb{E}_\delta + \frac{1}{\rho_o}\,\chi^{(em)}_{\alpha\beta}\mathbb{B}_\alpha\mathbb{E}_\beta - \frac{1}{2}c(\theta-\theta_o)^2 + \frac{1}{\rho_o}\lambda^{(m)}_\alpha\mathbb{B}_\alpha(\theta-\theta_o) + \\
& + \frac{1}{2\rho_o}L^{(m)}_{\alpha\beta}\mathbb{B}_\alpha\mathbb{B}_\beta(\theta-\theta_o) + \frac{1}{\rho_o}\lambda^{(e)}_\alpha\mathbb{E}_\alpha(\theta-\theta_o) + \frac{1}{2\rho_o}L^{(e)}_{\alpha\beta}\mathbb{E}_\alpha\mathbb{E}_\beta(\theta-\theta_o) + \qquad (5.55) \\
& + \frac{1}{\rho_o}[\varepsilon^{(m)}_{\beta\gamma\delta}\mathbb{B}_\beta + \frac{1}{2}b^{(m)}_{\alpha\beta\gamma\delta}\mathbb{B}_\alpha\mathbb{B}_\beta + \varepsilon^{(e)}_{\beta\gamma\delta}\mathbb{E}_\beta + \frac{1}{2}b^{(e)}_{\alpha\beta\gamma\delta}\mathbb{E}_\alpha\mathbb{E}_\beta - \rho_o\nu_{\gamma\delta}(\theta-\theta_o)]E_{\gamma\delta} + \\
& + \frac{1}{2\rho_o}c_{\alpha\beta\gamma\delta}E_{\alpha\beta}E_{\gamma\delta}\ .
\end{aligned}
$$

Here, instead of $C_{\alpha\beta}$ we have used

(5.56) $$E_{\alpha\beta} := \frac{1}{2}(C_{\alpha\beta} - \delta_{\alpha\beta})\ ,$$

as deformation measure. According to $(5.32)^4$, this can be developed as

(5.57) $$E_{\alpha\beta} = \bar{E}_{\alpha\beta} + e_{\alpha\beta}\ ,$$

where

(5.58) $$\bar{E}_{\alpha\beta} = \frac{1}{2}(\bar{C}_{\alpha\beta} - \delta_{\alpha\beta}), \qquad e_{\alpha\beta} = \frac{1}{2}(\bar{F}_{\gamma\beta}u_{\gamma,\alpha} + \bar{F}_{\gamma\alpha}u_{\gamma,\beta})\ .$$

Neglecting the deformations in the $\underline{\xi}$-state the latter may be approximated by

(5.59) $$e_{\alpha\beta} = \frac{1}{2}(R_{\gamma\beta}u_{\gamma,\alpha} + R_{\gamma\alpha}u_{\gamma,\beta})\ .$$

The polynomial representation (5.55) for the free energy is such that elastic and thermal effects are essentially linear and that interactions with three fields are excluded. Yet, some magnetic and electric nonlinearities (fourth order terms) are nevertheless present. The reasons for this are twofold:

i) For ferromagnetic materials and for cubic crystals the fourth order anisotropy effects (represented by ${}_4\chi^{(m)}_{\alpha\beta\gamma\delta}$) are often rather important and may, frequently, even be more important than the second order terms; for cubic crystals fourth order terms are the only nontrivial magnetic anisotropy effects, in general.

ii) For materials with central symmetry (e.g. cubic crystals or isotropic bodies) first and third order coefficients in (5.55) vanish. In this case second and fourth order terms like $L_{\alpha\beta}$ and $b_{\alpha\beta\gamma\delta}$ must be dominant.

We further note that it is possible to extend the above functional form of the free energy to include still higher order effects without the result that this would fundamentally influence the subsequent analysis. This will not be done here, and for the time being we also refrain from assigning names to the various coefficients occurring in (5.55), simply because the results of Section 4.7 have shown that unique interpretations are not possible. The coefficients occurring in (5.55) also satisfy certain symmetry requirements, which are readily obtained from (5.55). Because they are so obvious, we shall not explicitly state them here.

It is now straightforward to derive explicit expressions for entropy, polarization, magnetization and stress by substituting (5.55) into (5.19) and performing the respective differentiations. For the entropy this leads to

$$\begin{aligned}(5.60)\qquad \eta &= +c(\theta-\theta_o) - \frac{1}{\rho_o}\lambda_\alpha^{(m)}\mathbb{B}_\alpha - \frac{1}{2\rho_o}L_{\alpha\beta}^{(m)}\mathbb{B}_\alpha\mathbb{B}_\beta + \\ &\quad - \frac{1}{\rho_o}\lambda_\alpha^{(e)}\mathbb{E}_\alpha - \frac{1}{2\rho_o}L_{\alpha\beta}^{(e)}\mathbb{E}_\alpha\mathbb{E}_\beta + \nu_{\alpha\beta}E_{\alpha\beta} = \\ &= c(\bar\theta-\theta_o) - \frac{1}{\rho_o}\lambda_\alpha^{(m)}\bar{\mathbb{B}}_\alpha - \frac{1}{2\rho_o}L_{\alpha\beta}^{(m)}\bar{\mathbb{B}}_\alpha\bar{\mathbb{B}}_\beta - \frac{1}{\rho_o}\lambda_\alpha^{(e)}\bar{\mathbb{E}}_\alpha + \\ &\quad - \frac{1}{2\rho_o}L_{\alpha\beta}^{(e)}\bar{\mathbb{E}}_\alpha\bar{\mathbb{E}}_\beta + \nu_{\alpha\beta}\bar{E}_{\alpha\beta} + c\theta - \frac{1}{\rho_o}(\lambda_\alpha^{(m)} + L_{\alpha\beta}^{(m)}\bar{\mathbb{B}}_\beta)b_\alpha + \\ &\quad - \frac{1}{\rho_o}(\lambda_\alpha^{(e)} + L_{\alpha\beta}^{(e)}\bar{\mathbb{E}}_\beta)e_\beta + \nu_{\alpha\beta}e_{\alpha\beta} =: \bar\eta + s\ ,\end{aligned}$$

whereby $0(\varepsilon^2)$-terms are neglected and where the entropy of the intermediate state is given by

$$\begin{aligned}(5.61)\qquad \bar\eta &= c(\bar\theta-\theta_o) - \frac{1}{\rho_o}\lambda_\alpha^{(m)}\bar{\mathbb{B}}_\alpha - \frac{1}{2\rho_o}L_{\alpha\beta}^{(m)}\bar{\mathbb{B}}_\alpha\bar{\mathbb{B}}_\beta - \frac{1}{\rho_o}\lambda_\alpha^{(e)}\bar{\mathbb{E}}_\alpha - \frac{1}{2\rho_o}L_{\alpha\beta}^{(e)}\bar{\mathbb{E}}_\alpha\bar{\mathbb{E}}_\beta + \\ &\quad + \nu_{\alpha\beta}\bar{E}_{\alpha\beta}\ ,\end{aligned}$$

and the perturbed entropy s may be written as

$$(5.62)\qquad s = s^1_{\alpha\beta}u_{\alpha,\beta} + s^2_\alpha b_\alpha + s^3_\alpha e_\alpha + s^4\theta\ ,$$

with coefficients which can easily be derived from (5.60) and $(5.58)^2$ as

$$\begin{aligned}(5.63)\qquad s^1_{\alpha\beta} &= \bar{F}_{\alpha\gamma}\nu_{\beta\gamma}, \qquad s^2_\alpha = -\frac{1}{\rho_o}(\lambda_\alpha^{(m)} + L_{\alpha\beta}^{(m)}\bar{\mathbb{B}}_\beta) \\ s^3_\alpha &= -\frac{1}{\rho_o}(\lambda_\alpha^{(e)} + L_{\alpha\beta}^{(e)}\bar{\mathbb{E}}_\beta), \qquad s^4 = c\ .\end{aligned}$$

In case deformations of the intermediate state are neglected we simply set

$$\bar{F}_{\alpha\beta} = R_{\alpha\beta} \qquad \text{and} \qquad (\bar{\cdot}) = (\cdot)^o\ .$$

In a completely analogous way also the constitutive equations for polarization, magnetization and the Piola-Kirchhoff stress can be deduced. One obtains for the

polarization

(5.64) $$\mathbb{P}_\alpha = \bar{\mathbb{P}}_\alpha + p_\alpha = \bar{\mathbb{P}}_\alpha + \overset{1}{p}_{\alpha\beta\gamma} u_{\beta,\gamma} + \overset{2}{p}_{\alpha\beta} b_\beta + \overset{3}{p}_{\alpha\beta} e_\beta + \overset{4}{p}_\alpha \theta ,$$

where

(5.65) $$\begin{aligned} \bar{\mathbb{P}}_\alpha &= -(\varepsilon^{(e)}_{\alpha\gamma\delta} + b^{(e)}_{\alpha\beta\gamma\delta}\bar{\mathbb{E}}_\beta)\bar{E}_{\gamma\delta} - ({}_2\chi^{(e)}_{\alpha\beta} + {}_4\chi^{(e)}_{\alpha\beta\gamma\delta}\bar{\mathbb{E}}_\gamma\bar{\mathbb{E}}_\delta)\bar{\mathbb{E}}_\beta + \\ &\quad - \chi^{(em)}_{\beta\alpha}\bar{\mathbb{B}}_\beta - (\lambda^{(e)}_\alpha + L^{(e)}_{\alpha\beta}\bar{\mathbb{E}}_\beta)(\bar{\theta} - \theta_o) , \end{aligned}$$

and

(5.66) $$\begin{aligned} \overset{1}{p}_{\alpha\beta\gamma} &= -\bar{F}_{\beta\delta}(\varepsilon^{(e)}_{\alpha\gamma\delta} + b^{(e)}_{\alpha\varepsilon\gamma\delta}\bar{\mathbb{E}}_\varepsilon) , \\ \overset{2}{p}_{\alpha\beta} &= -\chi^{(em)}_{\beta\alpha} , \\ \overset{3}{p}_{\alpha\beta} &= -{}_2\chi^{(e)}_{\alpha\beta} - 3\,{}_4\chi^{(e)}_{\alpha\beta\gamma\delta}\bar{\mathbb{E}}_\gamma\bar{\mathbb{E}}_\delta - b^{(e)}_{\alpha\beta\gamma\delta}\bar{E}_{\gamma\delta} - L^{(e)}_{\alpha\beta}(\bar{\theta} - \theta_o) , \\ \overset{4}{p}_{\alpha\beta} &= -\lambda^{(e)}_\alpha - L^{(e)}_{\alpha\beta}\bar{\mathbb{E}}_\beta ; \end{aligned}$$

magnetization

(5.67) $$\mathbb{M}_\alpha = \bar{\mathbb{M}}_\alpha + m_\alpha = \bar{\mathbb{M}}_\alpha + \overset{1}{m}_{\alpha\beta\gamma} u_{\beta,\gamma} + \overset{2}{m}_{\alpha\beta} b_\beta + \overset{3}{m}_{\alpha\beta} e_\beta + \overset{4}{m}_\alpha \theta ,$$

where

(5.68) $$\begin{aligned} \bar{\mathbb{M}}_\alpha &= -(\varepsilon^{(m)}_{\alpha\gamma\delta} + b^{(m)}_{\alpha\beta\gamma\delta}\bar{\mathbb{B}}_\beta)\bar{E}_{\gamma\delta} - ({}_2\chi^{(m)}_{\alpha\beta} + {}_4\chi^{(m)}_{\alpha\beta\gamma\delta}\bar{\mathbb{B}}_\gamma\bar{\mathbb{B}}_\delta)\bar{\mathbb{B}}_\beta + \\ &\quad - \chi^{(em)}_{\alpha\beta}\bar{\mathbb{E}}_\beta - (\lambda^{(m)}_\alpha + L^{(m)}_{\alpha\beta}\bar{\mathbb{B}}_\beta)(\bar{\theta} - \theta_o) , \end{aligned}$$

and

(5.69) $$\begin{aligned} \overset{1}{m}_{\alpha\beta\gamma} &= -\bar{F}_{\beta\delta}(\varepsilon^{(m)}_{\alpha\gamma\delta} + b^{(m)}_{\alpha\varepsilon\gamma\delta}\bar{\mathbb{B}}_\varepsilon) , \\ \overset{2}{m}_{\alpha\beta} &= -{}_2\chi^{(m)}_{\alpha\beta} - 3\,{}_4\chi^{(m)}_{\alpha\beta\gamma\delta}\bar{\mathbb{B}}_\gamma\bar{\mathbb{B}}_\delta - b^{(m)}_{\alpha\beta\gamma\delta}\bar{E}_{\gamma\delta} - L^{(m)}_{\alpha\beta}(\bar{\theta} - \theta_o) , \\ \overset{3}{m}_{\alpha\beta} &= -\chi^{(em)}_{\alpha\beta} , \\ \overset{4}{m}_\alpha &= -\lambda^{(m)}_\alpha - L^{(m)}_{\alpha\beta}\bar{\mathbb{B}}_\beta , \end{aligned}$$

Piola-Kirchhoff stress tensor, which according to $(5.19)^4$ and (4.24) is given by

(5.70) $$T_{i\beta} = \rho_o \frac{\partial\Psi}{\partial E_{\beta\gamma}} F_{i\gamma} ,$$

so that

$$(5.71)\qquad T_{i\beta} = \bar{T}_{i\beta} + \delta_{i\alpha} t_{\alpha\beta} = \bar{T}_{i\beta} + \delta_{i\alpha}(\overset{1}{t}_{\alpha\beta\gamma\delta} u_{\gamma,\delta} + \overset{2}{t}_{\alpha\beta\gamma} b_\gamma + \overset{3}{t}_{\alpha\beta\gamma} e_\gamma + \overset{4}{t}_{\alpha\beta}\theta) ,$$

where

$$(5.72)\qquad \begin{aligned} \bar{T}_{i\beta} = \delta_{i\alpha}\bar{F}_{\alpha\gamma}[c_{\beta\gamma\delta\varepsilon}\bar{E}_{\delta\varepsilon} &+ (\varepsilon^{(m)}_{\delta\beta\gamma} + \tfrac{1}{2} b^{(m)}_{\delta\varepsilon\beta\gamma}\bar{\mathbb{B}}_\varepsilon)\bar{\mathbb{B}}_\delta + \\ &+ (\varepsilon^{(e)}_{\delta\beta\gamma} + \tfrac{1}{2} b^{(e)}_{\delta\varepsilon\beta\gamma}\bar{\mathbb{E}}_\varepsilon)\bar{\mathbb{E}}_\delta - \rho_o \nu_{\beta\gamma}(\bar{\theta} - \theta_o)] , \end{aligned}$$

and

$$\begin{aligned} \overset{1}{t}_{\alpha\beta\gamma\delta} = \bar{F}_{\alpha\mu}\bar{F}_{\gamma\varepsilon} c_{\beta\mu\varepsilon\delta} + \delta_{\alpha\gamma}[c_{\beta\delta\mu\nu}\bar{E}_{\mu\nu} &+ (\varepsilon^{(m)}_{\varepsilon\beta\delta} + \tfrac{1}{2} b^{(m)}_{\varepsilon\mu\beta\delta}\bar{\mathbb{B}}_\mu)\bar{\mathbb{B}}_\varepsilon + \\ &+ (\varepsilon^{(e)}_{\varepsilon\beta\delta} + \tfrac{1}{2} b^{(e)}_{\varepsilon\mu\beta\delta}\bar{\mathbb{E}}_\mu)\bar{\mathbb{E}}_\varepsilon - \rho_o \nu_{\beta\delta}(\bar{\theta} - \theta_o)] , \end{aligned}$$

$$(5.73)\qquad \overset{2}{t}_{\alpha\beta\gamma} = \bar{F}_{\alpha\delta}(\varepsilon^{(m)}_{\gamma\beta\delta} + b^{(m)}_{\gamma\varepsilon\beta\delta}\bar{\mathbb{B}}_\varepsilon) ,$$

$$\overset{3}{t}_{\alpha\beta\gamma} = \bar{F}_{\alpha\delta}(\varepsilon^{(e)}_{\gamma\beta\delta} + b^{(e)}_{\gamma\varepsilon\beta\delta}\bar{\mathbb{E}}_\varepsilon) ,$$

$$\overset{4}{t}_{\alpha\beta} = -\rho_o \bar{F}_{\alpha\gamma}\nu_{\beta\gamma} .$$

The above derivation is perfectly general and applies whether deformations in the intermediate state are small or large. If they are small, all coefficients characterized by a lower case letter (say $\overset{1}{p}_{\alpha\beta\gamma}$, $\overset{4}{t}_{\alpha\beta}$ etc.) may further be simplified by replacing all quantities with an overhead bar by the corresponding rigid body quantities. The error introduced into the balance laws of the perturbed quantities is thereby of order $O(E\varepsilon)$.
Note that the term containing $c_{\beta\mu\varepsilon\delta}$ in the expression for $\overset{1}{t}_{\alpha\beta\gamma\delta}$ is, in practice, always much larger than the remaining terms. In fact, this property is the justification for the assumption that the deformations due to the electromagnetic and due to the thermal fields are small. It seems to be reasonable, therefore, to approximate $\overset{1}{t}_{\alpha\beta\gamma\delta}$ by

$$(5.74)\qquad \overset{1}{t}_{\alpha\beta\gamma\delta} \cong R_{\alpha\mu}R_{\gamma\varepsilon}c_{\beta\mu\varepsilon\delta} .$$

Here, we have written $R_{\alpha\mu}$ for $\bar{F}_{\alpha\mu}$, because the error introduced by dropping the remaining terms in $(5.73)^1$ is of the same order as the one obtained by the replacement $\bar{F}_{\alpha\beta} \to R_{\alpha\beta}$.

We conclude this Section with the presentation of the decomposition of the generalized versions of Ohm's and Fourier's law, as given in (5.20). The coefficients in these equations are functions of the variables $E_{\alpha\beta}$, $\mathbb{E}_\alpha$, $\mathbb{B}_\alpha$, θ and $\theta_{,\alpha}$.

We shall exclude a possible dependency of these constitutive relations on the free charge $\mathbb{Q}$, and shall, furthermore, restrict ourselves to a linear dependence of current and heat flux on electric field and temperature gradient. We then have

$$(5.75)\qquad \Lambda_{\alpha\beta} = \Lambda_{\alpha\beta}(E_{\gamma\delta},\mathbb{B}_\gamma,\theta) ,$$

where $\Lambda_{\alpha\beta}$ stands for $\sigma_{\alpha\beta}$, $\kappa_{\alpha\beta}$ and $\beta_{\alpha\beta}$, respectively. To be more specific we shall choose the following polynomial expansions

$$(5.76)\qquad \Lambda_{\alpha\beta} = \Lambda^{(r)}_{\alpha\beta} + \Lambda^{(d)}_{\alpha\beta\gamma\delta}E_{\gamma\delta} + {}_3\Lambda^{(m)}_{\alpha\beta\gamma}\mathbb{B}_\gamma + {}_4\Lambda^{(m)}_{\alpha\beta\gamma\delta}\mathbb{B}_\gamma\mathbb{B}_\delta + \Lambda^{(t)}_{\alpha\beta}(\theta-\theta_o) .$$

Here we have included second order terms in $\mathbb{B}_\alpha$, but have deleted third order terms, because they vanish in a material with point symmetry anyhow. Decomposing the representations (5.20) for electric current and energy flux gives

$$(5.77)\qquad \begin{aligned} \bar{\mathbb{J}}_\alpha &= \bar{\sigma}_{\alpha\beta}\bar{\mathbb{E}}_\beta + \bar{\beta}_{\beta\alpha}\frac{\bar{\theta}_{,\beta}}{\bar{\theta}} , \\ \bar{Q}_\alpha &= -\bar{\kappa}_{\alpha\beta}\bar{\theta}_{,\beta} + \bar{\beta}_{\alpha\beta}\bar{\mathbb{E}}_\beta , \end{aligned}$$

as governing equations in the intermediate state and

$$(5.78)\qquad \begin{aligned} j_\alpha &= \overset{1}{j}_{\alpha\beta\gamma}u_{\beta,\gamma} + \overset{2}{j}_{\alpha\beta}b_\beta + \overset{3}{j}_{\alpha\beta}e_\beta + \overset{4}{j}_\alpha\theta + \overset{5}{j}_{\alpha\beta}\theta_{,\beta} , \\ q_\alpha &= \overset{1}{q}_{\alpha\beta\gamma}u_{\beta,\gamma} + \overset{2}{q}_{\alpha\beta}b_\beta + \overset{3}{q}_{\alpha\beta}e_\beta + \overset{4}{q}_\alpha\theta + \overset{5}{q}_{\alpha\beta}\theta_{,\beta} , \end{aligned}$$

as the approximate equations in the perturbed state, in which

$$(5.79)\qquad \begin{aligned} \overset{1}{j}_{\alpha\beta\gamma} &= \bar{F}_{\beta\delta}\Big(\sigma^{(d)}_{\alpha\varepsilon\gamma\delta}\bar{\mathbb{E}}_\varepsilon + \beta^{(d)}_{\varepsilon\alpha\gamma\delta}\frac{\bar{\theta}_{,\varepsilon}}{\bar{\theta}}\Big) , \\ \overset{2}{j}_{\alpha\beta} &= [{}_3\sigma^{(m)}_{\alpha\gamma\beta} + 2{}_4\sigma^{(m)}_{\alpha\gamma\beta\delta}\bar{\mathbb{B}}_\delta]\bar{\mathbb{E}}_\gamma + [{}_3\beta^{(m)}_{\gamma\alpha\beta} + 2{}_4\beta^{(m)}_{\gamma\alpha\beta\delta}\bar{\mathbb{B}}_\delta]\frac{\bar{\theta}_{,\gamma}}{\bar{\theta}} , \\ \overset{3}{j}_{\alpha\beta} &= \sigma^{(r)}_{\alpha\beta} + \sigma^{(d)}_{\alpha\beta\gamma\delta}\bar{E}_{\gamma\delta} + {}_3\sigma^{(m)}_{\alpha\beta\gamma}\bar{\mathbb{B}}_\gamma + {}_4\sigma^{(m)}_{\alpha\beta\gamma\delta}\bar{\mathbb{B}}_\gamma\bar{\mathbb{B}}_\delta + \sigma^{(t)}_{\alpha\beta}(\bar{\theta}-\theta_o) , \\ \overset{4}{j}_\alpha &= -[\beta^{(r)}_{\beta\alpha} + \beta^{(d)}_{\beta\alpha\gamma\delta}\bar{E}_{\gamma\delta} + {}_3\beta^{(m)}_{\beta\alpha\gamma}\bar{\mathbb{B}}_\gamma + {}_4\beta^{(m)}_{\beta\alpha\gamma\delta}\bar{\mathbb{B}}_\gamma\bar{\mathbb{B}}_\delta + \beta^{(t)}_{\beta\alpha}(\bar{\theta}-\theta_o)]\frac{\bar{\theta}_{,\beta}}{\bar{\theta}^2} + \\ &\quad + \sigma^{(t)}_{\alpha\beta}\bar{\mathbb{E}}_\beta + \beta^{(t)}_{\beta\alpha}\frac{\bar{\theta}_{,\beta}}{\bar{\theta}} , \\ \overset{5}{j}_{\alpha\beta} &= \frac{1}{\bar{\theta}}[\beta^{(r)}_{\beta\alpha} + \beta^{(d)}_{\beta\alpha\gamma\delta}\bar{E}_{\gamma\delta} + {}_3\beta^{(m)}_{\beta\alpha\gamma}\bar{\mathbb{B}}_\gamma + {}_4\beta^{(m)}_{\beta\alpha\gamma\delta}\bar{\mathbb{B}}_\gamma\bar{\mathbb{B}}_\delta + \beta^{(t)}_{\beta\alpha}(\bar{\theta}-\theta_o)] , \end{aligned}$$

and

$$
\begin{aligned}
q^1_{\alpha\beta\gamma} &= \bar{F}_{\beta\delta}(-\kappa^{(d)}_{\alpha\epsilon\gamma\delta}\bar{\theta}_{,\epsilon} + \beta^{(d)}_{\alpha\epsilon\gamma\delta}\bar{\mathbb{E}}_{\epsilon}) , \\
q^2_{\alpha\beta} &= -[{}_3\kappa^{(m)}_{\alpha\gamma\beta} + 2{}_4\kappa^{(m)}_{\alpha\gamma\beta\delta}\bar{\mathbb{B}}_{\delta}]\bar{\theta}_{,\gamma} + [{}_3\beta^{(m)}_{\alpha\gamma\beta} + 2{}_4\beta^{(m)}_{\alpha\gamma\beta\delta}\bar{\mathbb{B}}_{\delta}]\bar{\mathbb{E}}_{\gamma} , \\
q^3_{\alpha\beta} &= \beta^{(r)}_{\alpha\beta} + \beta^{(d)}_{\alpha\beta\gamma\delta}\bar{E}_{\gamma\delta} + {}_3\beta^{(m)}_{\alpha\beta\gamma}\bar{\mathbb{B}}_{\gamma} + {}_4\beta^{(m)}_{\alpha\beta\gamma\delta}\bar{\mathbb{B}}_{\gamma}\bar{\mathbb{B}}_{\delta} + \beta^{(t)}_{\alpha\beta}(\bar{\theta}-\theta_o) , \qquad (5.80)\\
q^4_{\alpha} &= -\kappa^{(t)}_{\alpha\beta}\bar{\theta}_{,\beta} + \beta^{(t)}_{\alpha\beta}\bar{\mathbb{E}}_{\beta} , \\
q^5_{\alpha\beta} &= -[\kappa^{(r)}_{\alpha\beta} + \kappa^{(d)}_{\alpha\beta\gamma\delta}\bar{E}_{\gamma\delta} + {}_3\kappa^{(m)}_{\alpha\beta\gamma}\bar{\mathbb{B}}_{\gamma} + {}_4\kappa^{(m)}_{\alpha\beta\gamma\delta}\bar{\mathbb{B}}_{\gamma}\bar{\mathbb{B}}_{\delta} + \kappa^{(t)}_{\alpha\beta}(\bar{\theta}-\theta_o)] .
\end{aligned}
$$

Specific forms of the coefficients, yet only for isotropic materials have been derived by Pipkin and Rivlin, [15,16], and by Borghesani and Morro, [17,18]. These authors also discuss the physical significance of the various terms in (5.79) and (5.80).

At this point we have thus completed the presentation of the linearized field equations. If (5.62), (5.64), (5.67), (5.71) and (5.78) are substituted into the balance laws (5.31) and (5.44), what we obtain is a system of 11 independent field equations for the 11 unknowns b_α, e_α, q, u_α and θ. This system could still further by simplified by introducing electromagnetic potentials φ and a_α by (see also (4.131))

$$b_\alpha = e_{\alpha\beta\gamma}a_{\gamma,\beta} , \qquad e_\alpha = \varphi_{,\alpha} - \dot{a}_\alpha , \qquad (5.81)$$

where a_α should also satisfy a gauge condition; for the Lorentz gauge

$$a_{\alpha,\alpha} - \mu_o\varepsilon_o\dot{\varphi} = 0 . \qquad (5.82)$$

For small deformations all quantities referred to the intermediate state that appear in these perturbation equations can be replaced by the corresponding variables in the rigid body state. To complete the description the form of the constitutive equations in this state will also be given. They are

$$
\begin{aligned}
\eta^o &= c(\theta^o-\theta_o) - \frac{1}{\rho_o}\lambda^{(m)}_{\alpha}\mathbb{B}^o_{\alpha} - \frac{1}{2\rho_o}L^{(m)}_{\alpha\beta}\mathbb{B}^o_{\alpha}\mathbb{B}^o_{\beta} - \frac{1}{\rho_o}\lambda^{(e)}_{\alpha}\mathbb{E}^{(o)}_{\alpha} - \frac{1}{2\rho_o}L^{(e)}_{\alpha\beta}\mathbb{E}^o_{\alpha}\mathbb{E}^o_{\beta} , \\
\mathbb{P}^o_{\alpha} &= -({}_2\chi^{(e)}_{\alpha\beta} + {}_4\chi^{(e)}_{\alpha\beta\gamma\delta}\mathbb{E}^o_{\gamma}\mathbb{E}^o_{\delta})\mathbb{E}^o_{\beta} - \chi^{(em)}_{\beta\alpha}\mathbb{B}^o_{\beta} - (\lambda^{(e)}_{\alpha} + L^{(e)}_{\alpha\beta}\mathbb{E}^o_{\beta})(\theta^o-\theta_o) , \\
\mathbb{M}^{Lo}_{\alpha} &= -({}_2\chi^{(m)}_{\alpha\beta} + {}_4\chi^{(m)}_{\alpha\beta\gamma\delta}\mathbb{B}^o_{\gamma}\mathbb{B}^o_{\delta})\mathbb{B}^o_{\beta} - \chi^{(em)}_{\alpha\beta}\mathbb{E}^o_{\beta} - (\lambda^{(m)}_{\alpha} + L^{(m)}_{\alpha\beta}\mathbb{B}^o_{\beta})(\theta^o-\theta_o) , \qquad (5.83)\\
\mathbb{J}^o_{\alpha} &= \sigma^o_{\alpha\beta}\mathbb{E}^o_{\beta} + \beta^o_{\beta\alpha}\frac{\theta^o_{,\beta}}{\theta^o} , \\
Q^o_{\alpha} &= -\kappa^o_{\alpha\beta}\theta^o_{,\beta} + \beta^o_{\alpha\beta}\mathbb{E}^o_{\beta} ,
\end{aligned}
$$

where the coefficients $\sigma^o_{\alpha\beta}$, $\beta^o_{\alpha\beta}$ and $\kappa^o_{\alpha\beta}$ follow from

$$\Lambda^o_{\alpha\beta} = \Lambda_{\alpha\beta}(0,\mathbb{B}^o_\gamma,\theta^o) = \Lambda^{(r)}_{\alpha\beta} + {}_3\Lambda^{(m)}_{\alpha\beta\gamma}\mathbb{B}^o_\gamma + {}_4\Lambda^{(m)}_{\alpha\beta\gamma\delta}\mathbb{B}^o_\gamma\mathbb{B}^o_\delta + \Lambda^{(t)}_{\alpha\beta}(\theta^o - \theta_o) \; . \tag{5.84}$$

When these constitutive relations are substituted into the balance equations (5.47)-(5.49) and (5.51), the rigid body problem is reduced to a problem for the 14 unknowns $\mathbb{E}^o_\alpha$, $\mathbb{B}^o_\alpha$, $\mathbb{Q}^o$, θ^o, Ξ_α and Ω_α. Again this problem could even further be reduced by introducing electromagnetic potentials according to

$$\mathbb{B}^o_\alpha = e_{\alpha\beta\gamma}A^o_{\gamma,\beta}, \qquad \mathbb{E}^o_\alpha = \phi^o_{,\alpha} - \dot{A}^o_\alpha \tag{5.85}$$

with

$$A^o_{\alpha,\alpha} - \mu_o\varepsilon_o\dot{\Phi}^o = 0 \; . \tag{5.86}$$

All there remains, therefore, are now the boundary and jump conditions.

4. Decomposition of the Jump and Boundary Conditions

The decomposition of the jump conditions is a very complex problem, in general, and this is the reason why we restricted ourselves to singular surfaces of the second order already in Section 5.2.1. This restriction will be maintained here too. We shall be even more restrictive and shall assume simultaneously that there are no propagating singular surfaces in the intermediate state. Hence, in this state $\bar{W}_N = 0$. An acceleration wave, if present, can then only exist in the perturbed state; the speed of propagation in this perturbed state then forms the total wave speed and may without confusion be called W_N. The above special situation does not necessarily require, however, that W_N be small. In fact, W_N is a wave speed the numerical value of which follows from the material properties. In what follows it is advantageous to distinguish between the two different kinds of singular surfaces, namely

i) a propagating singular surface $\Sigma^{(i)}$ (acceleration wave) with velocity W_N;
ii) a material surface $\Sigma^{(ii)}$, for which $W_N = 0$.

Since $\Sigma^{(i)}$ does not exist in the intermediate state, all quantities in this state must be continuous at $\Sigma^{(i)}$, i.e.

$$[\![\bar{\mathbb{E}}_\alpha]\!] = 0, \text{ etc, at } \Sigma^{(i)} \; .$$

The linearization of the jump conditions is rather simple in a Lagrangian formulation, and from (5.21), (5.23) and (5.29) we immediately obtain

$$[\![b_\alpha]\!]N_\alpha = 0, \qquad e_{\alpha\beta\gamma}[\![e_\beta]\!]N_\gamma + [\![b_\alpha]\!]W_N = 0 ,$$

$$[\![d^a_\alpha + p_\alpha]\!]N_\alpha = 0, \qquad e_{\alpha\beta\gamma}[\![h^a_\beta - m_\beta]\!]N_\gamma - [\![(d^a_\alpha + p_\alpha)]\!]W_N = 0 ,$$

$$(5.87) \qquad [\![j_\alpha]\!]N_\alpha - [\![q]\!]W_N = 0 ,$$

$$[\![t_{\alpha\beta}]\!]N_\beta = \bar{F}^{-1}_{\gamma\alpha}\{\bar{\mathbb{E}}_\gamma[\![p_\beta]\!]N_\beta + \bar{\mathbb{B}}_\beta([\![m_\beta]\!]N_\gamma - [\![m_\gamma]\!]N_\beta) + e_{\beta\gamma\delta}\bar{\mathbb{B}}_\beta[\![p_\delta]\!]W_N\} ,$$

$$[\![q_\alpha]\!]N_\alpha = \{\rho_o[\![u]\!] + \bar{\mathbb{M}}_\alpha[\![b_\alpha]\!] - \bar{\mathbb{E}}_\alpha[\![p_\alpha]\!]\}W_N ,$$

all valid on $\Sigma^{(i)}$. In the above u denotes the perturbed internal energy, i.e.

$$(5.88) \qquad u = \{\Psi - \overline{(\Psi)}\} + \bar{\eta}\theta + \bar{\theta}s - \frac{1}{\rho_o}\bar{\mathbb{E}}_\alpha p_\alpha - \frac{1}{\rho_o}\bar{\mathbb{P}}_\alpha e_\alpha .$$

On the other hand, on $\Sigma^{(ii)}$ the jumps in $\bar{\mathbb{E}}_\alpha$, $\bar{\mathbb{B}}_\alpha$ etc. need not be zero, but W_N is. In this case the linearized jump conditions read

$$[\![b_\alpha]\!]N_\alpha = 0, \qquad e_{\alpha\beta\gamma}[\![e_\beta]\!]N_\gamma = 0 ,$$

$$[\![d^a_\alpha + p_\alpha]\!]N_\alpha = 0, \qquad e_{\alpha\beta\gamma}[\![h^a_\beta - m_\beta]\!]N_\gamma = 0, \qquad [\![j_\alpha]\!]N_\alpha = 0 ,$$

$$(5.89) \qquad \begin{aligned}[\![t_{\alpha\beta}]\!]N_\beta = &-\bar{F}^{-1}_{\gamma\delta}\bar{F}^{-1}_{\varepsilon\alpha}\{\langle\!\langle\bar{\mathbb{E}}_\gamma\rangle\!\rangle[\![\bar{\mathbb{P}}_\beta]\!]N_\beta + \langle\!\langle\bar{\mathbb{B}}_\beta\rangle\!\rangle([\![\bar{\mathbb{M}}_\beta]\!]N_\gamma - [\![\bar{\mathbb{M}}_\gamma]\!]N_\beta)\}u_{\delta,\varepsilon} + \\ &+ \bar{F}^{-1}_{\gamma\alpha}\{\langle\!\langle\bar{\mathbb{E}}_\gamma\rangle\!\rangle[\![p_\beta]\!]N_\beta + [\![\bar{\mathbb{P}}_\beta]\!]N_\beta\langle\!\langle e_\gamma\rangle\!\rangle + \langle\!\langle\bar{\mathbb{B}}_\beta\rangle\!\rangle([\![m_\beta]\!]N_\gamma - [\![m_\gamma]\!]N_\beta) + \\ &+ ([\![\bar{\mathbb{M}}_\beta]\!]N_\gamma - [\![\bar{\mathbb{M}}_\gamma]\!]N_\beta)\langle\!\langle b_\beta\rangle\!\rangle\} ,\end{aligned}$$

$$[\![q_\alpha]\!]N_\alpha = 0 , \qquad \text{on } \Sigma^{(ii)} ,$$

whereas the jump conditions in the intermediate state follow from (5.21), (5.23) and (5.29) by invoking $W_N = 0$ and writing all quantities with an overhead bar.

This completes the linearization of the jump conditions. It should be noted that they can still be simplified (formally just a little bit, but for practical calculations enormously) when $O(E\varepsilon)$-terms are neglected, i.e. when the $\overline{(\cdot)}$-fields are replaced by the rigid body fields $(\cdot)^o$. This step is a trivial one and will therefore be deleted here.

5.3 LINEARISATION OF THE OTHER MODELS AND COMPARISON

In the preceding Section we presented the decomposition and linearization procedure for the Lorentz model. This would suffice, in general, because the proofs in Chapters 3 and 4 showed that all formulations are equivalent anyhow. This is true, but since various formulations are used in the literature and because one should be able to compare these with our presentation, we shall in this Section collect the balance laws and jump conditions for all five models. The comparison of the constitutive equations will be postponed until Sections 5.4 and 5.5, however.

To perform the comparison of the various models, recall that the results of Chapter 4 showed us that, in the Lagrangian formulation there are essentially only two different formulations. The first is the Chu-formulation (with its two different stress models) the other one is the Lorentz model which with a different stress tensor also embraces the statistical model. The Maxwell-Minkowski model can be connected to either one of these two basic models. In the previous Chapter the Lagrangian formulation of it was developed with the aid of the Chu-magnetization, rather than the Lorentz-magnetization, because most formulas took a much simpler form this way.

In what follows we shall discuss therefore models (I,II,III) and (IV,V) as the two basic different groups, although the separation of the various models could equally well be made according to (I,II) and (III,IV,V).

We begin our comparison by stating the decoupled electromagnetic equations. They are obtained from the equations (4.125)-(4.130) (compare also the systems (5.30), (5.31) of the previous Section).

In the Chu-formulation they read

$$\bar{\mathbb{B}}^a_{\alpha,\alpha} = -\mu_o \bar{\mathbb{M}}^C_{\alpha,\alpha} ,$$

$$\dot{\bar{\mathbb{B}}}^a_\alpha + e_{\alpha\beta\gamma} \bar{\mathbb{E}}_{\gamma,\beta} = -\mu_o \dot{\bar{\mathbb{M}}}^C_\alpha ,$$

$$(5.90) \qquad \bar{\mathbb{D}}^a_{\alpha,\alpha} = \bar{\mathbb{Q}} - \bar{\mathbb{P}}_{\alpha,\alpha},$$

$$-\dot{\bar{\mathbb{D}}}^a_\alpha + e_{\alpha\beta\gamma} \bar{\mathbb{H}}_{\gamma,\beta} = \bar{\mathbb{J}}_\alpha + \dot{\bar{\mathbb{P}}}_\alpha ,$$

$$\dot{\bar{\mathbb{Q}}} + \bar{\mathbb{J}}_{\alpha,\alpha} = 0 ,$$

and

$$b^{a}_{\alpha,\alpha} = -\mu_o m^{C}_{\alpha,\alpha} ,$$

$$\dot{b}^{a}_{\alpha} + e_{\alpha\beta\gamma} e_{\gamma,\beta} = -\mu_o \dot{m}^{C}_{\alpha} ,$$

$$(5.91) \qquad d^{a}_{\alpha,\alpha} = q - p_{\alpha,\alpha} ,$$

$$-\dot{d}^{a}_{\alpha} + e_{\alpha\beta\gamma} h_{\gamma,\beta} = j_{\alpha} + \dot{p}_{\alpha} ,$$

$$\dot{q} + j_{\alpha,\alpha} = 0 ,$$

in which $\bar{\mathbb{B}}^{a}_{\alpha}$, $\bar{\mathbb{D}}^{a}_{\alpha}$, b^{a}_{α} and d^{a}_{α} are auxiliary fields which, with the aid of (4.9) and (5.32), become

$$(5.92) \qquad \bar{\mathbb{D}}^{a}_{\alpha} = \varepsilon_o \bar{J}\bar{C}^{-1}_{\alpha\beta}\bar{\mathbb{E}}_{\beta}, \qquad \bar{\mathbb{B}}^{a}_{\alpha} = \mu_o \bar{J}\bar{C}^{-1}_{\alpha\beta}\bar{\mathbb{H}}_{\beta} ,$$

and

$$(5.93) \qquad \begin{aligned} d^{a}_{\alpha} &= \varepsilon_o \bar{J}[\bar{C}^{-1}_{\alpha\beta} e_{\beta} + (\bar{C}^{-1}_{\alpha\beta}\bar{F}^{-1}_{\gamma\delta} - \bar{C}^{-1}_{\beta\gamma}\bar{F}^{-1}_{\alpha\delta} - \bar{C}^{-1}_{\alpha\gamma}\bar{F}^{-1}_{\beta\delta})\bar{\mathbb{E}}_{\beta} u_{\delta,\gamma}] , \\ b^{a}_{\alpha} &= \mu_o \bar{J}[\bar{C}^{-1}_{\alpha\beta} h_{\beta} + (\bar{C}^{-1}_{\alpha\beta}\bar{F}^{-1}_{\gamma\delta} - \bar{C}^{-1}_{\beta\gamma}\bar{F}^{-1}_{\alpha\delta} - \bar{C}^{-1}_{\alpha\gamma}\bar{F}^{-1}_{\beta\delta})\bar{\mathbb{H}}_{\beta} u_{\delta,\gamma}] . \end{aligned}$$

Note the symmetry of the formulas (5.93). Its origin is the complete symmetry of the Maxwell equations in this formulation.

In the statistical and the Lorentz model the decomposition of the Maxwell equations was already made in Section 5.2.2. For completeness and for the purpose of comparison they will be repeated here. In the intermediate state they are

$$\bar{\mathbb{B}}_{\alpha,\alpha} = 0 ,$$

$$\dot{\bar{\mathbb{B}}}_{\alpha} + e_{\alpha\beta\gamma}\bar{\mathbb{E}}_{\gamma,\beta} = 0 ,$$

$$(5.94) \qquad \bar{\mathbb{D}}^{a}_{\alpha,\alpha} = \bar{\mathbb{Q}} - \bar{\mathbb{P}}_{\alpha,\alpha} ,$$

$$-\dot{\bar{\mathbb{D}}}^{a}_{\alpha} + e_{\alpha\beta\gamma}\bar{\mathbb{H}}^{a}_{\gamma,\beta} = \bar{\mathbb{J}}_{\alpha} + \dot{\bar{\mathbb{P}}}_{\alpha} + e_{\alpha\beta\gamma}\bar{\mathbb{M}}^{L}_{\gamma,\beta} ,$$

$$\dot{\bar{\mathbb{Q}}} + \bar{\mathbb{J}}_{\alpha,\alpha} = 0 ,$$

whereas in the perturbed state they become

$$b_{\alpha,\alpha} = 0 ,$$

$$\dot{b}_{\alpha} + e_{\alpha\beta\gamma} e_{\gamma,\beta} = 0 , \qquad \text{(over)}$$

$$(5.95)\quad d^a_{\alpha,\alpha} = q - p_{\alpha,\alpha} ,$$

$$-\dot{d}^a_\alpha + e_{\alpha\beta\gamma}h^a_{\gamma,\beta} = j_\alpha + \dot{p}_\alpha + e_{\alpha\beta\gamma}m^L_{\gamma,\beta} ,$$

$$\dot{q} + j_{\alpha,\alpha} = 0 ,$$

with $\bar{\mathbb{D}}^a_\alpha$, $\bar{\mathbb{H}}^a_\alpha$, d^a_α and h^a_α as given in (5.34) and (5.35):

$$\bar{\mathbb{D}}^a_\alpha = \varepsilon_o \overline{J}\overline{C}^{-1}_{\alpha\beta}\bar{\mathbb{E}}_\beta ,$$

(5.96)

$$\bar{\mathbb{H}}^a_\alpha = \frac{1}{\mu_o \bar{J}} \bar{C}_{\alpha\beta}\bar{\mathbb{B}}_\beta - \frac{\varepsilon_o}{\bar{J}} e_{\beta\gamma\delta}\bar{C}_{\alpha\beta}\bar{F}_{\varepsilon\gamma}\dot{\xi}_\varepsilon\bar{\mathbb{E}}_\delta$$

and

$$d^a_\alpha = \varepsilon_o \bar{J}\{\bar{C}^{-1}_{\alpha\beta}e_\beta + (\bar{C}^{-1}_{\alpha\beta}\bar{F}^{-1}_{\gamma\delta} - \bar{C}^{-1}_{\beta\gamma}\bar{F}^{-1}_{\alpha\delta} - \bar{C}^{-1}_{\alpha\gamma}\bar{F}^{-1}_{\beta\delta})\bar{\mathbb{E}}_\beta u_{\delta,\gamma}\} ,$$

(5.97)

$$h^a_\alpha = \frac{1}{\bar{J}}\{\frac{1}{\mu_o} \bar{C}_{\alpha\beta}b_\beta - \varepsilon_o e_{\beta\gamma\delta}\bar{F}_{\varepsilon\gamma}\dot{\xi}_\varepsilon\bar{C}_{\alpha\beta}e_\delta +$$

$$+ [(\frac{1}{\mu_o}\bar{\mathbb{B}}_\beta - e_{\beta\nu\mu}\bar{F}_{\varepsilon\nu}\dot{\xi}_\varepsilon\bar{\mathbb{E}}_\mu)(\bar{F}_{\delta\alpha}\delta_{\beta\gamma} + \bar{F}_{\delta\beta}\delta_{\alpha\gamma} - \bar{C}_{\alpha\beta}\bar{F}_{\gamma\delta}) +$$

$$- \varepsilon_o e_{\beta\gamma\mu}\bar{\mathbb{E}}_\mu\bar{C}_{\alpha\beta}\dot{\xi}_\delta]u_{\delta,\gamma} - \varepsilon_o e_{\beta\gamma\delta}\bar{\mathbb{E}}_\delta\bar{C}_{\alpha\beta}\bar{F}_{\varepsilon\gamma}\dot{u}_\varepsilon\} .$$

Note that in the above sets of Maxwell equations we have retained the upper indices C and L for the magnetizations.

Clearly, because polarization and magnetization are based on different conceptions, we cannot expect the formulas (5.96) and (5.97) to be symmetric. This difference becomes formally particularly apparent if the equations in the intermediate state are approximated by those in the rigid body state. Then the rigid body motion does, formally, not enter the Chu-formulation; but it does show up in the Lorentz-formulation.

We conclude this listing of the various forms of the Maxwell equations with the Maxwell-Minkowski formulation. The equations are

$$\bar{\mathbb{B}}_{\alpha,\alpha} = 0 ,$$

$$\dot{\bar{\mathbb{B}}}_\alpha + e_{\alpha\beta\gamma}\bar{\mathbb{E}}_{\gamma,\beta} = 0 ,$$

$$(5.98)\quad \bar{\mathbb{D}}_{\alpha,\alpha} = \bar{\mathbb{Q}} ,$$

$$-\dot{\bar{\mathbb{D}}}_\alpha + e_{\alpha\beta\gamma}\bar{\mathbb{H}}_{\gamma,\beta} = \bar{\mathbb{J}}_\alpha ,$$

$$\dot{\bar{\mathbb{Q}}} + \mathbb{J}_{\alpha,\alpha} = 0 ,$$

$$
\begin{aligned}
&b_{\alpha,\alpha} = 0 \ , \\
&\dot{b}_\alpha + e_{\alpha\beta\gamma} e_{\gamma,\beta} = 0 \ , \\
(5.99) \qquad &d_{\alpha,\alpha} = q \ , \\
&-\dot{d}_\alpha + e_{\alpha\beta\gamma} h_{\gamma,\beta} = j_\alpha \ , \\
&\dot{q} + j_{\alpha,\alpha} = 0 \ ,
\end{aligned}
$$

and must be supplemented by the relations

$$
(5.100) \qquad
\begin{aligned}
\bar{\mathbb{D}}_\alpha &= \varepsilon_o \bar{J} \bar{C}^{-1}_{\alpha\beta} \bar{\mathbb{E}}_\beta + \bar{\mathbb{P}}_\alpha \ , \\
\bar{\mathbb{B}}_\alpha &= \mu_o \bar{J} \bar{C}^{-1}_{\alpha\beta} \bar{\mathbb{H}}_\beta + \mu_o \bar{\mathbb{M}}^C_\alpha \ ,
\end{aligned}
$$

and

$$
(5.101) \qquad
\begin{aligned}
d_\alpha &= \varepsilon_o \bar{J}[\bar{C}^{-1}_{\alpha\beta} e_\beta + (\bar{C}^{-1}_{\alpha\beta}\bar{F}^{-1}_{\gamma\delta} - \bar{C}^{-1}_{\beta\gamma}\bar{F}^{-1}_{\alpha\delta} - \bar{C}^{-1}_{\alpha\gamma}\bar{F}^{-1}_{\beta\delta})\bar{\mathbb{E}}_\beta u_{\delta,\gamma}] + p_\alpha \ , \\
b_\alpha &= \mu_o \bar{J}[\bar{C}^{-1}_{\alpha\beta} e_\beta + (\bar{C}^{-1}_{\alpha\beta}\bar{F}^{-1}_{\gamma\delta} - \bar{C}^{-1}_{\beta\gamma}\bar{F}^{-1}_{\alpha\delta} - \bar{C}^{-1}_{\alpha\gamma}\bar{F}^{-1}_{\beta\delta})\bar{\mathbb{H}}_\beta u_{\delta,\gamma}] + \mu_o m^C_\alpha \ .
\end{aligned}
$$

Apart from the use of different auxiliary fields the most essential difference in these formulations lies in the choice of the magnetization: $\mathbb{M}^L_\alpha$ against $\mathbb{M}^C_\alpha$, which are related to each other according to (see (4.65))

$$
(5.102) \qquad
\begin{aligned}
\bar{\mathbb{M}}^C_\alpha &= \bar{J}\bar{C}^{-1}_{\alpha\beta}\bar{\mathbb{M}}^L_\beta \ , \\
m^C_\alpha &= \bar{J}\bar{C}^{-1}_{\alpha\beta} m^L_\beta + \bar{J}(\bar{C}^{-1}_{\alpha\beta}\bar{F}^{-1}_{\gamma\delta} - \bar{C}^{-1}_{\beta\gamma}\bar{F}^{-1}_{\alpha\delta} - \bar{C}^{-1}_{\alpha\gamma}\bar{F}^{-1}_{\beta\delta})\bar{\mathbb{M}}^L_\beta u_{\delta,\gamma} \ .
\end{aligned}
$$

Before we pass on to the mechanical balance laws, we would like to draw the readers attention to the basic differences in the Gauss and Faraday laws of the three formulations above. Because these laws are homogeneous in the Lorentz and in the Maxwell-Minkowski formulations electromagnetic potentials could easily and straightforwardly be introduced as was shown at the end of Section 5.2.4 (cf. (5.81), (5.82) or (4.131)). In the Chu-formulation, on the other hand, these equations are inhomogeneous. Therefore, in this formulation the potentials A_α and Φ cannot be introduced in the same way as in Section 5.2.4. They must rather be defined as

$$
(5.103) \qquad
\begin{aligned}
\bar{\mathbb{B}}^a_\alpha &= e_{\alpha\beta\gamma}\bar{A}_{\gamma,\beta} - \mu_o \bar{\mathbb{M}}^C_\alpha, \qquad \bar{\mathbb{E}}_\alpha = \bar{\Phi}_{,\alpha} - \dot{\bar{A}}_\alpha \ , \\
b^a_\alpha &= e_{\alpha\beta\gamma} a_{\gamma,\beta} - \mu_o m^C_\alpha, \qquad e_\alpha = \varphi_{,\alpha} - \dot{a}_\alpha \ ,
\end{aligned}
$$

with the gauche conditions

$$(5.104)\qquad \bar{A}_{\alpha,\alpha} - \varepsilon_o \mu_o \dot{\bar{\Phi}} = 0, \qquad a_{\alpha,\alpha} - \varepsilon_o \mu_o \dot{\phi} = 0 .$$

The above listed electromagnetic equations must still be supplemented by two sets of constitutive relations for e.g. polarization and magnetization. They are all given in Chapter 4, and since they are directly derivable from a free energy functional the linearization is trivial and will, therefore, be omitted here.

As was done for the Maxwell equations, we also wish to recapitulate the mechanical balance laws. If body forces of electromagnetic origin are not specified the equations can easily be taken over from (5.43) and (5.44). The momentum equations are

$$(5.105)\qquad \begin{aligned} \rho_o \ddot{\xi}_\alpha &= \delta_{i\alpha}(\bar{T}_{i\beta,\beta} + \rho_o \bar{F}^e_i) , \\ \rho_o \ddot{u}_\alpha &= t_{\alpha\beta,\beta} + \rho_o f^e_\alpha , \end{aligned}$$

whereas the energy equations in the intermediate and in the perturbed state can be written as

$$(5.106)\qquad \begin{aligned} \rho_o \bar{\theta}\dot{\bar{\eta}} &= \bar{\mathbb{J}}_\alpha \bar{\mathbb{E}}_\alpha - \bar{Q}_{\alpha,\alpha} \\ \rho_o \dot{\bar{\eta}}\theta + \rho_o \bar{\theta}\dot{s} &= j_\alpha \bar{\mathbb{E}}_\alpha + \bar{\mathbb{J}}_\alpha e_\alpha - q_{\alpha,\alpha} . \end{aligned}$$

All that is needed to complete the above equations for the various formulations is to prescribe the body force; for, all remaining quantities are either independent fields or else are given by constitutive relations. The correspondence conditions of the various formulations of the latter were treated in Section 4.7 and will again be taken up in the following Sections. Incidentally, that angular momentum is satisfied identically by satisfying objectivity requirements in the constitutive relations is reason for us not to list the balance law of moment of momentum. Moreover, it is noted that the reduced energy equation is the same in all formulations.

Let us now list the body force expressions of the various formulations.

In the Chu-formulation we have, from (4.11)[1],

for <u>model I</u>:

$$
(5.107)\quad
\begin{aligned}
{}^{I}(\rho_o\bar{F}^e_i) &= \delta_{i\delta}[\bar{F}^{-1}_{\alpha\delta}\{(\bar{\mathbb{Q}}-\bar{\mathbb{P}}_{\beta,\beta})\bar{\mathbb{E}}_\alpha + e_{\alpha\beta\gamma}(\bar{\mathbb{J}}_\beta+\dot{\bar{\mathbb{P}}}_\beta)\bar{\mathbb{B}}^a_\gamma - \mu_o\bar{\mathbb{M}}^C_{\beta,\beta}\bar{\mathbb{H}}_\alpha\} + \\
&\quad + \{\bar{F}^{-1}_{\beta\delta}(\bar{\mathbb{P}}_\alpha\bar{\mathbb{E}}_\beta + \mu_o\bar{\mathbb{M}}^C_\alpha\bar{\mathbb{H}}_\beta)\}_{,\alpha}] , \\
{}^{I}(\rho_o f^e_\alpha) &= {}^{II}(\rho_o f^e_\alpha) + [\bar{F}^{-1}_{\beta\alpha}(\bar{\mathbb{P}}_\varepsilon e_\beta + p_\varepsilon\bar{\mathbb{E}}_\beta + \mu_o\bar{\mathbb{M}}^C_\varepsilon h_\beta + \mu_o m^C_\varepsilon\bar{\mathbb{H}}_\beta) + \\
&\quad - \bar{F}^{-1}_{\beta\gamma}\bar{F}^{-1}_{\delta\alpha}(\bar{\mathbb{P}}_\varepsilon\bar{\mathbb{E}}_\beta + \mu_o\bar{\mathbb{M}}^C_\varepsilon\bar{\mathbb{H}}_\beta)u_{\gamma,\delta}]_{,\varepsilon} ,
\end{aligned}
$$

where ${}^{II}(\rho_o f^e_\alpha)$ is listed in $(5.108)^2$. On the other hand,
for <u>model II</u>: (from $(4.12)^1$)

$$
(5.108)\quad
\begin{aligned}
{}^{II}(\rho_o\bar{F}^e_i) &= \delta_{i\delta}\bar{F}^{-1}_{\alpha\delta}[(\bar{\mathbb{Q}}-\bar{\mathbb{P}}_{\beta,\beta})\bar{\mathbb{E}}_\alpha + e_{\alpha\beta\gamma}(\bar{\mathbb{J}}_\beta+\dot{\bar{\mathbb{P}}}_\beta)\bar{\mathbb{B}}^a_\gamma - \mu_o\bar{\mathbb{M}}^C_{\beta,\beta}\bar{\mathbb{H}}_\alpha] , \\
{}^{II}(\rho_o f^e_\alpha) &= \bar{F}^{-1}_{\beta\alpha}\{(q - p_{\varepsilon,\varepsilon})\bar{\mathbb{E}}_\beta + (\bar{\mathbb{Q}}-\bar{\mathbb{P}}_{\varepsilon,\varepsilon})e_\beta + e_{\beta\gamma\delta}(\bar{\mathbb{J}}_\gamma+\dot{\bar{\mathbb{P}}}_\gamma)b^a_\delta + \\
&\quad + e_{\beta\gamma\delta}(j_\gamma+\dot{p}_\gamma)\bar{\mathbb{B}}^a_\delta - \mu_o\bar{\mathbb{M}}^C_{\varepsilon,\varepsilon}h_\beta - \mu_o m^C_{\varepsilon,\varepsilon}\bar{\mathbb{H}}_\beta\} + \\
&\quad - \bar{F}^{-1}_{\beta\gamma}\bar{F}^{-1}_{\delta\alpha}\{(\bar{\mathbb{Q}}-\bar{\mathbb{P}}_{\varepsilon,\varepsilon})\bar{\mathbb{E}}_\beta + e_{\beta\mu\nu}(\bar{\mathbb{J}}_\mu+\dot{\bar{\mathbb{P}}}_\mu)\bar{\mathbb{B}}^a_\nu - \mu_o\bar{\mathbb{M}}^C_{\varepsilon,\varepsilon}\bar{\mathbb{H}}_\beta\}u_{\gamma,\delta} .
\end{aligned}
$$

In actual calculations one only needs the force expression for one single model, because the stress tensors of the two models are related by (4.22), or

$$
(5.109)\quad
\begin{aligned}
{}^{I}\bar{T}_{i\alpha} &= {}^{II}\bar{T}_{i\alpha} - \delta_{i\delta}\bar{F}^{-1}_{\beta\delta}(\bar{\mathbb{P}}_\alpha\bar{\mathbb{E}}_\beta + \mu_o\bar{\mathbb{M}}^C_\alpha\bar{\mathbb{H}}_\beta) , \\
{}^{I}t_{\alpha\beta} &= {}^{II}t_{\alpha\beta} - \bar{F}^{-1}_{\gamma\alpha}(p_\beta\bar{\mathbb{E}}_\gamma + \bar{\mathbb{P}}_\beta e_\gamma + \mu_o m^C_\beta\bar{\mathbb{H}}_\gamma + \mu_o\bar{\mathbb{M}}^C_\beta h_\gamma) + \\
&\quad + \bar{F}^{-1}_{\varepsilon\gamma}\bar{F}^{-1}_{\delta\alpha}(\bar{\mathbb{P}}_\beta\bar{\mathbb{E}}_\varepsilon + \mu_o\bar{\mathbb{M}}^C_\beta\bar{\mathbb{H}}_\varepsilon)u_{\gamma,\delta} .
\end{aligned}
$$

When the stress tensor and the body force of model I are substituted into the momentum equation what results is the body force of model II and a term which agrees with ${}^{II}T_{i\alpha}$.

In the <u>Maxwell-Minkowski model</u> the body force is given by $(4.97)^1$. When decomposed this becomes

$$
\begin{aligned}
{}^{III}(\rho_o\bar{F}^e_i) &= \delta_{i\delta}[\bar{F}^{-1}_{\alpha\delta}\{\bar{\mathbb{Q}}\bar{\mathbb{E}}_\alpha + e_{\alpha\beta\gamma}\bar{\mathbb{J}}_\beta\bar{\mathbb{B}}_\gamma + \bar{\mathbb{P}}_\beta\bar{\mathbb{E}}_{\beta,\alpha} + \mu_o\bar{\mathbb{M}}^C_\beta\bar{\mathbb{H}}_{\beta,\alpha} + \\
&\quad + e_{\alpha\beta\gamma}(\bar{\mathbb{D}}_\beta\dot{\bar{\mathbb{B}}}_\gamma + \dot{\bar{\mathbb{D}}}_\beta\bar{\mathbb{B}}_\gamma)\} + \bar{F}^{-1}_{\beta\delta,\alpha}(\bar{\mathbb{P}}_\alpha\bar{\mathbb{E}}_\beta + \mu_o\bar{\mathbb{M}}^C_\alpha\bar{\mathbb{H}}_\beta)] ,
\end{aligned}
$$

(over)

$${}^{III}(\rho_o f^e_\alpha) = \bar{\bar{F}}^{-1}_{\beta\alpha}\{q\bar{\mathbb{E}}_\beta + \bar{\mathbb{Q}}e_\beta + e_{\beta\gamma\delta}(\bar{\mathbb{J}}_\gamma b_\delta + j_\gamma\bar{\mathbb{B}}_\delta) +$$

(5.110) $$+ \bar{\mathbb{P}}_\gamma e_{\gamma,\beta} + p_\gamma\bar{\mathbb{E}}_{\gamma,\beta} + \mu_o\bar{\mathbb{M}}^C_\gamma h_{\gamma,\beta} + \mu_o m^C_\gamma\bar{\mathbb{H}}_{\gamma,\beta} + e_{\beta\gamma\delta}(\bar{\mathbb{D}}_\gamma\dot{b}_\delta + \dot{\bar{\mathbb{D}}}_\gamma b_\delta + d_\gamma\dot{\bar{\mathbb{B}}}_\delta + \dot{d}_\gamma\bar{\mathbb{B}}_\delta)\} +$$

$$+ \bar{\bar{F}}^{-1}_{\beta\alpha,\gamma}\{\bar{\mathbb{P}}_\gamma e_\beta + p_\gamma\bar{\mathbb{E}}_\beta + \mu_o\bar{\mathbb{M}}^C_\gamma h_\beta + \mu_o m^C_\gamma\bar{\mathbb{H}}_\beta\} +$$

$$- \bar{\bar{F}}^{-1}_{\beta\gamma}\bar{\bar{F}}^{-1}_{\delta\alpha}\{\mathbb{Q}\bar{\mathbb{E}}_\beta + e_{\beta\mu\nu}\bar{\mathbb{J}}_\mu\bar{\mathbb{B}}_\nu + \bar{\mathbb{P}}_\varepsilon\bar{\mathbb{E}}_{\varepsilon,\beta} + \mu_o\bar{\mathbb{M}}^C_\varepsilon\bar{\mathbb{H}}_{\varepsilon,\beta} + e_{\beta\mu\nu}(\bar{\mathbb{D}}_\mu\dot{\bar{\mathbb{B}}}_\nu + \dot{\bar{\mathbb{D}}}_\mu\bar{\mathbb{B}}_\nu)\}u_{\gamma,\delta} +$$

$$- (\bar{\bar{F}}^{-1}_{\beta\gamma}\bar{\bar{F}}^{-1}_{\delta\alpha}u_{\gamma,\delta})_{,\varepsilon}(\bar{\mathbb{P}}_\varepsilon\bar{\mathbb{E}}_\beta + \mu_o\bar{\mathbb{M}}^C_\varepsilon\bar{\mathbb{H}}_\beta) .$$

We recall that the stress tensors in the models I and III are identical, hence

(5.111) $${}^{I}\bar{T}_{i\alpha} = {}^{III}\bar{T}_{i\alpha} \qquad \text{and} \qquad {}^{I}t_{\alpha\beta} = {}^{III}t_{\alpha\beta} .$$

The relationships between $({}^{II}\bar{T}_{i\alpha}, {}^{II}t_{\alpha\beta})$ and $({}^{III}\bar{T}_{i\alpha}, {}^{III}t_{\alpha\beta})$ can then easily be read off from (5.109).

Next, we list the body forces for the statistical and the Lorentz-formulations. In the Lorentz-formulation (model V) they are already given in (5.45) and (5.46). For reasons of comparison they will be repeated here:

$${}^{V}(\rho_o\bar{F}^e_i) = \delta_{i\delta}\bar{\bar{F}}^{-1}_{\alpha\delta}\{(\bar{\mathbb{Q}} - \bar{\mathbb{P}}_{\beta,\beta})\mathbb{E}_\alpha + e_{\alpha\beta\gamma}(\bar{\mathbb{J}}_\beta + \dot{\bar{\mathbb{P}}}_\beta)\bar{\mathbb{B}}_\gamma + (\bar{\mathbb{M}}^L_{\alpha,\beta} - \bar{\mathbb{M}}^L_{\beta,\alpha})\bar{\mathbb{B}}_\beta\} ,$$

(5.112)
$$\begin{aligned}{}^{V}(\rho_o f^e_\alpha) = {} & \bar{\bar{F}}^{-1}_{\beta\alpha}\{(\bar{\mathbb{Q}} - \bar{\mathbb{P}}_{\varepsilon,\varepsilon})e_\beta + (q - p_{\varepsilon,\varepsilon})\bar{\mathbb{E}}_\beta + e_{\beta\gamma\delta}(\bar{\mathbb{J}}_\gamma + \dot{\bar{\mathbb{P}}}_\gamma)b_\delta + \\ & + e_{\beta\gamma\delta}(j_\gamma + \dot{p}_\gamma)\bar{\mathbb{B}}_\delta + \bar{\mathbb{B}}_\gamma(m^L_{\beta,\gamma} - m^L_{\gamma,\beta}) + b_\gamma(\bar{\mathbb{M}}^L_{\beta,\gamma} - \bar{\mathbb{M}}^L_{\gamma,\beta})\} + \\ & - \bar{\bar{F}}^{-1}_{\beta\gamma}\bar{\bar{F}}^{-1}_{\delta\alpha}\{(\bar{\mathbb{Q}} - \bar{\mathbb{P}}_{\varepsilon,\varepsilon})\bar{\mathbb{E}}_\beta + e_{\beta\mu\nu}(\mathbb{J}_\mu + \dot{\bar{\mathbb{P}}}_\mu)\bar{\mathbb{B}}_\nu + \bar{\mathbb{B}}_\varepsilon(\bar{\mathbb{M}}^L_{\beta,\varepsilon} - \bar{\mathbb{M}}^L_{\varepsilon,\beta})\}u_{\gamma,\delta} .\end{aligned}$$

On the other hand, in model IV we have (cf. (4.74)[1])

$${}^{IV}(\rho_o\bar{F}^e_i) = {}^{V}(\rho_o\bar{F}^e_i) + \delta_{i\delta}\{\bar{\bar{F}}^{-1}_{\alpha\delta}(\bar{\mathbb{E}}_\alpha\bar{\mathbb{P}}_\beta - \bar{\mathbb{M}}^L_\alpha\bar{\mathbb{B}}_\beta + \delta_{\alpha\beta}\bar{\mathbb{M}}^L_\gamma\bar{\mathbb{B}}_\gamma)\}_{,\beta} ,$$

(5.113) $${}^{IV}(\rho_o f^e_\alpha) = {}^{V}(\rho_o f^e_\alpha) + \{\bar{\bar{F}}^{-1}_{\beta\alpha}[\bar{\mathbb{E}}_\beta p_\varepsilon + e_\beta\bar{\mathbb{P}}_\varepsilon - \bar{\mathbb{M}}^L_\beta b_\varepsilon - m^L_\beta\bar{\mathbb{B}}_\varepsilon + \delta_{\beta\varepsilon}(\bar{\mathbb{M}}^L_\gamma b_\gamma + m^L_\gamma\bar{\mathbb{B}}_\gamma)] +$$

$$- \bar{\bar{F}}^{-1}_{\beta\gamma}\bar{\bar{F}}^{-1}_{\delta\alpha}[\bar{\mathbb{E}}_\beta\bar{\mathbb{P}}_\varepsilon - \bar{\mathbb{M}}^L_\beta\bar{\mathbb{B}}_\varepsilon + \delta_{\beta\varepsilon}\bar{\mathbb{M}}^L_\mu\bar{\mathbb{B}}_\mu]u_{\gamma,\delta}\}_{,\varepsilon} .$$

These forces need not be calculated, however, because the divergence term on the right-hand side of (5.113) will with opposite sign also show up in the stress tensor of the statistical formulation; hence, the momentum equation remains unchanged. Indeed, (cf. (4.77))

$$ {}^{IV}\bar{T}_{i\alpha} = {}^{V}\bar{T}_{i\alpha} - \delta_{i\delta}\bar{F}^{-1}_{\beta\delta}(\bar{\mathbb{P}}_{\alpha}\bar{\mathbb{E}}_{\beta} - \bar{\mathbb{B}}_{\alpha}\bar{\mathbb{M}}^{L}_{\beta} + \delta_{\alpha\beta}\bar{\mathbb{B}}_{\gamma}\bar{\mathbb{M}}^{L}_{\gamma}) \,, $$

$$ (5.114) \qquad {}^{IV}t_{\alpha\beta} = {}^{V}t_{\alpha\beta} - \bar{F}^{-1}_{\gamma\alpha}[p_{\beta}\bar{\mathbb{E}}_{\gamma} + \bar{\mathbb{P}}_{\beta}e_{\gamma} - b_{\beta}\bar{\mathbb{M}}^{L}_{\gamma} - \bar{\mathbb{B}}_{\beta}m^{L}_{\gamma} + \delta_{\beta\gamma}(b_{\delta}\bar{\mathbb{M}}^{L}_{\delta} + \bar{\mathbb{B}}_{\delta}m^{L}_{\delta})] + $$

$$ + \bar{F}^{-1}_{\varepsilon\gamma}\bar{F}^{-1}_{\delta\alpha}[\bar{\mathbb{P}}_{\beta}\bar{\mathbb{E}}_{\varepsilon} - \bar{\mathbb{B}}_{\beta}\bar{\mathbb{M}}^{L}_{\varepsilon} + \delta_{\beta\varepsilon}\bar{\mathbb{B}}_{\mu}\bar{\mathbb{M}}^{L}_{\mu}]u_{\gamma,\delta} \,. $$

To find the link between the groups (I,II,III) and (IV,V) we also need the decomposed versions of the relation (4.136) between ${}^{V}T_{i\alpha}$ and ${}^{II}T_{i\alpha}$. This relation gives

$$ {}^{V}\bar{T}_{i\alpha} = {}^{II}\bar{T}_{i\alpha} + \delta_{i\delta}\bar{F}^{-1}_{\beta\delta}[\bar{\mathbb{B}}^{a}_{\alpha}\bar{\mathbb{H}}_{\beta} - \bar{\mathbb{B}}_{\alpha}\bar{\mathbb{H}}^{a}_{\beta} - \frac{1}{2}\delta_{\alpha\beta}(\bar{\mathbb{B}}^{a}_{\gamma}\bar{\mathbb{H}}_{\gamma} - \bar{\mathbb{B}}_{\gamma}\bar{\mathbb{H}}^{a}_{\gamma})] \,, $$

$$ (5.115) \qquad \begin{aligned} {}^{V}t_{\alpha\beta} = {}^{II}t_{\alpha\beta} + \bar{F}^{-1}_{\gamma\alpha}[b^{a}_{\beta}\bar{\mathbb{H}}_{\gamma} + \bar{\mathbb{B}}^{a}_{\beta}h_{\gamma} - b_{\beta}\bar{\mathbb{H}}^{a}_{\gamma} - \bar{\mathbb{B}}_{\beta}h^{a}_{\gamma} + \\ - \frac{1}{2}\delta_{\beta\gamma}(b^{a}_{\delta}\bar{\mathbb{H}}_{\delta} + \bar{\mathbb{B}}^{a}_{\delta}h_{\delta} - b_{\delta}\bar{\mathbb{H}}^{a}_{\delta} - \bar{\mathbb{B}}_{\delta}h^{a}_{\delta})] + \\ - \bar{F}^{-1}_{\varepsilon\gamma}\bar{F}^{-1}_{\delta\alpha}[\bar{\mathbb{B}}^{a}_{\beta}\bar{\mathbb{H}}_{\varepsilon} - \bar{\mathbb{B}}_{\beta}\bar{\mathbb{H}}^{a}_{\varepsilon} - \frac{1}{2}\delta_{\beta\varepsilon}(\bar{\mathbb{B}}^{a}_{\mu}\bar{\mathbb{H}}_{\mu} - \bar{\mathbb{B}}_{\mu}\bar{\mathbb{H}}^{a}_{\mu})]u_{\gamma,\delta} \,. \end{aligned} $$

With the aid of the formulas $(5.34)^2$, $(5.35)^2$, $(5.92)^2$ and $(5.93)^2$ the auxiliary fields $\bar{\mathbb{B}}^{a}_{\alpha}$, $\bar{\mathbb{H}}^{a}_{\alpha}$, b^{a}_{α} and h^{a}_{α} can be eliminated.
If we wished to do so, we could also give the decoupled constitutive equations for the stresses ${}^{I}T_{i\alpha}, \ldots, {}^{V}T_{i\alpha}$. We shall not do it here, and we restrict ourselves to recalling that only ${}^{II}T_{i\alpha}$ and ${}^{V}T_{i\alpha}$ are directly derivable from a free energy; hence, their linearization is trivial. However, the free energies occurring in these relations are not identical; in fact they are related according to (4.144), which when decoupled yields

$$ \overline{{}^{V}(\rho_{o}\Psi)} = \overline{{}^{II}(\rho_{o}\bar{\Psi})} - \frac{\mu_{o}}{2\bar{J}}\bar{C}_{\alpha\beta}\bar{\mathbb{M}}^{C}_{\alpha}\bar{\mathbb{M}}^{C}_{\beta} \,, $$

$$ (5.116) \qquad \begin{aligned} \{{}^{V}(\rho_{o}\Psi) - \overline{{}^{V}(\rho_{o}\Psi)}\} = \{{}^{II}(\rho_{o}\bar{\Psi}) - \overline{{}^{II}(\rho_{o}\bar{\Psi})}\} - \frac{\mu_{o}}{\bar{J}}\bar{C}_{\alpha\beta}\bar{\mathbb{M}}^{C}_{\alpha}m^{C}_{\beta} + \\ + \frac{\mu_{o}}{\bar{J}}\bar{\mathbb{M}}^{C}_{\alpha}\bar{\mathbb{M}}^{C}_{\beta}[\bar{F}^{-1}_{\gamma\delta}\bar{C}_{\alpha\beta} - \delta_{\alpha\delta}\bar{F}_{\gamma\beta} - \delta_{\beta\delta}\bar{F}_{\gamma\alpha}]u_{\gamma,\delta} \,. \end{aligned} $$

We conclude this Section with a survey of the decomposed jump and boundary conditions. As before, the singular surfaces will be assumed to be of <u>order 2</u>. Since the deduction of the respective conditions is straightforward and in fact analogous to the methods illustrated in Section 5.2.4, we shall present the final results only. In accord with the assumptions laid down in Section 5.2.4 $\Sigma^{(i)}$ will denote the propagating surface of acceleration waves on which the intermediate fields $(\bar{\cdot})$ do not suffer a jump ($[\![(\bar{\cdot})]\!] = 0$, on $\Sigma^{(i)}$). On a material surface $\Sigma^{(ii)}$, on the other hand, $W_N = 0$, but the intermediate fields may jump there.

We start with the jump conditions for the electromagnetic fields. One obtains:

for <u>model I</u> and <u>model II</u>

(5.117)
$$[\![\bar{\mathbb{B}}^a_\alpha + \mu_o \bar{\mathbb{M}}^C_\alpha]\!] N_\alpha = 0, \qquad [\![\bar{\mathbb{D}}^a_\alpha + \bar{\mathbb{P}}_\alpha]\!] N_\alpha = 0,$$
$$e_{\alpha\beta\gamma}[\![\bar{\mathbb{E}}_\beta]\!] N_\gamma = 0, \quad e_{\alpha\beta\gamma}[\![\bar{\mathbb{H}}_\beta]\!] N_\gamma = 0, \quad [\![\bar{\mathbb{J}}_\alpha]\!] N_\alpha = 0, \quad \text{on } \Sigma^{(ii)}$$

and

(5.118)
$$[\![b^a_\alpha + \mu_o m^C_\alpha]\!] N_\alpha = 0, \qquad [\![d^a_\alpha + p_\alpha]\!] N_\alpha = 0,$$
$$e_{\alpha\beta\gamma}[\![e_\beta]\!] N_\gamma + [\![(b^a_\alpha + \mu_o m^C_\alpha)]\!] W_N = 0,$$
$$e_{\alpha\beta\gamma}[\![h_\beta]\!] N_\gamma - [\![(d^a_\alpha + p_\alpha)]\!] W_N = 0,$$
$$[\![j_\alpha]\!] N_\alpha - [\![q]\!] W_N = 0, \qquad \text{on } \Sigma^{(i)} \text{ and } \Sigma^{(ii)},$$

for <u>model III</u>

(5.119)
$$[\![\bar{\mathbb{B}}_\alpha]\!] N_\alpha = 0, \qquad [\![\bar{\mathbb{D}}_\alpha]\!] N_\alpha = 0,$$
$$e_{\alpha\beta\gamma}[\![\bar{\mathbb{E}}_\beta]\!] N_\gamma = 0, \quad e_{\alpha\beta\gamma}[\![\bar{\mathbb{H}}_\beta]\!] N_\gamma = 0, \quad [\![\bar{\mathbb{J}}_\alpha]\!] N_\alpha = 0, \quad \text{on } \Sigma^{(ii)},$$

and

(5.120)
$$[\![b_\alpha]\!] N_\alpha = 0, \qquad [\![d_\alpha]\!] N_\alpha = 0,$$
$$e_{\alpha\beta\gamma}[\![e_\beta]\!] N_\gamma + [\![b_\alpha]\!] W_N = 0,$$
$$e_{\alpha\beta\gamma}[\![h_\beta]\!] N_\gamma - [\![d_\alpha]\!] W_N = 0,$$
$$[\![j_\alpha]\!] N_\alpha - [\![q]\!] W_N = 0, \qquad \text{on } \Sigma^{(i)} \text{ and } \Sigma^{(ii)},$$

for <u>model IV</u> and <u>model V</u>

(5.121)
$$[\![\bar{\mathbb{B}}_\alpha]\!] N_\alpha = 0, \qquad [\![\bar{\mathbb{D}}^a_\alpha + \bar{\mathbb{P}}_\alpha]\!] N_\alpha = 0,$$
$$e_{\alpha\beta\gamma}[\![\bar{\mathbb{E}}_\beta]\!] N_\gamma = 0, \quad e_{\alpha\beta\gamma}[\![\bar{\mathbb{H}}^a_\beta - \bar{\mathbb{M}}^L_\beta]\!] N_\gamma = 0, \quad [\![\bar{\mathbb{J}}_\alpha]\!] N_\alpha = 0, \quad \text{on } \Sigma^{(ii)},$$

and

$$[\![b_\alpha]\!]N_\alpha = 0, \qquad [\![d^a_\alpha + p_\alpha]\!]N_\alpha = 0 ,$$

$$e_{\alpha\beta\gamma}[\![e_\beta]\!]N_\gamma + [\![b_\alpha]\!]W_N = 0 ,$$

(5.122) $$e_{\alpha\beta\gamma}[\![h^a_\beta - m^L_\beta]\!]N_\gamma - [\![d^a_\alpha + p_\alpha]\!]N_\alpha = 0 ,$$

$$[\![j_\alpha]\!]N_\alpha - [\![q]\!]W_N = 0 , \qquad \text{on } \Sigma^{(i)} \text{ and } \Sigma^{(ii)} .$$

These equations can easily be specified for $\Sigma^{(i)}$ and $\Sigma^{(ii)}$.

Yet, for the jump conditions of momentum and energy of matter and fields we must distinguish between $\Sigma^{(i)}$ and $\Sigma^{(ii)}$. We first list those valid at the wave surface $\Sigma^{(i)}$; they only need be given for the perturbed relations. One obtains: for <u>model I</u> and <u>model III</u>

$$[\![{}^{I}t_{\alpha\beta}]\!]N_\beta = [\![{}^{III}t_{\alpha\beta}]\!]N_\beta =$$

(5.123) $$= \bar{F}^{-1}_{\gamma\alpha}\{-\bar{\mathbb{P}}_\beta[\![e_\gamma]\!]N_\beta - \mu_o\bar{\mathbb{M}}^C_\beta[\![h_\gamma]\!]N_\beta + e_{\beta\gamma\delta}\bar{\mathbb{B}}^a_\beta[\![p_\delta]\!]W_N\} ,$$

$$[\![q_\alpha]\!]N_\alpha = \{\rho_o[\![{}^{II}u]\!] - \bar{\mathbb{E}}_\alpha[\![p_\alpha]\!] - \bar{\mathbb{H}}_\alpha[\![\mu_o m^C_\alpha]\!]\}W_N , \qquad \text{on } \Sigma^{(i)} .$$

We note that in the first of the above conditions the symmetry between the electric and the magnetic fields, which is characteristic for the Chu-formulation, is destroyed, because in the derivation the term

$$e_{\beta\gamma\delta}\langle\!\langle \mathbb{D}^a_\beta\rangle\!\rangle[\![\mu_o\mathbb{M}^C_\delta]\!]W_N$$

was dropped which is proportional to c^{-2}. Furthermore, for model III we have to replace in $(5.123)^1$ $\bar{\mathbb{B}}^a_\beta$ by:

$$\bar{\mathbb{B}}_\beta - \mu_o\bar{\mathbb{M}}^C_\beta .$$

For <u>model II</u> we have

$$[\![{}^{II}t_{\alpha\beta}]\!]N_\beta = \bar{F}^{-1}_{\gamma\alpha}\{\bar{\mathbb{E}}_\gamma[\![p_\beta]\!]N_\beta + \bar{\mathbb{H}}_\gamma[\![\mu_o m^C_\beta]\!]N_\beta + e_{\beta\gamma\delta}\bar{\mathbb{B}}^a_\beta[\![p_\delta]\!]W_N\} ,$$

(5.124)

$$[\![q_\alpha]\!]N_\alpha = \{\rho_o[\![{}^{II}u]\!] - \bar{\mathbb{E}}_\alpha[\![p_\alpha]\!] - \bar{\mathbb{H}}_\alpha[\![\mu_o m^C_\alpha]\!]\}W_N, \qquad \text{on } \Sigma^{(i)} .$$

Moreover, for <u>model IV</u>

$$[\![{}^{IV}t_{\alpha\beta}]\!]N_\beta = \bar{F}^{-1}_{\gamma\alpha}\{-\bar{\mathbb{P}}_\beta[\![e_\gamma]\!]N_\beta - \bar{\mathbb{M}}^L_\beta[\![b_\beta]\!]N_\gamma + e_{\beta\gamma\delta}\bar{\mathbb{B}}_\beta[\![p_\delta]\!]W_N\} ,$$

(5.125)

$$[\![q_\alpha]\!]N_\alpha = \{\rho_o[\![{}^{V}u]\!] + \bar{\mathbb{M}}_\alpha[\![b_\alpha]\!] - \bar{\mathbb{E}}_\alpha[\![p_\alpha]\!]\}W_N , \qquad \text{on } \Sigma^{(i)} .$$

and for model V:

$$[\![{}^{V}t_{\alpha\beta}]\!]N_\beta = \bar{F}^{-1}_{\gamma\alpha}\{\bar{\mathbb{E}}_\gamma[\![p_\beta]\!]N_\beta + \bar{\mathbb{B}}_\beta([\![m^L_\beta]\!]N_\gamma - [\![m^L_\gamma]\!]N_\beta) + e_{\beta\gamma\delta}\bar{\mathbb{B}}_\beta[\![p_\delta]\!]W_N\} ,$$

(5.126)

$$[\![q_\alpha]\!]N_\alpha = \{\rho_o[\![{}^{V}u]\!] + \bar{\mathbb{M}}^L_\alpha[\![b_\alpha]\!] - \bar{\mathbb{E}}_\alpha[\![p_\alpha]\!]\}W_N , \qquad \text{on } \Sigma^{(i)} .$$

Since in all of the above jump conditions for the energy q_α is the same function (needless to say once more that in model V q_α stands for q^S_α rather than q^L_α) the right-hand sides of these conditions must be identical also. That this is indeed the case can most easily be seen from (4.143) which in perturbed form reads

$$\rho_o{}^{V}u = \rho_o{}^{II}u - \mu_o h_\alpha \bar{\mathbb{M}}^C_\alpha - \mu_o \bar{\mathbb{H}}_\alpha m^C_\alpha - \frac{\mu_o}{J}\bar{C}_{\alpha\beta}\bar{\mathbb{M}}^C_\alpha m^C_\beta +$$

(5.127)

$$+ \frac{\mu_o}{2J}\bar{\mathbb{M}}^C_\alpha\bar{\mathbb{M}}^C_\beta[\bar{F}^{-1}_{\gamma\delta}\bar{C}_{\alpha\beta} - \delta_{\alpha\delta}\bar{F}_{\gamma\beta} - \delta_{\beta\delta}\bar{F}_{\gamma\alpha}]u_{\gamma,\delta} .$$

Finally, we list the boundary conditions for momentum and energy on material surface $\Sigma^{(ii)}$. On these, W_N vanishes but the intermediate fields may jump instead. Consequently,

for <u>model I</u> and <u>model III</u>

$$[\![{}^{I}\bar{T}_{i\alpha}]\!]N_\alpha = [\![{}^{III}\bar{T}_{i\alpha}]\!]N_\alpha =$$

$$= \delta_{i\delta}\bar{F}^{-1}_{\alpha\delta}\{-\langle\bar{\mathbb{P}}_\beta\rangle[\![\bar{\mathbb{E}}_\alpha]\!]N_\beta - \langle\mu_o\bar{\mathbb{M}}^C_\beta\rangle[\![\bar{\mathbb{H}}_\alpha]\!]N_\beta ,$$

(5.128) $$[\![{}^{I}t_{\alpha\beta}]\!]N_\beta = [\![{}^{III}t_{\alpha\beta}]\!]N_\beta = \bar{F}^{-1}_{\gamma\alpha}\{-\langle\bar{\mathbb{P}}_\beta\rangle[\![e_\gamma]\!]N_\beta - \langle p_\beta\rangle[\![\bar{\mathbb{E}}_\gamma]\!]N_\beta +$$

$$- \langle\mu_o\bar{\mathbb{M}}^C_\beta\rangle[\![h_\gamma]\!]N_\beta - \langle\mu_o m^C_\beta\rangle[\![\bar{\mathbb{H}}_\gamma]\!]N_\beta\} +$$

$$+ \bar{F}^{-1}_{\gamma\delta}\bar{F}^{-1}_{\varepsilon\alpha}\{\langle\bar{\mathbb{P}}_\beta\rangle[\![\bar{\mathbb{E}}_\gamma]\!]N_\beta + \langle\mu_o\bar{\mathbb{M}}^C_\beta\rangle[\![\bar{\mathbb{H}}_\gamma]\!]N_\beta\}u_{\delta,\varepsilon} , \qquad \text{on } \Sigma^{(ii)} ,$$

in which for model III $\mathbb{B}^a_\alpha$ and b^a_α must again be replaced by

$$(\bar{\mathbb{B}}_\alpha - \mu_o\bar{\mathbb{M}}^C_\alpha) \qquad \text{and} \qquad (b_\alpha - \mu_o m^C_\alpha), \text{ respectively} .$$

For <u>model II</u>

$$[\![{}^{II}\bar{T}_{i\alpha}]\!]N_\alpha = \delta_{i\delta}\bar{F}^{-1}_{\alpha\delta}\{\langle\bar{\mathbb{E}}_\alpha\rangle[\![\bar{\mathbb{P}}_\beta]\!]N_\beta + \langle\bar{\mathbb{H}}_\alpha\rangle[\![\mu_o\bar{\mathbb{M}}^C_\beta]\!]N_\beta\} ,$$

$$[\![{}^{II}t_{\alpha\beta}]\!]N_\beta = \bar{F}^{-1}_{\gamma\alpha}\{\langle\bar{\mathbb{E}}_\gamma\rangle[\![p_\beta]\!]N_\beta + \langle e_\gamma\rangle[\![\bar{\mathbb{P}}_\beta]\!]N_\beta + \langle\bar{\mathbb{H}}_\gamma\rangle[\![\mu_o m^C_\beta]\!]N_\beta +$$

(5.129)

$$+ \langle h_\gamma\rangle[\![\mu_o\bar{\mathbb{M}}^C_\beta]\!]N_\beta\} - \bar{F}^{-1}_{\gamma\delta}\bar{F}^{-1}_{\varepsilon\alpha}\{\langle\bar{\mathbb{E}}_\gamma\rangle[\![\bar{\mathbb{P}}_\beta]\!]N_\beta + \langle\bar{\mathbb{H}}_\gamma\rangle[\![\mu_o\bar{\mathbb{M}}^C_\beta]\!]N_\beta\}u_{\delta,\varepsilon},$$

$$\text{on } \Sigma^{(ii)}$$

for <u>model IV</u>

$$\llbracket {}^{IV}\bar{T}_{i\alpha} \rrbracket N_\alpha = \delta_{i\delta} \bar{F}^{-1}_{\alpha\delta} \{ -\langle \bar{\mathbb{P}}_\beta \rangle \llbracket \bar{\mathbb{E}}_\alpha \rrbracket N_\beta - \langle \bar{\mathbb{M}}^L_\beta \rangle \llbracket \bar{\mathbb{B}}_\beta \rrbracket N_\alpha \} \; ,$$

(5.130) $$\llbracket {}^{IV}t_{\alpha\beta} \rrbracket N_\beta = \bar{F}^{-1}_{\gamma\alpha} \{ -\langle \bar{\mathbb{P}}_\beta \rangle \llbracket e_\gamma \rrbracket N_\beta - \langle p_\beta \rangle \llbracket \bar{\mathbb{E}}_\gamma \rrbracket N_\beta - \langle \bar{\mathbb{M}}^L_\beta \rangle \llbracket b_\beta \rrbracket N_\gamma - \langle m^L_\beta \rangle \llbracket \bar{\mathbb{B}}_\beta \rrbracket N_\gamma \} +$$

$$+ \bar{F}^{-1}_{\gamma\delta} \bar{F}^{-1}_{\varepsilon\alpha} \{ \langle \bar{\mathbb{P}}_\beta \rangle \llbracket \bar{\mathbb{E}}_\gamma \rrbracket N_\beta + \langle \bar{\mathbb{M}}^L_\beta \rangle \llbracket \bar{\mathbb{B}}_\beta \rrbracket N_\gamma \} u_{\delta,\varepsilon} , \qquad \text{on } \Sigma^{(ii)}$$

and for <u>model V</u>

$$\llbracket {}^{V}\bar{T}_{i\alpha} \rrbracket N_\alpha = \delta_{i\delta} \bar{F}^{-1}_{\alpha\delta} \{ \langle \bar{\mathbb{E}}_\alpha \rangle \llbracket \bar{\mathbb{P}}_\beta \rrbracket N_\beta + \langle \bar{\mathbb{B}}_\beta \rangle (\llbracket \bar{\mathbb{M}}^L_\beta \rrbracket N_\alpha - \llbracket \bar{\mathbb{M}}^L_\alpha \rrbracket N_\beta) \} \; ,$$

(5.131)
$$\llbracket {}^{V}t_{\alpha\beta} \rrbracket N_\beta = \bar{F}^{-1}_{\gamma\alpha} \{ \langle \bar{\mathbb{E}}_\gamma \rangle \llbracket p_\beta \rrbracket N_\beta + \langle e_\gamma \rangle \llbracket \bar{\mathbb{P}}_\beta \rrbracket N_\beta +$$

$$+ \langle \bar{\mathbb{B}}_\beta \rangle (\llbracket m^L_\beta \rrbracket N_\gamma - \llbracket m^L_\gamma \rrbracket N_\beta) + \langle b_\beta \rangle (\llbracket \bar{\mathbb{M}}^L_\beta \rrbracket N_\gamma - \llbracket \bar{\mathbb{M}}^L_\gamma \rrbracket N_\beta) \} +$$

$$- \bar{F}^{-1}_{\gamma\delta} \bar{F}^{-1}_{\varepsilon\alpha} \{ \langle \bar{\mathbb{E}}_\gamma \rangle \llbracket \bar{\mathbb{P}}_\beta \rrbracket N_\beta + \langle \bar{\mathbb{B}}_\beta \rangle (\llbracket \bar{\mathbb{M}}^L_\beta \rrbracket N_\gamma - \llbracket \bar{\mathbb{M}}^L_\gamma \rrbracket N_\beta) \} u_{\delta,\varepsilon} , \quad \text{on } \Sigma^{(ii)} \; .$$

The boundary conditions of energy on $\Sigma^{(ii)}$ are equal in all formulations and simply read

(5.132) $$\llbracket \bar{Q}_\alpha \rrbracket N_\alpha = \llbracket q_\alpha \rrbracket N_\alpha = 0 \; , \qquad \text{on } \Sigma^{(ii)} \; .$$

Needless to say here that, with the use of the relations (5.109), (5.114), (5.115) and (5.127), all these jump conditions and boundary conditions can be transformed into each other.

5.4 THE MEANING OF INTERCHANGING DEPENDENT AND INDEPENDENT CONSTITUTIVE VARIABLES IN ONE FORMULATION

In Chapters 3 and 4 it was demonstrated that all theories of magnetizable and polarizable thermoelastic materials are fully equivalent. The equivalence requirements could be stated as interrelationships between the free energies and heat flux vectors of the formulations being compared. It was further demonstrated in Section 4.7 of Chapter 4 that full equivalence of different formulations can be destroyed simply by comparing energy expressions, which are too restrictive to allow complete matching. In particular, using in each theory a polynomial expansion of the free energy in terms of its variables and truncating these expansions at a certain prescribed level, destroyed complete agreement. One can, of course, insist in polynomial representations, but must then accept that equivalence of two formulations is only possible to within terms not being omitted in the expansion process.

A similar situation prevails if one tries to compare theories, which are based on one single model (say the Lorentz model), in which different dependent and independent constitutive variables are used. The situation is similar as in the nonlinear theory of elasticity, in which there also exist two different constitutive formulations, one in which the free energy is a function of the strains and a second one, whose energy function is obtained from the former through a Legendre transformation. This latter energy, which is called enthalpy (or complementary energy) is a function of the stresses rather than the strains. It gives the strains as functions of the stresses. Inverting the stress-strain relationship, that is expressing stress as a function of strain then effectively amounts to finding the free energy from the enthalpy. If these inversions can be performed, the theories based on enthalpy and free energy, respectively, are fully equivalent. It is known that this step is not a trivial one, except in the linear elasticity theory, in which it amounts to writing the elasticity coefficients in terms of the compliances.

In an electromechanical interaction theory changes of dependent and independent variables are possible also among the electromagnetic constitutive variables. Indeed, it is not the change of strain and stress as dependent and independent constitutive variables, that ordinarily gives rise to different constitutive approaches, but rather that of electromagnetic variables.

We showed in Chapter 2 already that there are four different constitutive theories in just one formulation, not counting that stress and strain and temperature and entropy could also be interchanged as dependent and independent variables. In what follows we shall only be dealing with two possibilities, namely the Lorentz formulation, in which $(\mathbb{E}_\alpha, \mathbb{B}_\alpha)$ and $(\mathbb{P}_\alpha/\rho_o, \mathbb{M}_\alpha/\rho_o)$ are taken as the respective independent constitutive variables. Results for the Chu-formulation and other variable combinations will be stated, however.

In the Lorentz model and if $\mathbb{E}_\alpha$ and $\mathbb{B}_\alpha$ are selected we have

$$(5.133) \qquad \check{\Psi} = U - \eta\theta - \frac{1}{\rho_o}\mathbb{E}_\alpha\mathbb{P}_\alpha = \check{\Psi}(E_{\alpha\beta}, \mathbb{E}_\alpha, \mathbb{B}_\alpha, \theta) \ ,$$

and

$$(5.134) \qquad \eta = -\frac{\partial\check{\Psi}}{\partial\theta} \ , \quad \mathbb{P}_\alpha = -\rho_o\frac{\partial\check{\Psi}}{\partial\mathbb{E}_\alpha} \ , \quad \mathbb{M}^L_\alpha = -\rho_o\frac{\partial\check{\Psi}}{\partial\mathbb{B}_\alpha} \ , \quad {}_{V}T^P_{\alpha\beta} = \rho_o\frac{\partial\check{\Psi}}{\partial E_{\alpha\beta}} \ .$$

On the other hand, when $\mathbb{P}_\alpha/\rho_o$ and $\mathbb{M}^L_\alpha/\rho_o$ are the independent fields, we use as energy functional

$$(5.135) \qquad \hat{\Psi} = U - \eta\theta + \frac{1}{\rho_o}\mathbb{M}^L_\alpha\mathbb{B}_\alpha = \hat{\Psi}\left(E_{\alpha\beta}, \frac{\mathbb{P}_\alpha}{\rho_o}, \frac{\mathbb{M}^L_\alpha}{\rho_o}, \theta\right) \ ,$$

and from it we obtain

$$(5.136)\qquad \eta = -\frac{\partial\hat{\Psi}}{\partial\theta}\,,\quad \mathbb{E}_\alpha = \frac{\partial\hat{\Psi}}{\partial\mathbb{P}_\alpha/\rho_o}\,,\quad \mathbb{B}_\alpha = \frac{\partial\hat{\Psi}}{\partial\mathbb{M}^L_\alpha/\rho_o}\,,\quad V_T{}^P_{\alpha\beta} = \rho_o\frac{\partial\hat{\Psi}}{\partial E_{\alpha\beta}}\,.$$

In view of the definitions of $\check{\Psi}$ and $\hat{\Psi}$, (5.133) and (5.135), the two constitutive theories lead to identical results if

$$(5.137)\qquad \hat{\Psi} = \check{\Psi} + \frac{1}{\rho_o}\mathbb{E}_\alpha\mathbb{P}_\alpha + \frac{1}{\rho_o}\mathbb{B}_\alpha\mathbb{M}^L_\alpha\,,$$

or with the use of (5.134), if

$$(5.138)\qquad \hat{\Psi}\Big(E_{\alpha\beta},\frac{\mathbb{P}_\alpha}{\rho_o},\frac{\mathbb{M}^L_\alpha}{\rho_o},\theta\Big) = \check{\Psi}\Big(E_{\alpha\beta},\frac{\partial\hat{\Psi}}{\partial\mathbb{P}_\alpha/\rho_o},\frac{\partial\hat{\Psi}}{\partial\mathbb{M}^L_\alpha/\rho_o},\theta\Big) + \frac{\partial\hat{\Psi}}{\partial\mathbb{P}_\alpha/\rho_o}\frac{\mathbb{P}_\alpha}{\rho_o} + \frac{\partial\hat{\Psi}}{\partial\mathbb{M}^L_\alpha/\rho_o}\frac{\mathbb{M}^L_\alpha}{\rho_o}\,.$$

Of course there is also a dual relation to (5.138), namely

$$(5.139)\qquad \check{\Psi}(E_{\alpha\beta},\mathbb{E}_\alpha,\mathbb{B}_\alpha,\theta) = \hat{\Psi}\Big(E_{\alpha\beta},-\frac{\partial\check{\Psi}}{\partial\mathbb{E}_\alpha},-\frac{\partial\check{\Psi}}{\partial\mathbb{B}_\alpha},\theta\Big) + \mathbb{E}_\alpha\frac{\partial\check{\Psi}}{\partial\mathbb{E}_\alpha} + \mathbb{B}_\alpha\frac{\partial\check{\Psi}}{\partial\mathbb{B}_\alpha}$$

For given functionals $\hat{\Psi}$ and $\check{\Psi}$, equations (5.138) and (5.139) must be satisfied identically if the two constitutive theories are to be equivalent. If only one of the energy functionals, $\hat{\Psi}$ or $\check{\Psi}$, is given, then (5.138) or (5.139) are functional differential equations to determine the other. The solution to these remains an open problem, as we shall not attack it here. Nevertheless, in order to demonstrate that this problem is extremely complex in general, consider, as a more or less arbitrary example, the following energy functional

$$(5.140)\qquad \hat{\Psi} = \hat{\Psi}(E_{\alpha\beta},\mathbb{M}_\alpha) = f(M^2) + \frac{1}{2\rho_o}\{\lambda(E_{\alpha\alpha})^2 + 2GE_{\alpha\beta}E_{\alpha\beta}\}$$

which may be regarded as the most simple functional form for an isotropic nonlinearly magnetizable body. In (5.140)

$$(5.141)\qquad M^2 := \frac{1}{\rho_o^2}\mathbb{M}^L_\alpha\mathbb{M}^L_\alpha\,,$$

and $f(M^2)$ is a continuous, differentiable function. With the aid of (5.136) we may derive the following expression for $\mathbb{B}_\alpha$

$$(5.142)\qquad \mathbb{B}_\alpha = 2\,\frac{df}{dM^2}\,\frac{\mathbb{M}^L_\alpha}{\rho_o}\,.$$

Substituting this into the identity (5.137) allows the determination of $\check{\Psi}$,

$$(5.143)\qquad \check{\Psi}(E_{\alpha\beta},\mathbb{B}_\alpha) = f(M^2) - 2M^2\frac{df}{dM^2} + \frac{1}{2\rho_o}\{\lambda(E_{\alpha\alpha})^2 + 2GE_{\alpha\beta}E_{\alpha\beta}\} .$$

Yet, the right-hand side of (5.143) can be determined as an explicit function of $\mathbb{B}_\alpha$, or more precisely of $\mathbb{B}^2 = \mathbb{B}_\alpha\mathbb{B}_\alpha$, only if the relation (5.142) is invertible, that is only if M^2 is expressible as a function of $\mathbb{B}_\alpha$. This is rarely the case in general. However, the special choice

$$(5.144)\qquad f(M^2) = \frac{1}{2}\,{}_2\chi\int_0^{M^2}[1 + 2\frac{{}_4\chi}{{}_2\chi}\xi]^{1/2}d\xi \qquad (\cong \frac{1}{2}\,{}_2\chi M^2 + \frac{1}{4}\,{}_4\chi M^4 + \ldots) ,$$

which is a nonlinear representation that may be regarded as an extension of the usual functional dependencies of the free energy on magnetization, gives (for ${}_2\chi \neq 0$, ${}_4\chi \neq 0$)

$$(5.145)\qquad M^2 = \frac{{}_2\chi}{4\,{}_4\chi}\{-1 + [1 + \frac{8\,{}_4\chi}{{}_2\chi^3}\mathbb{B}^2]^{1/2}\} ,$$

so that $\check{\Psi}$ may be written as

$$(5.146)\qquad \begin{aligned}\check{\Psi} = \frac{1}{2}\,{}_2\chi\int_0^{M^2}[1 + 2\frac{{}_4\chi}{{}_2\chi}\xi]^{1/2}d\xi - {}_2\chi[1 + 2\frac{{}_4\chi}{{}_2\chi}M^2]^{1/2}M^2 + \\ + \frac{1}{2\rho_o}\{\lambda(E_{\alpha\alpha})^2 + 2GE_{\alpha\beta}E_{\alpha\beta}\} ,\end{aligned}$$

in which M^2 is given by (5.145).

Often the construction of a free energy of one formulation from that of another is a matter of shear patience or simply becomes impossible analytically. This does not mean that equivalence is not possible in these cases; it simply means that a free energy function of the second formulation is given only implicitly. Such is already the case in the above representation if we leave $f(M^2)$ unspecified.

The above construction of a free energy may appear to be rather artificial, because the representation (5.140) is of very limited practical applicability. Nevertheless, it is important from a mathematical point of view, because it explicitly demonstrates that the functional differential equations mentioned above do indeed admit exact solutions, at least for the demonstrated case. Conditions imposed on the free energy functions, that guarantee the existence of such solutions would be of value, and in particular, it would be valuable if, for instance, free energies could be constructed which would still be of physical relevance, but would not admit a solution of the functional differential equations. In that case non-equivalence of two theories would be demonstrated. We shall not go any deeper in this subject, but will mention one simple and physically important case, in which existence of solutions of the equations (5.138) or (5.139) can easily be established and for which equivalence of the theories compared is guaranteed. What we have in mind is the case for which the free energy functions can be expressed as Taylor series expansions about zero deformation, constant temperature and zero electromagnetic fields. The proof to this case will not be outlined here, but from the procedure explained below the reader should be able to construct his own proof.

With these few comments we shall now leave the subject of an exact determination of the functional $\check{\Psi}$ from $\hat{\Psi}$ and pass on to an approximate satisfaction of the relations (5.138) and (5.139). To this end we shall choose truncated polynomials as expressions for the free energies. If the degree of the polynomial representation of $\hat{\Psi}$ is known, then with (5.138) that of $\check{\Psi}$ can be determined. In general the order of truncation of $\check{\Psi}$ needed to obtain full equivalence is not the same as that for $\hat{\Psi}$. We shall not be so general and choose for both energy functionals the same polynomial expressions. In complexity, we shall be as general as we were in (5.55) and thus write

$$
\begin{aligned}
\check{\Psi} = {} & \frac{1}{2\rho_o}\,{}_2\check{\chi}^{(m)}_{\alpha\beta}\mathbb{B}_\alpha\mathbb{B}_\beta + \frac{1}{4\rho_o}\,{}_4\check{\chi}^{(m)}_{\alpha\beta\gamma\delta}\mathbb{B}_\alpha\mathbb{B}_\beta\mathbb{B}_\gamma\mathbb{B}_\delta + \frac{1}{\rho_o}\check{\chi}^{(em)}_{\alpha\beta}\mathbb{B}_\alpha\mathbb{E}_\beta + \\
& + \frac{1}{2\rho_o}\,{}_2\check{\chi}^{(e)}_{\alpha\beta}\mathbb{E}_\alpha\mathbb{E}_\beta + \frac{1}{4\rho_o}\,{}_4\check{\chi}^{(e)}_{\alpha\beta\gamma\delta}\mathbb{E}_\alpha\mathbb{E}_\beta\mathbb{E}_\gamma\mathbb{E}_\delta - \frac{1}{2}\check{c}(\theta-\theta_o)^2 + \\
& + \frac{1}{\rho_o}\check{\lambda}^{(m)}_\alpha\mathbb{B}_\alpha(\theta-\theta_o) + \frac{1}{2\rho_o}\check{L}^{(m)}_{\alpha\beta}\mathbb{B}_\alpha\mathbb{B}_\beta(\theta-\theta_o) + \frac{1}{\rho_o}\check{\lambda}^{(e)}_\alpha\mathbb{E}_\alpha(\theta-\theta_o) + \\
& + \frac{1}{2\rho_o}\check{L}^{(e)}_{\alpha\beta}\mathbb{E}_\alpha\mathbb{E}_\beta(\theta-\theta_o) + \Big\{\frac{1}{\rho_o}\check{\varepsilon}^{(m)}_{\alpha\beta\gamma}\mathbb{B}_\beta + \frac{1}{2\rho_o}\check{b}^{(m)}_{\alpha\beta\gamma\delta}\mathbb{B}_\alpha\mathbb{B}_\beta + \frac{1}{\rho_o}\check{\varepsilon}^{(e)}_{\beta\gamma\delta}\mathbb{E}_\beta + \\
& + \frac{1}{2\rho_o}\check{b}^{(e)}_{\alpha\beta\gamma\delta}\mathbb{E}_\alpha\mathbb{E}_\beta - \check{\nu}_{\gamma\delta}(\theta-\theta_o)\Big\}E_{\gamma\delta} + \frac{1}{2\rho_o}\check{c}_{\alpha\beta\gamma\delta}E_{\alpha\beta}E_{\gamma\delta}\,,
\end{aligned}
\tag{5.147}
$$

and

$$
\begin{aligned}
\hat{\Psi} = {} & \frac{\rho_o}{2}\,{}_2\hat{\chi}^{(m)}_{\alpha\beta}\frac{\mathbf{M}^L_\alpha}{\rho_o}\frac{\mathbf{M}^L_\beta}{\rho_o} + \frac{\rho_o^3}{4}\,{}_4\hat{\chi}^{(m)}_{\alpha\beta\gamma\delta}\frac{\mathbf{M}^L_\alpha}{\rho_o}\frac{\mathbf{M}^L_\beta}{\rho_o}\frac{\mathbf{M}^L_\gamma}{\rho_o}\frac{\mathbf{M}^L_\delta}{\rho_o} + \rho_o\hat{\chi}^{(em)}_{\alpha\beta}\frac{\mathbf{M}^L_\alpha}{\rho_o}\frac{\mathbb{P}_\beta}{\rho_o} + \\
& + \frac{\rho_o}{2}\,{}_2\hat{\chi}^{(e)}_{\alpha\beta}\frac{\mathbb{P}_\alpha}{\rho_o}\frac{\mathbb{P}_\beta}{\rho_o} + \frac{\rho_o^3}{4}\,{}_4\hat{\chi}^{(e)}_{\alpha\beta\gamma\delta}\frac{\mathbb{P}_\alpha}{\rho_o}\frac{\mathbb{P}_\beta}{\rho_o}\frac{\mathbb{P}_\gamma}{\rho_o}\frac{\mathbb{P}_\delta}{\rho_o} - \frac{1}{2}\hat{c}(\theta-\theta_o)^2 + \\
& + \hat{\lambda}^{(m)}_\alpha\frac{\mathbf{M}^L_\alpha}{\rho_o}(\theta-\theta_o) + \frac{\rho_o}{2}\hat{L}^{(m)}_{\alpha\beta}\frac{\mathbf{M}^L_\alpha}{\rho_o}\frac{\mathbf{M}^L_\beta}{\rho_o}(\theta-\theta_o) + \hat{\lambda}^{(e)}_\alpha\frac{\mathbb{P}_\alpha}{\rho_o}(\theta-\theta_o) +
\end{aligned}
\tag{5.148}
$$

(over)

$$+ \frac{\rho_o}{2} \hat{L}^{(e)}_{\alpha\beta} \frac{\mathbb{P}_\alpha}{\rho_o} \frac{\mathbb{P}_\beta}{\rho_o}(\theta - \theta_o) + \{\hat{\varepsilon}^{(m)}_{\beta\gamma\delta} \frac{M^L_\beta}{\rho_o} + \frac{\rho_o}{2} \hat{b}^{(m)}_{\alpha\beta\gamma\delta} \frac{M^L_\alpha}{\rho_o} \frac{M^L_\beta}{\rho_o} + \hat{\varepsilon}^{(e)}_{\beta\gamma\delta} \frac{\mathbb{P}_\beta}{\rho_o} +$$

$$+ \frac{\rho_o}{2} \hat{b}^{(e)}_{\alpha\beta\gamma\delta} \frac{\mathbb{P}_\alpha}{\rho_o} \frac{\mathbb{P}_\beta}{\rho_o} - \hat{\nu}_{\gamma\delta}(\theta - \theta_o)\} E_{\gamma\delta} + \frac{1}{2\rho_o} \hat{c}_{\alpha\beta\gamma\delta} E_{\alpha\beta} E_{\gamma\delta} .$$

With these representations full equivalence is not possible, because the transformations indicated by (5.138) and (5.139) lead to terms which have been omitted in the formulation of the respective energy functions. Nevertheless, except for these terms equivalence can be established. If (5.147) and (5.148) are used to exploit (5.138) and (5.139) in this approximate sense a series of identities can be derived for the phenomenological coefficients of the two formulations. We have done this, and the calculations for the derivation of the corresponding relations are very long. Unfortunately, the emerging identities are much too long, and conclusions that can be drawn from them in this full generality are very meager in order to justify to list them here. Nevertheless, one result derivable from these identities may be quoted. It reads: If the coefficients $_2\hat{\chi}^{(m)}_{\alpha\beta}$, $_2\hat{\chi}^{(e)}_{\alpha\beta}$, $\chi^{(em)}_{\alpha\beta}$ vanish, all remaining coefficients accounting for electromagnetic effects must also vanish if the theories are to be equivalent in the above mentioned sense. Otherwise stated, if $_2\hat{\chi}^{(m)}$, $_2\hat{\chi}^{(e)}$ and $\hat{\chi}^{(em)}$ vanish in one theory and the two theories are to be equivalent the free energies reduce to

$$\Psi = \frac{1}{2\rho_o} c_{\alpha\beta\gamma\delta} E_{\alpha\beta} E_{\gamma\delta} - \nu_{\alpha\beta}(\theta - \theta_o) E_{\alpha\beta} - \frac{1}{2} c(\theta - \theta_o)^2 ,$$

and no other terms. In a truly polarizable and magnetizable material at least one of the coefficients $_2\chi^{(m)}_{\alpha\beta}$, $_2\chi^{(e)}_{\alpha\beta}$ or $\chi^{(em)}_{\alpha\beta}$ must therefore be non-zero. In the following we shall exploit the equations (5.138) and (5.139) for an isotropic body, in which

$$\lambda^{(m)}_\alpha = \lambda^{(e)}_\alpha = 0, \qquad \varepsilon^{(m)}_{\alpha\beta\gamma} = \varepsilon^{(e)}_{\alpha\beta\gamma} = 0 ,$$

$$_2\chi^{(m)}_{\alpha\beta} = {_2\chi^{(m)}}\delta_{\alpha\beta}, \quad _2\chi^{(e)}_{\alpha\beta} = {_2\chi^{(e)}}\delta_{\alpha\beta}, \quad \chi^{(em)}_{\alpha\beta} = 0 ,$$

$$_4\chi^{(m)}_{\alpha\beta\gamma\delta} = {_4\chi^{(m)}} \frac{1}{3}(\delta_{\alpha\beta}\delta_{\gamma\delta} + \delta_{\alpha\gamma}\delta_{\beta\delta} + \delta_{\alpha\delta}\delta_{\beta\gamma}),$$

$$(5.149) \qquad _4\chi^{(e)}_{\alpha\beta\gamma\delta} = {_4\chi^{(e)}} \frac{1}{3}(\delta_{\alpha\beta}\delta_{\gamma\delta} + \delta_{\alpha\gamma}\delta_{\beta\delta} + \delta_{\alpha\delta}\delta_{\beta\gamma}),$$

$$L^{(m)}_{\alpha\beta} = L^{(m)}\delta_{\alpha\beta}, \quad L^{(e)}_{\alpha\beta} = L^{(e)}\delta_{\alpha\beta}, \quad \nu_{\alpha\beta} = \nu\delta_{\alpha\beta} ,$$

(over)

$$b^{(m)}_{\alpha\beta\gamma\delta} = b^{(m)}_1 \delta_{\alpha\beta}\delta_{\gamma\delta} + \frac{1}{2} b^{(m)}_2 (\delta_{\alpha\gamma}\delta_{\beta\delta} + \delta_{\alpha\delta}\delta_{\beta\gamma}),$$

$$b^{(e)}_{\alpha\beta\gamma\delta} = b^{(e)}_1 \delta_{\alpha\beta}\delta_{\gamma\delta} + \frac{1}{2} b^{(e)}_2 (\delta_{\alpha\gamma}\delta_{\beta\delta} + \delta_{\alpha\delta}\delta_{\beta\gamma}),$$

$$c_{\alpha\beta\gamma\delta} = \lambda\delta_{\alpha\beta}\delta_{\gamma\delta} + G(\delta_{\alpha\gamma}\delta_{\beta\delta} + \delta_{\alpha\delta}\delta_{\beta\gamma}) .$$

Substituting these expressions into (5.147) and (5.148) gives

$$(5.150)\quad \begin{aligned} \check{\Psi} = {} & \frac{1}{2\rho_o}\,{}_2\check{\chi}^{(m)}\mathbb{B}_\alpha\mathbb{B}_\alpha + \frac{1}{4\rho_o}\,{}_4\check{\chi}^{(m)}(\mathbb{B}_\alpha\mathbb{B}_\alpha)^2 + \frac{1}{2\rho_o}\,{}_2\check{\chi}^{(e)}\mathbb{E}_\alpha\mathbb{E}_\alpha + \frac{1}{4\rho_o}\,{}_4\check{\chi}^{(e)}(\mathbb{E}_\alpha\mathbb{E}_\alpha)^2 \\ & - \frac{1}{2}\check{c}(\theta-\theta_o)^2 + \frac{1}{2\rho_o}\check{L}^{(m)}\mathbb{B}_\alpha\mathbb{B}_\alpha(\theta-\theta_o) + \frac{1}{2\rho_o}\check{L}^{(e)}\mathbb{E}_\alpha\mathbb{E}_\alpha(\theta-\theta_o) + \\ & + \frac{1}{2\rho_o}(\check{b}^{(m)}_1\mathbb{B}_\alpha\mathbb{B}_\alpha + \check{b}^{(e)}_1\mathbb{E}_\alpha\mathbb{E}_\alpha)E_{\beta\beta} + \frac{1}{2\rho_o}(\check{b}^{(m)}_2\mathbb{B}_\alpha\mathbb{B}_\beta + \check{b}^{(e)}_2\mathbb{E}_\alpha\mathbb{E}_\beta)E_{\alpha\beta} + \\ & - \check{\nu}(\theta-\theta_o)E_{\alpha\alpha} + \frac{1}{2\rho_o}\{\check{\lambda}(E_{\alpha\alpha})^2 + 2\check{G}E_{\alpha\beta}E_{\alpha\beta}\} , \end{aligned}$$

and

$$(5.151)\quad \begin{aligned} \hat{\Psi} = {} & \frac{\rho_o}{2}\,{}_2\hat{\chi}^{(m)}\frac{\mathbb{M}^L_\alpha}{\rho_o}\frac{\mathbb{M}^L_\alpha}{\rho_o} + \frac{\rho_o^3}{4}\,{}_4\hat{\chi}^{(m)}\left(\frac{\mathbb{M}^L_\alpha}{\rho_o}\frac{\mathbb{M}^L_\alpha}{\rho_o}\right)^2 + \frac{\rho_o}{2}\,{}_2\hat{\chi}^{(e)}\frac{\mathbb{P}_\alpha}{\rho_o}\frac{\mathbb{P}_\alpha}{\rho_o} + \frac{\rho_o^3}{4}\,{}_4\hat{\chi}^{(e)}\left(\frac{\mathbb{P}_\alpha}{\rho_o}\frac{\mathbb{P}_\alpha}{\rho_o}\right)^2 \\ & - \frac{1}{2}\hat{c}(\theta-\theta_o)^2 + \frac{\rho_o}{2}\hat{L}^{(m)}\frac{\mathbb{M}^L_\alpha}{\rho_o}\frac{\mathbb{M}^L_\alpha}{\rho_o}(\theta-\theta_o) + \frac{\rho_o}{2}\hat{L}^{(e)}\frac{\mathbb{P}_\alpha}{\rho_o}\frac{\mathbb{P}_\alpha}{\rho_o}(\theta-\theta_o) + \\ & + \frac{\rho_o}{2}(\hat{b}^{(m)}_1\frac{\mathbb{M}^L_\alpha}{\rho_o}\frac{\mathbb{M}^L_\alpha}{\rho_o} + \hat{b}^{(e)}_1\frac{\mathbb{P}_\alpha}{\rho_o}\frac{\mathbb{P}_\alpha}{\rho_o})E_{\beta\beta} + \frac{\rho_o}{2}(\hat{b}^{(m)}_2\frac{\mathbb{M}^L_\alpha}{\rho_o}\frac{\mathbb{M}^L_\beta}{\rho_o} + \hat{b}^{(e)}_2\frac{\mathbb{P}_\alpha}{\rho_o}\frac{\mathbb{P}_\beta}{\rho_o})E_{\alpha\beta} + \\ & - \hat{\nu}(\theta-\theta_o)E_{\alpha\alpha} + \frac{1}{2\rho_o}\{\hat{\lambda}(E_{\alpha\alpha})^2 + 2\hat{G}E_{\alpha\beta}E_{\alpha\beta}\} . \end{aligned}$$

With these an exploitation of (5.138) stays within a reasonable effort. The approach is analogous to that demonstrated in Section 4.7 and at the beginning of this Section. One simply evaluates $\mathbb{E}_\alpha$ and $\mathbb{B}_\alpha$ according to (5.136),

$$(5.152)\quad \begin{aligned} \mathbb{B}_\alpha &= {}_2\hat{\chi}^{(m)}\mathbb{M}^L_\alpha + {}_4\hat{\chi}^{(m)}\mathbb{M}^L_\beta\mathbb{M}^L_\beta\mathbb{M}^L_\alpha + \hat{L}^{(m)}\mathbb{M}^L_\alpha(\theta-\theta_o) + \hat{b}^{(m)}_1\mathbb{M}^L_\alpha E_{\beta\beta} + \hat{b}^{(m)}_2\mathbb{M}^L_\beta E_{\alpha\beta} , \\ \mathbb{E}_\alpha &= {}_2\hat{\chi}^{(e)}\mathbb{P}_\alpha + {}_4\hat{\chi}^{(e)}\mathbb{P}_\beta\mathbb{P}_\beta\mathbb{P}_\alpha + \hat{L}^{(e)}\mathbb{P}_\alpha(\theta-\theta_o) + \hat{b}^{(e)}_1\mathbb{P}_\alpha E_{\beta\beta} + \hat{b}^{(e)}_2\mathbb{P}_\beta E_{\alpha\beta} , \end{aligned}$$

expresses with their use $\check{\Psi}$ as a function of $\mathbb{P}_\alpha/\rho_o$ and $\mathbb{M}^L_\alpha/\rho_o$ rather than $\mathbb{E}_\alpha$ and $\mathbb{B}_\alpha$ and substitutes the emerging relation together with (5.148) into (5.138). When this is done the following identities are obtained:

$$
\begin{aligned}
&{}_2\hat{\chi}^{(m)} = -\frac{1}{{}_2\check{\chi}^{(m)}} = \frac{\mu_o\mu}{\mu-\mu_o} = \frac{\mu}{\chi^{(m)}}\,, \qquad {}_2\hat{\chi}^{(e)} = -\frac{1}{{}_2\check{\chi}^{(e)}} = \frac{1}{\chi^{(e)}}\,,\\
&{}_4\hat{\chi}^{(m)} = \left(\frac{\mu}{\chi^{(m)}}\right)^4 {}_4\check{\chi}^{(m)}, \qquad {}_4\hat{\chi}^{(e)} = \left(\frac{1}{\chi^{(e)}}\right)^4 {}_4\check{\chi}^{(e)}\,,\\
(5.153)\quad &\hat{L}^{(m)} = \left(\frac{\mu}{\chi^{(m)}}\right)^2 \check{L}^{(m)}, \qquad \hat{L}^{(e)} = \left(\frac{1}{\chi^{(e)}}\right)^2 \check{L}^{(m)}\,, \qquad\qquad \text{(A)}\\
&\hat{b}^{(m)}_{1,2} = \left(\frac{\mu}{\chi^{(m)}}\right)^2 \check{b}^{(m)}_{1,2}, \qquad \hat{b}^{(e)}_{1,2} = \left(\frac{1}{\chi^{(e)}}\right)^2 \check{b}^{(e)}_{1,2}\,,\\
&\hat{c} = \check{c} = \frac{c_W}{\theta_o}\,, \qquad \hat{\nu} = \check{\nu} = \nu, \qquad \hat{\lambda} = \check{\lambda} = \lambda, \qquad \hat{G} = \check{G} = G\,,
\end{aligned}
$$

with obvious inversions, which we shall not write down. In the above relations, μ is called magnetic permeability, $\chi^{(m)}$ the magnetic susceptibility and $\chi^{(e)}$ the electric susceptibility; for an ideal medium (i.e. a rigid, isotropic body) they are defined by

$$
\underline{\mathbb{B}} = \mu\underline{\mathbb{H}}, \quad \underline{\mathbb{M}} = \chi^{(m)}\underline{\mathbb{H}}, \quad \underline{\mathbb{P}} = \chi^{(e)}\underline{\mathbb{E}}\ .
$$

Further, c_W is the specific heat, ν the thermoelastic constant, and λ and G are the Lamé constants. In an isotropic body, therefore, the classical phenomenological coefficients are equal and thus clearly defined. This is not so for all other coefficients, as can clearly be seen from (5.153). In fact, the transformations all involve the magnetic and electric susceptibility. The definitions of fourth order electromagnetic, magnetostrictive, electrostrictive, thermoelectric and thermomagnetic coefficients are not unique in this restricted isotropic theory, however. Care should therefore be observed with the use of specific names for these effects. Note also that the above described transformation becomes simply impossible whenever ${}_2\chi^{(m)}$ and/or ${}_2\chi^{(e)}$ are zero. But in this case, the outlined procedure does not lead to a polynomial expression for $\hat{\Psi}$ of the form (5.151).

In the above only two possible constitutive formulations were investigated, namely those in which $(\mathbb{E}_\alpha, \mathbb{B}_\alpha)$ and $(\mathbb{P}_\alpha/\rho_o, \mathbb{M}_\alpha/\rho_o)$, respectively, were the independent constitutive variables. Of course, similar calculations can also be performed, if other sets of independent variables are chosen. They also lead to results similar to (5.153) relating the phenomenological coefficients of the respective constitutive formulations. Below we shall present the transformations of all possibilities of the statistical and the Lorentz models (IV,V).

On the other hand, the Chu-formulations (models I,II) and the Maxwell-Minkowski model (III) have been shown to be equivalent if in each of them the same functional dependencies for the free energy are taken. Hence, similar changes can also be performed for these formulations.

In the remainder of this Section we shall list the transformation rules for all these changes of dependent and independent constitutive variables; but we shall restrict ourselves to free energies of the complexity (5.150). Thus we write

$$
\begin{aligned}
\Psi = {} & \frac{1}{2\rho_o}\,{}_2\chi^{(m)} V_\alpha V_\alpha + \frac{1}{4\rho_o}\,{}_4\chi^{(m)} (V_\alpha V_\alpha)^2 + \frac{1}{2\rho_o}\,{}_2\chi^{(e)} W_\alpha W_\alpha + \frac{1}{4\rho_o}\,{}_4\chi^{(e)} (W_\alpha W_\alpha)^2 + \\
& - \frac{1}{2} c(\theta - \theta_o)^2 + \frac{1}{2\rho_o} L^{(m)} V_\alpha V_\alpha (\theta - \theta_o) + \frac{1}{2\rho_o} L^{(e)} W_\alpha W_\alpha (\theta - \theta_o) - \nu(\theta - \theta_o) E_{\alpha\alpha} + \\
& + \frac{1}{2\rho_o}(b_1^{(m)} V_\alpha V_\alpha + b_1^{(\ )} W_\alpha W_\alpha) E_{\beta\beta} + \frac{1}{2\rho_o}(b_2^{(m)} V_\alpha V_\beta + b_2^{(e)} V_\alpha V_\beta) E_{\alpha\beta} + \\
& + \frac{1}{2\rho_o}\{\lambda (E_{\alpha\alpha})^2 + 2G E_{\alpha\beta} E_{\alpha\beta}\} .
\end{aligned}
\tag{5.154}
$$

Here, (V_α, W_α) stands for the respective pairs of electromagnetic variables which will be chosen as the independent variables. In particular, the following choices will be made:

<u>in the models I,II,III</u>	<u>in the models IV,V</u>
$(V_\alpha, W_\alpha) = (\mathbb{E}_\alpha, \mathbb{H}_\alpha)$: $(\check{\check{\cdot}})$	$(V_\alpha, W_\alpha) = (\mathbb{E}_\alpha, \mathbb{B}_\alpha)$: $(\check{\cdot})$
$(\mathbb{P}_\alpha, \mathbb{M}^C_\alpha)$: $(\hat{\hat{\cdot}})$	$(\mathbb{P}_\alpha, \mathbb{M}^L_\alpha)$: $(\hat{\cdot})$
$(\mathbb{E}_\alpha, \mathbb{M}^C_\alpha)$: $(\approx)$	$(\mathbb{E}_\alpha, \mathbb{M}^L_\alpha)$: $(\tilde{\cdot})$
$(\mathbb{P}_\alpha, \mathbb{H}_\alpha)$: $(\overset{++}{\cdot})$	$(\mathbb{P}_\alpha, \mathbb{B}_\alpha)$: $(\overset{+}{\cdot})$

Correspondingly, Ψ in (5.154) is the energy functional of the formulation at hand and must be characterized for each of these. The symbols used are also indicated in the above table; clearly, they must also be applied in all coefficients on the right-hand side of (5.154). Recall, further, that $\check{\check{\Psi}}$ and $\hat{\hat{\Psi}}$, for instance, are not the same energy functionals, but are related to each other by Legendre transformations.

We shall now list all the equivalence relations for the various formulations:

(A): The transformations from the $(\check{\cdot})$- to the $(\hat{\cdot})$-formulation have been listed in (5.153) and are labeled with the symbol (A).

(B): The transformations between the constitutive theories $(\hat{\cdot})$ and $(\tilde{\cdot})$ are based on the energy functionals

$$
\hat{\psi} = U - \eta\theta + \frac{\mathbb{M}^L_\alpha}{\rho_o}\mathbb{B}_\alpha , \qquad \tilde{\psi} = U - \eta\theta + \frac{\mathbb{M}^L_\alpha}{\rho_o}\mathbb{B}_\alpha - \frac{\mathbb{P}_\alpha}{\rho_o}\mathbb{E}_\alpha , \tag{5.155}
$$

and equivalence follows if the following identities hold

$$
\begin{aligned}
&{}_2\tilde{\chi}^{(m)} = {}_2\hat{\chi}^{(m)} = \frac{\mu}{\chi^{(m)}}\,, &&{}_2\tilde{\chi}^{(e)} = -\frac{1}{{}_2\hat{\chi}^{(e)}} = -\chi^{(e)}\,,\\
&{}_4\tilde{\chi}^{(m)} = {}_4\hat{\chi}^{(m)}, &&{}_4\tilde{\chi}^{(e)} = (\chi^{(e)})^4\,{}_4\hat{\chi}^{(e)}\,,\\
&\tilde{L}^{(m)} = \hat{L}^{(m)}, &&\tilde{L}^{(e)} = (\chi^{(e)})^2\hat{L}^{(e)}\,,\\
&\tilde{b}^{(m)}_{1,2} = \hat{b}^{(m)}_{1,2}, &&\tilde{b}^{(e)}_{1,2} = (\chi^{(e)})^2\hat{b}^{(e)}_{1,2}\,,\\
&\tilde{c} = \hat{c} = \frac{c_W}{\theta_o}\,, \quad \tilde{\nu} = \hat{\nu} = \nu, && \tilde{\lambda} = \hat{\lambda} = \lambda, \quad \tilde{G} = \hat{G} = G\,.
\end{aligned}
\tag{5.156}
$$

(B)

(C): To relate the $(\tilde{\cdot})$- and $(\overset{+}{\cdot})$- formulations one must start with the energy functionals

$$
\tilde{\psi} = U - \eta\theta + \frac{\mathbb{M}^L_\alpha}{\rho_o}\mathbb{B}_\alpha - \frac{\mathbb{P}_\alpha}{\rho_o}\mathbb{E}_\alpha, \qquad \overset{+}{\psi} = U - \eta\theta\,, \tag{5.157}
$$

and then obtains the equivalence relations:

$$
\begin{aligned}
&{}_2\overset{+}{\chi}^{(m)} = -\frac{1}{{}_2\tilde{\chi}^{(m)}} = -\frac{\chi^{(m)}}{\mu}\,, &&{}_2\overset{+}{\chi}^{(e)} = -\frac{1}{{}_2\tilde{\chi}^{(e)}} = \frac{1}{\chi^{(e)}}\,,\\
&{}_4\overset{+}{\chi}^{(m)} = \left(\frac{\chi^{(m)}}{\mu}\right)^4 {}_4\tilde{\chi}^{(m)}, &&{}_4\overset{+}{\chi}^{(e)} = \left(\frac{1}{\chi^{(e)}}\right)^4 {}_4\tilde{\chi}^{(e)}\,,\\
&\overset{+}{L}^{(m)} = \left(\frac{\chi^{(m)}}{\mu}\right)^2\tilde{L}^{(m)}, &&\overset{+}{L}^{(e)} = \left(\frac{1}{\chi^{(e)}}\right)^2\tilde{L}^{(e)}\,,\\
&\overset{+}{b}^{(m)}_{1,2} = \left(\frac{\chi^{(m)}}{\mu}\right)^2\tilde{b}^{(m)}_{1,2}, &&\overset{+}{b}^{(e)}_{1,2} = \left(\frac{1}{\chi^{(e)}}\right)^2\tilde{b}^{(e)}_{1,2}\,,\\
&\overset{+}{c} = \tilde{c} = \frac{c_W}{\theta_o}\,, \quad \overset{+}{\nu} = \tilde{\nu} = \nu, && \overset{+}{\lambda} = \tilde{\lambda} = \lambda, \quad \overset{+}{G} = \tilde{G} = G\,.
\end{aligned}
\tag{5.158}
$$

(C)

(D): The energy functionals in the $(\overset{+}{\cdot})$- and $(\overset{\vee}{\cdot})$-formulations have already been stated before, namely in (5.133) and $(5.157)^2$. Exploiting the identity

$$
\overset{\vee}{\psi} = \overset{+}{\psi} - \frac{\mathbb{P}_\alpha}{\rho_o}\mathbb{E}_\alpha\,,
$$

thus leads to

$$
\begin{aligned}
&{}_2\overset{\vee}{\chi}{}^{(m)} = {}_2\overset{+}{\chi}{}^{(m)} = -\frac{\mu}{\chi^{(m)}}\,, && {}_2\overset{\vee}{\chi}{}^{(e)} = -\frac{1}{{}_2\overset{+}{\chi}{}^{(e)}} = -\chi^{(e)}\,,\\
&{}_4\overset{\vee}{\chi}{}^{(m)} = {}_4\overset{+}{\chi}{}^{(m)}, && {}_4\overset{\vee}{\chi}{}^{(e)} = (\chi^{(e)})^4\,{}_4\overset{+}{\chi}{}^{(e)}\,,\\
&\overset{\vee}{L}{}^{(m)} = \overset{+}{L}{}^{(m)}, && \overset{\vee}{L}{}^{(e)} = (\chi^{(e)})^2\overset{+}{L}{}^{(e)}\,,\\
&\overset{\vee}{b}{}^{(m)}_{1,2} = \overset{+}{b}{}^{(m)}_{1,2}, && \overset{\vee}{b}{}^{(e)}_{1,2} = (\chi^{(e)})^2\overset{+}{b}{}^{(e)}_{1,2}\,,\\
&\overset{\vee}{c} = \overset{+}{c} = \frac{c_W}{\theta_o}\,, \quad \overset{\vee}{\nu} = \overset{+}{\nu} = \nu, && \overset{\vee}{\lambda} = \overset{+}{\lambda} = \lambda, \quad \overset{\vee}{G} = \overset{+}{G} = G\,.
\end{aligned}
\tag{5.159 (D)}
$$

With these relations the constitutive theories of the models IV and V are all compared. Indeed, performing the transformations (A),(B),(C) and (D) must lead to an identity.

It remains to compare the constitutive theories for the models I,II and III.

(AA): We begin with the transformations for the ($\overset{\vee}{\ }$)- and ($\hat{\ }$)-formulations. In these (see (4.25) and (5.54)) we have

$$
\overset{\vee}{\psi} = U - \eta\theta - \frac{\mathbb{P}_\alpha}{\rho_o}\mathbb{E}_\alpha - \frac{\mu_o \mathbb{M}^C_\alpha}{\rho_o}\mathbb{H}_\alpha, \qquad \hat{\psi} = U - \eta\theta\,. \tag{5.160}
$$

Note, however, that here $U = {}^{II}U$, whereas in all the foregoing relations one must read ${}^{V}U$ for U. Comparing for both formulations the energy functionals of the complexity (5.154) leads to the following relations:

$$
\begin{aligned}
&{}_2\hat{\chi}^{(m)} = -\frac{\mu_o^2}{{}_2\overset{\vee}{\chi}{}^{(m)}} = \frac{\mu_o}{\chi^{(m)}}\,, && {}_2\hat{\chi}^{(e)} = -\frac{1}{{}_2\overset{\vee}{\chi}{}^{(e)}} = \frac{1}{\chi^{(e)}}\,,\\
&{}_4\hat{\chi}^{(m)} = \Big(\frac{1}{\chi^{(m)}}\Big)^4\,{}_4\overset{\vee}{\chi}{}^{(m)}, && {}_4\hat{\chi}^{(e)} = \Big(\frac{1}{\chi^{(e)}}\Big)^4\,{}_4\overset{\vee}{\chi}{}^{(e)}\,,\\
&\hat{L}^{(m)} = \Big(\frac{1}{\chi^{(m)}}\Big)^2\overset{\vee}{L}{}^{(m)}, && \hat{L}^{(e)} = \Big(\frac{1}{\chi^{(e)}}\Big)^2\overset{\vee}{L}{}^{(e)}\,,\\
&\hat{b}^{(m)}_{1,2} = \Big(\frac{1}{\chi^{(m)}}\Big)^2\overset{\vee}{b}{}^{(m)}_{1,2}, && \hat{b}^{(e)}_{1,2} = \Big(\frac{1}{\chi^{(e)}}\Big)^2\overset{\vee}{b}{}^{(e)}_{1,2}\,,\\
&\hat{c} = \overset{\vee}{c} = \frac{c_W}{\theta_o}\,, \quad \hat{\nu} = \overset{\vee}{\nu} = \nu, && \hat{\lambda} = \overset{\vee}{\lambda} = \lambda, \quad \hat{G} = \overset{\vee}{G} = G\,.
\end{aligned}
\tag{5.161 (AA)}
$$

(BB),(CC),(DD): It is now obvious how the transformations must look like for the remaining formulations. For the electrical coefficients the transformations (AA),(BB),(CC),(DD) are identical to the corresponding ones indicated with a single letter, whereas for the magnetic coefficients we must only replace the factor $(\mu/\chi^{(m)})$ in the one-letter transformations by $(1/\chi^{(m)})$ to obtain the double-letter transformations.

What remains is then the transformation of one formulation out of the group (IV,V) to one of (I,II,III). This problem is already solved in Section 4.7 in which the transformation from the $(\check{\cdot})$- to the $(\check{\check{\cdot}})$-formulation is discussed. (There, $\check{\check{\Psi}}$ was denoted by $\bar{\Psi}$.) For completeness we shall repeat the results here. We shall characterize this transformation, which makes the link between the two distinct groups (IV,V) and (I,II,III) and, therefore, completes the comparison of all possible constitutive theories, with the symbol($\mathbb{A}$):

$$
\begin{aligned}
&{}_2\check{\check{\chi}}^{(m)} = \mu_o\mu\,{}_2\check{\chi}^{(m)} = -\mu_o\chi^{(m)}, &&{}_2\check{\check{\chi}}^{(e)} = {}_2\check{\chi}^{(e)} = -\chi^{(e)}, \\
&{}_2\check{\check{\chi}}^{(m)} = \mu^4\,{}_4\check{\chi}^{(m)}, &&{}_4\check{\check{\chi}}^{(e)} = {}_4\check{\chi}^{(e)}, \\
&\check{\check{L}}^{(m)} = \mu^2\check{L}^{(m)}, &&\check{\check{L}}^{(e)} = \check{L}^{(e)}, \\
&\check{\check{b}}^{(m)}_{\alpha\beta\gamma\delta} = \mu^2\check{b}^{(m)}_{\alpha\beta\gamma\delta} + \mu_o\chi^{(m)}(2+\chi^{(m)})n_{\alpha\beta\gamma\delta}, &&\check{\check{b}}^{(e)}_{\alpha\beta\gamma\delta} = \check{b}^{(e)}_{\alpha\beta\gamma\delta}, \\
&\check{\check{c}} = \check{c} = \frac{c_W}{\theta_o}, \quad \check{\check{\nu}} = \check{\nu} = \nu, \quad \check{\check{\lambda}} = \check{\lambda} = \lambda, \quad \check{\check{G}} = \check{G} = G .
\end{aligned}
\tag{5.162}\ (\mathbb{A})
$$

This completes the explicit comparison of all possible constitutive theories in the formulations (IV,V) and (I,II,III) as far as the free energy goes. A complete comparison must, however, also include the constitutive relations for the electric current and the heat flux vector. They must in all formulations be the same functions. For relationships of the complexity of Section 5.2.3 this means that relations must hold such that (see (5.75))

$$
\Lambda_{\alpha\beta} = \check{\Lambda}_{\alpha\beta}(E_{\gamma\delta},\mathbb{B}_\gamma,\theta) = \hat{\Lambda}(E_{\gamma\delta},\frac{M^L_\gamma}{\rho_o},\theta) = \hat{\Lambda}(E_{\gamma\delta},-\frac{\partial\check{\check{\Psi}}}{\partial\mathbb{B}_\gamma},\theta) . \tag{5.163}
$$

These relations can be elaborated in an analogous way as above and thus lead to relations between coefficients as, for instance, $\check{\Lambda}^{(d)}_{\alpha\beta\gamma\delta}$ and $\hat{\Lambda}^{(d)}_{\alpha\beta\gamma\delta}$ etc. These are, however, at most second order effects and for most practical problems linear relationships of the form

$$
\begin{aligned}
&Q_\alpha\ (= Q^S_\alpha) = -\kappa_{\alpha\beta}\theta_{,\beta} + \beta_{\alpha\beta}\mathbb{E}_\beta , \\
&J_\alpha = \frac{1}{\theta}\beta_{\beta\alpha}\theta_{,\beta} + \sigma_{\alpha\beta}\mathbb{E}_\beta ,
\end{aligned}
\tag{5.164}
$$

suffice. In (5.164), now $\kappa_{\alpha\beta}$, $\beta_{\alpha\beta}$ and $\sigma_{\alpha\beta}$ are constants which correspond to the tensor of heat conduction, the tensor of electrical conductivity and the tensor of the thermoelectric effect. Since the temperature gradient and the Lagrangian electromagnetic field are the same variables in all formulations, it suffices for a full equivalence of all theories to choose for $\kappa_{\alpha\beta}$, $\beta_{\alpha\beta}$, $\sigma_{\alpha\beta}$ the same numerical values.

5.5 DISCUSSION

In this Chapter, the governing equations of thermoelastic polarizable and magnetizable solids were subjected to a decomposition procedure, which allowed a separation of the general problem, described mathematically by a set of nonlinear partial differential equations, into two simpler ones. One of these involves as unknowns only small quantities, so that all products of these could justly be omitted. The emerging set of equations was linearized this way. We saw that linearization by itself turned out to be a fairly complex problem if not attacked with the proper formulation. The Lagrangian description treated in Chapter 4 provided the appropriate vehicle to avoid all complexities which arose in the Eulerian description.

The decomposition of the govering field equations into equations valid for the intermediate state and the perturbation state was performed for the Lorentz model only, because all theories were proved to be equivalent. Hence, and if one so desires, all problems can be solved just with this formulation. Yet, because all formulations are used in the current literature, the transformations of the phenomenological parameters from one theory to another one should be known, if the link between all these theories should explicitly be possible. With the results derived in Sections 4.7 and 5.4 this can easily be achieved for the restricted constitutive class investigated there. All necessary informations to perform such transformations are contained in Figure 5.1, in which the two circles stand for the models (I,II,III) and (IV,V), respectively. At the periphery, we show those electromagnetic field variables, which are considered as independent electromagnetic fields of the constitutive theory. The symbols used in these constitutive theories are also indicated in the figure. In Section 5.4 we derived the transformation rules necessary for the respective constitutive formulations to lead to equivalent electromagnetic theories. The transformations as listed in Section 5.4 are indicated in Figure 5.1 by an arrow on the periphery connecting the two formulations they apply to. The arrow is also characterized by a symbol (A),(B),...,(DD), indicating the transformations rules derived in Section 5.4 and valid for the transfer of a formulation into the neightboring

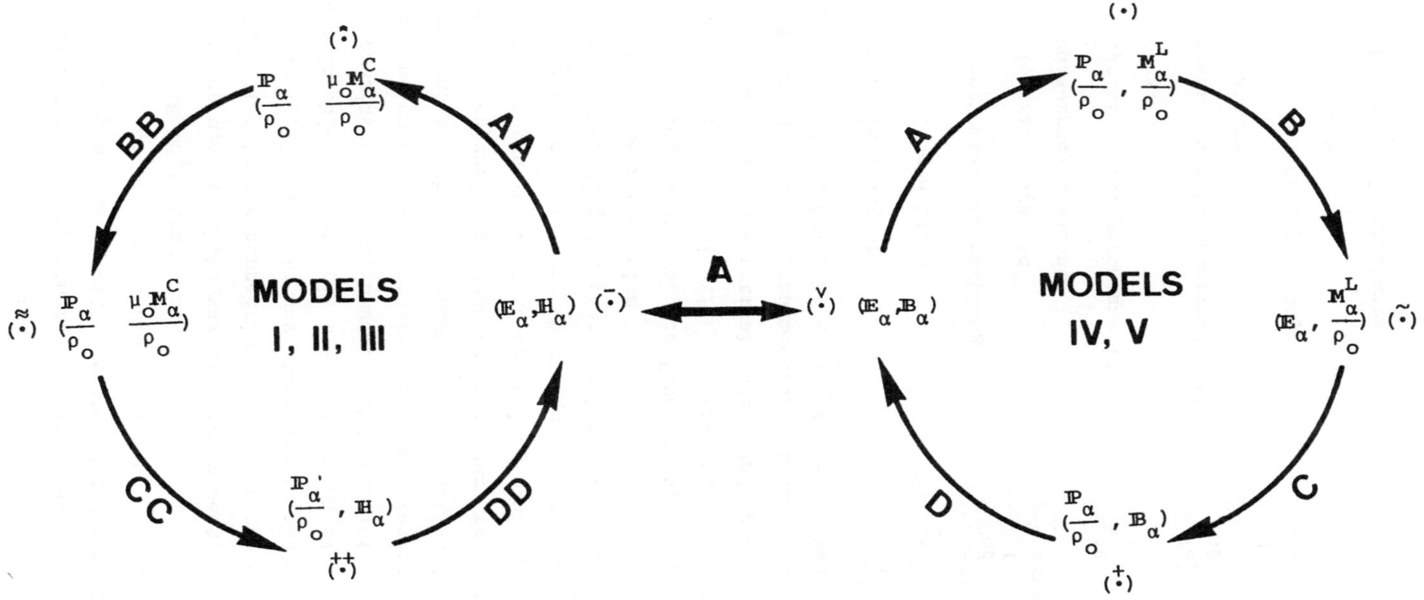

Figure 5.1. Scheme for the transformation of a constitutive formulation of any model into any other one.

one on the periphery of the circle. It is evident that to every constitutive theory within a circle any other one corresponds to. The missing piece to obtain full correspondence of all presented formulations is the double arrow connecting both circles and characterized by the symbol ($\mathbb{A}$). The transformation from the $(\mathbb{E}_\alpha,\mathbb{H}_\alpha)$-formulation in the models (I,II,III) to the $(\mathbb{E}_\alpha,\mathbb{B}_\alpha)$-formulation in the models (IV,V) was derived in Section 4.7 and recapitulated in Section 5.4. It allows a connection between anyone of the constitutive theories in the models (I,II,III) with any other one in the models (IV,V). With this result the main objective of this tractate is achieved.

We must emphasize once more, however, that conditions of equivalence were derived above for free energy functionals, which were regarded to be truncated polynomial expansions and that equivalence was established to within terms omitted in the respective expansion procedures. Moreover, for reasons of transparency in the presentation and in order to avoid long manipulations, we restricted ourselves to isotropic solids for which equivalence conditions turned out to be simple formulas relating the phenomenological coefficients. Of course, the calculations can be performed also for an anisotropic solid, but they are very long and, furthermore, equivalence formulas are complex.

Apart from its practical significance the study of the equivalence conditions has led us to a deeper theoretical understanding at least in the following regard: It is known that the phenomenological constants of the free energy as given in (5.147) and (5.148) bear standard names. These are as follows:

magnetic constants	${}_2\chi^{(m)}_{\alpha\beta}$
electric constants	${}_2\chi^{(e)}_{\alpha\beta}$
electromagnetic coupling	$\chi^{(em)}_{\alpha\beta}$
electric, magnetic anisotropy constants	${}_4\chi^{(e)}_{\alpha\beta\gamma\delta}$, ${}_4\chi^{(m)}_{\alpha\beta\gamma\delta}$
pyroelectric, pyromagnetic constants	$\lambda^{(e)}_\alpha$, $\lambda^{(m)}_\alpha$
thermoelectric, thermomagnetic constants	$L^{(e)}_{\alpha\beta}$, $L^{(m)}_{\alpha\beta}$
piezoelectric, piezomagnetic constants	$\varepsilon^{(e)}_{\alpha\beta\gamma}$, $\varepsilon^{(m)}_{\alpha\beta\gamma}$
thermoelastic constants	$\nu_{\alpha\beta}$
thermal constant	c
elastic constants	$c_{\alpha\beta\gamma\delta}$

The association of names with a certain phenomenological constant is not unique, however. In particular, we know that dependent on the constitutive approach the magnetic constants are called magnetic permeability or magnetic susceptibility. Moreover, we saw that for an isotropic body, in which there are no electromagnetic coupling terms and no pyro- and piezoelectric and -magnetic effects, only the last three constants in the above list turned out to remain the same in all formulations. The transformation rules of all other constants involved the magnetic or the electric constants so that the definition of a particular effect depended on the choice of the constitutive variables and on the formulation. For anisotropic bodies the situation becomes even worse, and indeed, one can show that in a body, in which the third order constants $e_{\alpha\beta\gamma}$ do not all vanish, not even the elastic coefficients are unique. Similarly, in a material with nonvanishing λ_α's thermoelastic constants and the specific heats are not uniquely determinable either. Theoretically, these results are very important ones, because they say, for instance, that in an anisotropic piezomagnetic body the elasticity coefficients depend on the formulation used. Strictly this means that in such a material the elasticity coefficients cannot be determined from an experiment in the absence of electromagnetic fields. Practically, on the other hand, the corrections of the field free elasticity coefficients are so small that the effect of the piezomagnetic constants on them can always be neglected.
The constitutive theories developed in this monograph apply to all bodies which are called magnetizable and polarizable thermoelastic solids. By setting the appropriate phenomenological constants to zero, all special cases of it can also be derived. If, for instance, all coefficients bearing the superscript (m) vanish, the body is called polarizable-only or dielectric. If, on the other hand, all coefficients with a superscript (e) vanish, the body is magnetizable-only. This definition is independent of the fact whether electrical conduction is present or not. Often, dielectric substances are electric insulators, however.

Before we close, we would like to draw the readers attention on a few practical problems the theories presented in this book and in particular the decomposition procedure of this tractate may be applied to. An extensive monograph on such problems is F.C. Moon's "*Problems in Magneto-Solid Mechanics*", [18], so that it suffices if we point out some of the more interesting problems. In principle the problems can be classified into two groups: firstly dynamical problems such as magnetoelastic wave propagation and vibration problems, and secondly magnetically and electrically induced bifurcation problems. Both classes of problems have been attacked by several authors already, but so far this has only be done in the Eulerian description. Early applications of the vibration and wave propagation are by Dunkin and Eringen [19], Kaliski and Petykiewicz [20], Paria [21], Alers and Fleary [22] and M.F. McCarthy [23]. The linearization procedure of these authors is, however, only sketchy and, as far as dynamical equations go the theories are quasi-static. A first attempt to investigate the wave propagation problem with a linearized theory which proceeds along the lines of this article is due to Hutter, [3], [4]. But his equations are based on those of Hutter and Pao [8], which have been found not to be completely correct for the reasons explained in Section 5.1. Hence, the influence of (polarization and) magnetization on the propagation of waves in solids should still further be attacked, and experiments should be performed, which would corroborate or disprove the theoretical predictions. This has already be done to a certain extent by Moon and Chattopadhyay [24], but an extensive comparison with the results in [3], [4] was not attempted.

Of similar nature are the vibration problems treated by Tiersten [1] and van de Ven [2]. Both authors use a linearization procedure equivalent to the one presented above, but using the Eulerian description and only for a quasi-static theory (in [9], however, it was shown that the results for a dynamic but non-relativistic theory correspond to those of the quasi-static case). Tiersten determines the eigenmodes of magnetically saturated plates of cubic Yttrium-Iron-Garnet in a large static transverse magnetic field, perpendicular to which a small time-periodic field is superimposed on. Van de Ven, on the other hand, discussed the vibrations of circular cylinders.

First applications in the bifurcation of electro- and magnetoelasticity trace back to 1967 when Moon and Pao [5] presented experimental and first theoretical results on the buckling of a soft ferromagnetic plate in a homogeneous magnetic field. The discrepancies of the theory and of the experimental evidence initiated further work, notably by Wallerstein and Peach [25], Popelar [26], Dalrymple, Peach and Vliegelahn [27], Pao and Yeh [11] and Alblas [10], [28], but with the exception of Refs [11], [10], [28] these articles do not make use of a proper linearization scheme. Alblas [28], [10], includes in his analysis the buckling of circular rods based on a Liapunov approach. Since in all of the aforementioned articles the plate boundary conditions were not derived consistently, the entire matter was reinvestigated by van de Ven [6], who also compared the existing formulations of static magnetoelasticity, in particular the Maxwell-Minkowski and the Ampère-current model, which in the static theory agrees with the Lorentz model. It turns out that, ultimately all models lead to the same buckling values, as they must; yet, the reasons for this buckling can in each case be interpreted differently:

i) In the Maxwell-Minkowski formulation buckling is due to distributed magnetic surface forces at the upper and lower surface of the plate and a shear force per unit length of magnetic origin at the boundary of the plate;

ii) in the Ampèrian-current model, on the other hand, buckling originates from distributed surface moments of magnetic origin at the upper and lower surface of the plate; the boundary is now free of forces.

The difference in the loading could, of course, be traced back to differences in the constitutive equations for the shear forces and bending moments in the plate.

In spite of the increased refinements in the formulations of the buckling of clamped rectangular plates the large discrepancies between theoretical prediction and experimental evidence was not removed; improvements were only very small. The causes for this fact are not certain, and the problem is still an open one, but there are indications that edge effects might alter the results correspondingly.

Other interesting bifurcation problems are the buckling of superconducting rings and coils which carry a large electric current. A first step towards a solution was done by Moon and Chattopadhyay [29], but further investigations are still needed to improve their solution. A very recent contribution to this field of research is due to Alblas and Grysa [30]. These problems are of extreme practical interest in future reactor technology.

REFERENCES

[1] Tiersten, H.F., *Thickness Vibrations of Saturated Magnetoelastic Plates*, J.A.P. 36 (1965), 2250-2259.

[2] Ven, A.A.F. van de, *On the Vibrations of a Nonconducting Magnetically Saturated Cylinder in a Magnetic Field*, Proc. of Vibr. Probl. 11 (1970), 89-102.

[3] Hutter, K., *Wave Propagation and Attenuation in Paramagnetic and Soft Ferromagnetic Materials*, Int. J. Eng. Sc. 13 (1975), 1067-1084.

[4] Hutter, K., *Wave Propagation and Attenuation in Paramagnetic and Soft Ferromagnetic Materials, Part II*, Int. J. Eng. Sc. 14 (1976), 883-894.

[5] Moon, F.C. and Y.H. Pao, *Magnetoelastic Buckling of a Thin Plate*, J.A.M. 35 (1968), 53-58.

[6] Ven, A.A.F. van de, *Magnetoelastic Buckling of Thin Plates in a Uniform Transverse Magnetic Field,* J. of Elasticity, 8 (1978), 297-312.

[7] Toupin, R.A., *A Dynamical Theory of Elastic Dielectrics,* Int. J. Eng. Sc., 1 (1963), 101-126.

[8] Hutter, K. and Y.H. Pao, *A Dynamic Theory for Magnetizable Elastic Solids with Thermal and Electrical Conduction,* J. of Elasticity, 4 (1974), 89-114.

[9] Ven, A.A.F. van de, *Interaction of Electromagnetic and Elastic Fields in Solids,* Dr. of Science Thesis, University of Technology Eindhoven, The Netherlands, 1975.

[10] Alblas, J.B., *General Theory of Electro- and Magneto-Elasticity,* in Electromagnetic Interactions in Elastic Solids, ed. by H. Parkus, Springer, Wien, 1978.

[11] Pao, Y.H. and C.S. Yeh, *A Linear Theory of Soft Ferromagnetic Elastic Solids,* Int. J. of Eng. Sc., 11 (1973), 415-436.

[12] Tiersten, H.F., *Coupled Magnetomechanical Equations for Magnetically Saturated Insulators,* J. of Math. Phys., 5 (1964), 1298-1318.

[13] Baumhauer, J.C. and H.F. Tiersten, *Nonlinear Electroelastic Equations for Small Fields Superposed on a Bias,* J. Acoust. Soc. Am., 54 (1973), 1017-1025.

[14] Jordan, N.F. and A.C. Eringen, *On the Static Nonlinear Theory of Electromagnetic Thermoelastic Solids,* Int. J. Eng. Sc., 2 (1964), 59-114.

[15] Pipkin, A.S. and R.S. Rivlin, *Electrical Conduction in Deformed Isotropic Materials,* J. Math. Phys., 1 (1960), 127-130.

[16] Pipkin, A.S. and R.S. Rivlin, *Galvanomagnetic and Thermomagnetic Effects in Isotropic Materials,* J. Math. Phys., 1 (1960). 542-546.

[17] Eringen, A.C. and E.S. Suhubi, *Elastodynamics,* Vol. I, Academic Press, New York and London, 1974.

[18] Moon, F.C., *Problems in Magneto-Solid Mechanics,* Mechanics Today, Vol. 4, 307-390, ed. S. Nemat-Nasser, Pergamon Press Inc., New York, 1978.

[19] Dunkin, J.W. and A.C. Eringen, *Propagation Of Waves in an Electromagnetic Elastic Solid,* Int. J. Eng. Sc., 1 (1963), 461-495.

[20] Kaliski, S. and J. Petykiewicz, *Dynamical Equations of Motion and Solving Functions for Elastic and Inelastic Anisotropic Bodies in the Magnetic Field,* Proc. of Vibr. Probl. 1 (1959/1960), 17-35.

[21] Paria, G., *Magneto-Elasticity and Magneto-Thermoelasticity,* Adv. in Appl. Mech., 10 (1967), 73-112.

[22] Alers, G.A. and P.A. Fleary, *Modification of the Velocity of Sound in Metals by Magnetic Fields,* Phys. Rev., 129 (1963), 2425.

[23] McCarthy, M.F., *The Propagation and Growth of Plane Acceleration Waves in a Perfectly Electrically Conducting Elastic Material in a Magnetic Field,* Int. J. Eng. Sc. 4 (1966), 361-381.

[24] Moon, F.C. and S. Chattopadhyay, *Magnetically Induced Stress Waves in a Conducting Solid-Theory and Experiment,* J.A.M., 41 (1974), 641-645.

[25] Wallerstein, D.V. and M.O. Peach, *Magnetoelastic Buckling of Beams and Thin Plates of Magnetically Soft Material,* J.A.M., 39 (1972), 451-455.

[26] Popelar, C.H., *Postbuckling Analysis of a Magnetoelastic Beam,* J.A.M., 39 (1972), 207-211.

[27] Dalrymple, J.M., M.O. Peach and G.L. Vliegelahn, *Magnetoelastic Buckling of Thin Magnetically Soft Plates in Cylindrical Mode,* J.A.M., 41 (1974), 145-150.

[28] Alblas, J.B., *A General Theory of Magneto-Elastic Stability,* to appear in Vekua 70-th Anniversary Volume, 1977.

[29] Moon, F.C., and S. Chattopadhyay, *Elastic Stability of a Thermonuclear Reactor Coil,* Proc. 5th Symposium on Engineering Problems of Fusion Research, November 1973, Princeton, N.J. IEEE Nuclear and Plasma Sci. Soc., N.Y., Publ. No. 73, CH 0843-3-NPS, (April 1974), 544-578.

[30] Alblas, J.B. and K.W. Grysa, *Magneto-Elastic Stability of an Unconstrained Assembly of Coils,* September 1978, to appear in Archives of Mechanics.

APPENDIX A : ON OBJECTIVITY

In this Appendix we briefly state, how one can show what transformation properties the various electromagnetic field variables introduced in the main body of this article enjoy. To this end, we shall use three and four dimensional notation. Let x^A be the (contravariant) four vector (x_i,t), consisting of the position vector x_i $(i = 1,2,3)$ of a particle and time t, and let $x^{*A} = x^A(x^B)$ be any C^1-transformation $(x_i,t) \rightarrow (x_i^*,t^*)$. A covariant four tensor Ψ_{AB} and a contravariant four tensor Ψ^{AB} are then quantities, which under such transformations transform according to

$$\text{(A.1)} \qquad \Psi^*_{AB} = \frac{\partial x^C}{\partial x^{*A}}\frac{\partial x^D}{\partial x^{*B}}\Psi_{CD}, \qquad \Psi^{*AB} = \frac{\partial x^{*A}}{\partial x^C}\frac{\partial x^{*B}}{\partial x^D}\Psi^{CD}$$

Likewise, a contravariant four vector transforms under general transformations $(x_i,t) \rightarrow (x_i^*,t^*)$ according to

$$\text{(A.2)} \qquad \sigma^{*A} = \frac{\partial x^{*A}}{\partial x^B}\sigma^B .$$

Of special interest to us are <u>Euclidean</u> transformations given by

$$\text{(A.3)} \qquad x_i^* = 0_{ij}(t)x_j + c_i(t), \qquad x_i = 0_{ji}(t)(x_j^* - c_j(t)),$$
$$t^* = t, \qquad t = t^*$$

and in what follows we would like to explore some consequences implied by them.

i) Let

$$\sigma^A = (J_i,\mathcal{Q})$$

be a contravariant vector. Then a routine calculation shows that under Euclidean transformations

$$\text{(A.4)} \qquad \sigma^{*A} = (J_i^*,\mathcal{Q}^*) = (0_{ij}(J_j - \dot{x}_j\mathcal{Q}) + \mathcal{Q}(\dot{x}_i^*),\mathcal{Q}^*) ,$$

or

$$\text{(A.4a)} \qquad (J_i - \mathcal{Q}\dot{x}_i)^* = 0_{ij}(t)(J_j - \mathcal{Q}\dot{x}_j),$$
$$\mathcal{Q}^* = \mathcal{Q} .$$

In other words, $\mathcal{Q}$ and $(J_i - \mathcal{Q}\dot{x}_i) = \mathcal{J}_i$ transform under Euclidean transformations as an objective scalar and an objective vector, respectively.

ii) Let Ψ_{AB} be a skew-symmetric covariant four-tensor with the components

$$\text{(A.5)} \qquad \Psi_{AB} = \begin{pmatrix} 0 & b_3 & -b_2 & e_1 \\ & 0 & b_1 & e_2 \\ (-) & & 0 & e_3 \\ & & & 0 \end{pmatrix}$$

(we choose to name these components b_i and e_i for suggestive reasons later-on). Performing a Euclidean transformation (A.3) shows that the three vectors $b_i := (b_1,b_2,b_3)$ and $e_i := (e_1,e_2,e_3)$ transforms as follows:

$$\text{(A.6)} \qquad \begin{aligned} b_i^* &= \det(0)0_{ik}b_k \, , \\ e_i^* + e_{ijk}(\dot{x}_j^*)b_k^* &= 0_{ij}(t)(e_j + e_{jk\ell}\dot{x}_k b_\ell) \, , \end{aligned}$$

or otherwise stated, b_i is an objective axial vector and $(e_i + e_{ijk}\dot{x}_j b_k)$ an objective vector under the Euclidean transformation group. To prove $(A.6)^1$ for instance, note that

$$\Psi_{ij}^* = 0_{ik}0_{j\ell}\Psi_{k\ell} \quad , \quad \text{where } \Psi_{k\ell} = e_{k\ell m}b_m \, ,$$

where e_{ijk} is the usual three dimensional permutation tensor. In much the same, though more complicated way one can also prove that

$$e_i^* = 0_{i\ell}\{e_\ell + e_{\ell mn}[\dot{0}_{jm}(x_j^* - c_j) - 0_{jm}\dot{c}_j]b_n\} \, ,$$

which with the aid of the identity

$$\dot{0}_{jm}(x_j^* - c_j) - 0_{jm}\dot{c}_j = \dot{x}_m - 0_{im}(\dot{x_i^*})$$

immediately implies (A.6).

A special application of (A.5) is the tensor whose components are $b_i := m_i$, $e_i = 0$. Then m_i must be an objective axial tensor under the Euclidean transformation group.

iii) Another covariant skew symmetric tensor of importance is

$$\text{(A.7)} \qquad \Psi_{AB} = \begin{pmatrix} 0 & m_3 & -m_2 & (\underline{m} \times \dot{\underline{x}})_1 \\ & 0 & m_1 & (\underline{m} \times \dot{\underline{x}})_2 \\ (-) & & 0 & (\underline{m} \times \dot{\underline{x}})_3 \\ & & & 0 \end{pmatrix}$$

and it can be shown by a straightforward calculation, that the three-vector $m_i := (m_1, m_2, m_3)$ is an objective axial vector under Euclidean transformations. (This is just a special case of ii).)

iv) Let Ψ^{AB} be a skew symmetric contravariant four-tensor with the components

$$\text{(A.8)} \qquad \Psi^{AB} = \begin{pmatrix} 0 & h_3 & -h_2 & -d_1 \\ & 0 & h_1 & -d_2 \\ (-) & & 0 & -d_3 \\ & & & 0 \end{pmatrix} .$$

A calculation identical to the one performed above shows that the vectors $d_i := (d_1, d_2, d_3)$ and $h_1 := (h_1, h_2, h_3)$ transform under the Euclidean group as

$$\text{(A.9)} \qquad \begin{aligned} d_i^* &= O_{ij} d_j , \\ h_i^* - e_{ijk} (\dot{x}_j^*) d_k^* &= \det(O) O_{ij} (h_j - e_{jk\ell} \dot{x}_k d_\ell) . \end{aligned}$$

Hence d_i and $(h_i - e_{ijk} \dot{x}_j d_k)$ are an objective vector and an objective axial vector under Euclidean transformations.

A special situation is again the case for which

$$h_i = m_i \quad \text{and} \quad d_i = 0$$

which immediately shows that m_i must be an objective axial vector.

v) As a last example, consider the contravariant skew-symmetric tensor

$$\text{(A.10)} \qquad \Psi^{AB} = \begin{pmatrix} 0 & (\underline{p} \times \underline{\dot{x}})_3 & -(\underline{p} \times \underline{\dot{x}})_2 & p_1 \\ & 0 & (\underline{p} \times \underline{\dot{x}})_1 & p_2 \\ & & 0 & p_3 \\ (-) & & & 0 \end{pmatrix}$$

Its transformation properties are most easily found in two steps. Firstly, we write

$$(\Psi^*)^{k4} = \frac{\partial x^{*k}}{\partial x^\ell} \frac{\partial x^{*4}}{\partial x^4} \Psi^{\ell 4}$$

and obtain with the aid of (A.10) and $(A.3)^1$,

$$\text{(A.11)} \qquad p_i^* = O_{ij} p_j$$

proving that p_i is an objective vector. On the other hand

$$(\Psi^*)^{ij} = \dot{p}_i^*(x_j^*) - \dot{p}_j^*(x_i^*) = \frac{\partial x^{*i}}{\partial x^k}\frac{\partial x^{*j}}{\partial x^\ell}\Psi^{k\ell} + \frac{\partial x^{*i}}{\partial x^4}\frac{\partial x^{*j}}{\partial x^\ell}\Psi^{4\ell} + \frac{\partial x^{*i}}{\partial x^k}\frac{\partial x^{*j}}{\partial x^4}\Psi^{k4}$$

and it is now an easy matter, using (A.10) and (A.3) to show that the expression on the far right and in the middle of this chain are the same if p_i is assumed to obey (A.11).

It is shown in theoretical electrodynamics that the Maxwell equations of deformable continua can be written in the form

$$\text{(A.12)} \qquad e^{ABCD}\frac{\partial \varphi_{CD}}{\partial x^B} = 0 \quad \text{and} \quad \frac{\partial \eta^{AB}}{\partial x^B} = \sigma^A ,$$

where φ_{CD} and η^{CD} are skew symmetric covariant and contravariant four-tensors, respectively and where σ^A is a contravariant vector. Furthermore, e_{ABCD} is the four-dimensional permutation tensor, which is antisymmetric with respect to any interchange of two indices and vanishes if any two indices are the same. Moreover, $e_{1234} = 1$, and lowering and rising of indices is achieved by the use of the metric tensor $g_{AB} = g^{AB}$, whose matrix is given by

$$\text{(A.13)} \qquad g_{AB} = \begin{vmatrix} 1 & 0 & 0 & 0 \\ 0 & 1 & 0 & 0 \\ 0 & 0 & 1 & 0 \\ 0 & 0 & 0 & -1 \end{vmatrix} .$$

Hence $e^{1234} = -1$, since $e_{1234} = +1$.

The equations (A.12) are general and hold in vacuo as well as in matter, but the contribution of matter is usually separated from that of vacuo; this separation is achieved by writing

$$\text{(A.14)} \qquad \varphi_{AB} = \Phi_{AB} - \mu_{AB}, \qquad \eta^{AB} = H^{AB} - \pi^{AB} ,$$

where μ^{AB} and π^{AB} are a covariant and a contravariant skew symmetric four tensor, respectively, which vanish in vacuo. Thus Φ_{AB} and H^{AB} are the vacuum fields. Note that in view of the transformation properties explained under i)-v) for general skew symmetric tensors, there will be no need to derive such properties for φ_{AB}, Φ_{AB}, μ_{AB}, η^{AB}, H^{AB} and π^{AB} anew. Before we list these tensors in the various descriptions, recall that the vacuum-fields Φ_{AB} and H^{AB} are related to each other through the equation

$$\text{(A.15)} \qquad \Phi_{AB} = \hat{\Phi}_{AB}(H^{CD}) ,$$

a relation which is sometimes called the Maxwell-Lorentz-aether relation. We shall see that it is not invariant under the general transformation $(x_i,t) \to (x_i^*,t^*)$.

We now list the various formulations and give the invariance properties their variables enjoy.

a) The Minkowski formulation. In this formulation one chooses $\mu_{AB} = 0$ and does not separate η^{AB} into two parts. Thus

$$\text{(A.16)} \qquad \varphi_{AB} = \begin{pmatrix} 0 & B_3 & -B_2 & E_1 \\ & 0 & B_1 & E_2 \\ (-) & & 0 & E_3 \\ & & & 0 \end{pmatrix}, \quad \eta^{AB} = \begin{pmatrix} 0 & H_3 & -H_2 & -D_1 \\ & 0 & H_1 & -D_2 \\ (-) & & 0 & -D_3 \\ & & & 0 \end{pmatrix}.$$

Hence, because of the properties ii) and iv) of skew symmetric tensors, we have under the Euclidean transformation group

D_i, $\mathcal{E}_i := E_i + e_{ijk}\dot{x}_j B_k$, transform as objective vectors ,

B_i, $\mathcal{H}_i := H_i - e_{ijk}\dot{x}_j D_k$, transform as objective axial vectors.

It is also a routine matter to show that (A.12) agrees with (2.35). Finally, the Maxwell Lorentz aether relation is formally introduced by writing $\eta^{AB} = H^{AB} - \pi^{AB}$ with

$$\text{(A.17)} \qquad H^{AB} = \begin{pmatrix} 0 & H_3^a & -H_2^a & -D_1^a \\ & 0 & H_1^a & -D_2^a \\ (-) & & 0 & -D_3^a \\ & & & 0 \end{pmatrix}, \quad \pi^{AB} = \begin{pmatrix} 0 & -M_3 & M_2 & P_1 \\ & 0 & -M_1 & P_2 \\ (-) & & 0 & P_3 \\ & & & 0 \end{pmatrix}$$

where H_i^a and D_i^a are auxiliary fields and M_i and P_i are the Minkowskian magnetization and polarization. Again, in view of the items ii) and iv) above

D_i^a, P_i , transform as objective vectors

(A.18) $\left.\begin{aligned} \mathcal{M}_i &:= M_i - e_{ijk}\dot{x}_j P_k \\ \mathcal{H}_i^a &:= H_i^a - e_{ijk}\dot{x}_j D_k^a \end{aligned}\right\}$ transform as objective axial vectors

under the Euclidean transformation group. Now, the Maxwell-Lorentz aether relation (A.15) is given by

$$\text{(A.19)} \qquad H_i^a = \frac{1}{\mu_o} B_i \qquad \text{and} \qquad D_i^a = \varepsilon_o E_i ,$$

and it is trivial to show that these are <u>not</u> invariant under Euclidean transformations.

b) The Statistical Model. Except for notation this model is identical with the Minkowski model. Hence, under Euclidean transformations

$$\text{(A.20)}\qquad \begin{array}{ll} D_i^a,\ E_i,\ P_i, & \text{transform as objective vectors,} \\ M_i,\ B_i,\ H_i, & \text{transform as objective axial vectors.} \end{array}$$

c) The Lorentz formulation. In this formulation one sets $\mu_{AB} = 0$, as was done in the previous ones. Furthermore, H^{AB} is given as in (A.17), but

$$\text{(A.21)}\qquad \pi^{AB} = \begin{pmatrix} 0 & -M_3^L + (\underline{P}^L \times \underline{\dot{x}})_3 & M_2^L - (\underline{P}^L \times \underline{\dot{x}})_2 & P_1^L \\ & 0 & -M_1^L + (\underline{P}^L \times \underline{\dot{x}})_1 & P_2^L \\ & & 0 & P_3^L \\ (-) & & & 0 \end{pmatrix}.$$

This tensor can easily be written as the sum of two tensors, one containing the polarization, the other containing the magnetization only. From iv) and v) and the previous results it then follows that

$$\begin{array}{ll} D_i^a,\ E_i,\ P_i^L = P_i\ , & \text{transform as objective vectors,} \\ M_i^L = M_i,\ B_i,\ H_i^a, & \text{transform as objective axial vectors,} \end{array}$$

under the Euclidean transformation group.

d) The Two-Dipole Model (Chu-formulation). This model is the only one with non-vanishing μ_{AB}. Indeed,

$$\text{(A.22)}\qquad \begin{array}{l} \Phi_{AB} = \begin{pmatrix} 0 & B_3^a & -B_2^a & E_1^C \\ & 0 & B_1^a & E_2^C \\ & & 0 & E_3^C \\ (-) & & & 0 \end{pmatrix},\quad \mu_{AB} = \begin{pmatrix} 0 & -M_3^C & M_2^C & (\underline{M}^C \times \underline{\dot{x}})_1 \\ & 0 & -M_1^C & (\underline{M}^C \times \underline{\dot{x}})_2 \\ (-) & & 0 & (\underline{M}^C \times \underline{\dot{x}})_3 \\ & & & 0 \end{pmatrix} \\[2ex] \eta^{AB} = \begin{pmatrix} 0 & H_3^C & -H_2^C & -D_1^a \\ & 0 & H_1^C & -D_2^a \\ (-) & & 0 & -D_3^a \\ & & & 0 \end{pmatrix},\quad \pi^{AB} = \begin{pmatrix} 0 & (\underline{P}^C \times \underline{\dot{x}})_3 & -(\underline{P}^C \times \underline{\dot{x}})_2 & P_1^C \\ & 0 & (\underline{P}^C \times \underline{\dot{x}})_1 & P_2^C \\ & & 0 & P_3^C \\ & (-) & & 0 \end{pmatrix}. \end{array}$$

Here, all variables are the Chu-variables and

$$B_i^a = \mu_o H_i^C \qquad \text{and} \qquad D_i^a = \varepsilon_o E_i^C$$

and the Maxwell-Lorentz aether relations. It follows from ii)-v) above that under Euclidean transformations

$$D_i^a,\ \mathcal{E}_i := E_i + e_{ijk}\dot{x}_j B_k^a,\ P_i^C, \quad \text{transform as objective vectors,}$$

$$B_i^a,\ \mathcal{H}_i := H_i^C - e_{ijk}\dot{x}_j D_k^a,\ M_i^C, \quad \text{transform as objective axial vectors.}$$

It is not difficult to show for each set of four tensors, introduced above that the three-dimensional Maxwell equations in the respective formulations are obtained. Moreover, as clearly seen from the above derivation E_i (or E_i^C) and H_i (or H_i^a or H_i^C) are not objective vectors under the Euclidean transformation group but that $\varepsilon_o E_i$, $\varepsilon_o E_i^C$, $\mu_o H_i$, $\mu_o H_i^a$ and $\mu_o H_i^C$ are, as can easily be seen by invoking the Maxwell-Lorentz aether relations in the expressions for D_i^a, B_i^a. Note also, that M_i^L and M_i^C are both objective. Of such properties we have freely made use in the main body of this article.

Finally we mention once more that it is through the Maxwell-Lorentz aether relations that the Maxwell equations are not invariant under general transformations $(x_i,t) \to (x_i^*,t^*)$. Their form is such that the Maxwell equations in vacuo (in which the Maxwell-Lorentz aether relations are substituted) are invariant only under a very restricted transformation group, the Lorentz group. This should not be confused with the basic fact that φ_{AB}, Φ_{AB}, μ_{AB}, η^{AB}, H^{AB} and π^{AB} are four-tensors, which must obey (A.1). Consequently, the transformation properties under the Euclidean group hold irrespective of the invariance properties of the Maxwell equations.

Of course, Euclidean transformations are one group only, for which the transformations (A.1) and (A.2) hold. In principle, any other transformation group can be investigated and of special interest are Lorentz transformations, because they leave the Maxwell equations including the Maxwell-Lorentz aether relation invariant. These transformations are well-known and so we do not elaborate on them.

APPENDIX B : SOME DETAILED CALCULATIONS OF THE MAXWELL-MINKOWSKI MODEL

In this Appendix we present a motivation for the choice of ρT and R, (2.76), and a derivation of equation (2.78).

We start by transforming the Poynting-vector

$$\underline{E}^M \times \underline{H}^M .$$

With the aid of (2.59) we show that

$$e_{ijk}E^M_j H^M_j = e_{ijk}(E_j - e_{j\ell m}\dot{x}_\ell B_m)(H_k + e_{kpq}\dot{x}_p D_q) =$$

$$= e_{ijk}E_jH_k + (E_jD_j + H_jB_j)\dot{x}_i - (E_jD_i + H_jB_i)\dot{x}_j + \tag{B.1}$$

$$+ e_{jk\ell}P_kB_\ell\dot{x}_j\dot{x}_i + O(V^2/c^2) .$$

Next, using the Maxwell equations (2.68) and (2.71), we derive

$$(e_{ijk}E_jH_k)_{,i} = -J_iE_i - \overset{*}{D}_iE_i - \overset{*}{B}_iH_i =$$

$$= -J_iE_i - (\varepsilon_o E_i\dot{E}_i + \mu_o H_i\dot{H}_i) - E_i\dot{P}_i - \mu_o H_i\dot{M}_i +$$

$$- (E_jD_j - H_jB_j)\dot{x}_{i,i} + (E_iD_j + H_iB_j)\dot{x}_{i,j} =$$

$$= -J_iE_i - \frac{d}{dt}[\frac{1}{2}(\varepsilon_o E_iE_i + \mu_o H_iH_i)] - E_i\overset{*}{P}_i - \mu_o H_i\overset{*}{M}_i + \tag{B.2}$$

$$- (E_iP_j + \mu_o H_iM_j)\dot{x}_{i,j} - (\varepsilon_o E_jE_j + \mu_o H_jH_j)\dot{x}_{i,i} +$$

$$+ (E_iD_j + H_iB_j)\dot{x}_{i,j} .$$

Substituting (B.1) and (B.2) into (2.75) yields

$$(\dot{\rho} + \rho\dot{x}_{i,i})(U + \frac{1}{2}\dot{x}_i\dot{x}_i + T) + \rho\dot{T} - \rho\dot{U} + q_{i,i} - \rho r^{ext} +$$

$$- J_iE_i - E_i\overset{*}{P}_i - \mu_o H_i\overset{*}{M}_i - [R_i - (P_jE_j + \mu_o H_jM_j)\dot{x}_i +$$

$$+ e_{jk\ell}P_kB_\ell\dot{x}_j\dot{x}_i]_{,i} - [t_{ij} + E_iP_j + \mu_o H_iM_j]\dot{x}_{i,j} + \tag{B.3}$$

$$+ [\rho\ddot{x}_i - \rho F_i^{ext} - t_{ij,j} - (E_iD_j + H_iB_j)_{,j} + \frac{1}{2}(\varepsilon_o E_kE_k + \mu_o H_kH_k)_{,i}]\dot{x}_i .$$

For this relation to be invariant under rigid-body translations, the term proportional to $\dot{x}_i\dot{x}_j$, i.e.:

$$e_{jk\ell}P_kB_\ell\dot{x}_i\dot{x}_j \ ,$$

must be compensated by R_i. Moreover, $\underline{R}$ must vanish with vanishing $\underline{P}$ and $\underline{M}$. Both requirements are satisfied by assuming $(2.76)^2$ which reads

(B.4) $$R_i = (P_jE_j + \mu_oM_jH_j)\dot{x}_i + e_{jk\ell}P_kB_\ell\dot{x}_j\dot{x}_i \ .$$

Furthermore, if T would not be an objective scalar under rigid-body translations (with velocity $\underline{b}(t)$) the term ρT would lead to a term proportional to $\dot{\underline{b}}(t)$. Since this would be the only term of this kind, the above assumption leads to a contradiction and, hence, ρT must be objective. In that case, there is no distinction possible between T and U, and so T may be absorbed by U, or in other words, we may take

(B.5) $$\rho T = 0 \ ,$$

as was done in $(2.76)^1$.

We now substitute (B.4) and (B.5) into (B.3) and use the relation

(B.6) $$(E_iD_j + H_iB_j)_{,j} - \tfrac{1}{2}(\varepsilon_oE_kE_k + \mu_oH_kH_k)_{,i} = \\ = QE_i + e_{ijk}J_jB_k + P_jE_{j,i} + \mu_oM_jH_{j,i} + e_{ijk}(D_j\overset{*}{B}_k + \overset{*}{D}_jB_k) \ ,$$

which follows from (2.68) and (2.71). This then leads to (2.78).

AUTHOR INDEX

SUBJECT INDEX

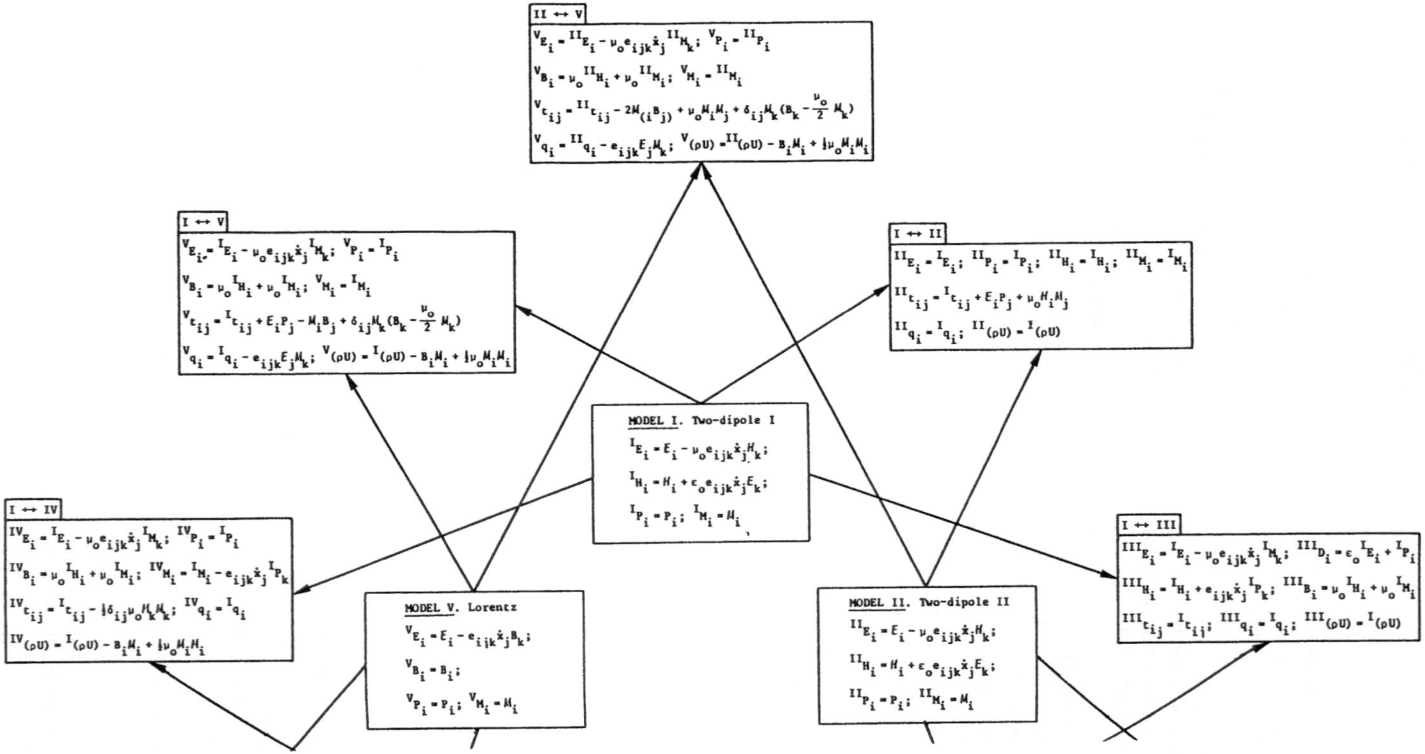
II ↔ V
${}^{V}E_i = {}^{II}E_i - \mu_o e_{ijk}\dot{x}_j {}^{II}M_k$; ${}^{V}P_i = {}^{II}P_i$
${}^{V}B_i = \mu_o {}^{II}H_i + \mu_o {}^{II}M_i$; ${}^{V}M_i = {}^{II}M_i$
${}^{V}t_{ij} = {}^{II}t_{ij} - 2M_{(i}B_{j)} + \mu_o M_i M_j + \delta_{ij} M_k (B_k - \frac{\mu_o}{2} M_k)$
${}^{V}q_i = {}^{II}q_i - e_{ijk}E_j M_k$; ${}^{V}(\rho U) = {}^{II}(\rho U) - B_i M_i + \frac{1}{2}\mu_o M_i M_i$
I ↔ V
${}^{V}E_i = {}^{I}E_i - \mu_o e_{ijk}\dot{x}_j {}^{I}M_k$; ${}^{V}P_i = {}^{I}P_i$
${}^{V}B_i = \mu_o {}^{I}H_i + \mu_o {}^{I}M_i$; ${}^{V}M_i = {}^{I}M_i$
${}^{V}t_{ij} = {}^{I}t_{ij} + E_i P_j - M_i B_j + \delta_{ij} M_k (B_k - \frac{\mu_o}{2} M_k)$
${}^{V}q_i = {}^{I}q_i - e_{ijk}E_j M_k$; ${}^{V}(\rho U) = {}^{I}(\rho U) - B_i M_i + \frac{1}{2}\mu_o M_i M_i$
I ↔ II
${}^{II}E_i = {}^{I}E_i$; ${}^{II}P_i = {}^{I}P_i$; ${}^{II}H_i = {}^{I}H_i$; ${}^{II}M_i = {}^{I}M_i$
${}^{II}t_{ij} = {}^{I}t_{ij} + E_i P_j + \mu_o H_i M_j$
${}^{II}q_i = {}^{I}q_i$; ${}^{II}(\rho U) = {}^{I}(\rho U)$
MODEL I. Two-dipole I
${}^{I}E_i = E_i - \mu_o e_{ijk}\dot{x}_j H_k$;
${}^{I}H_i = H_i + \epsilon_o e_{ijk}\dot{x}_j E_k$;
${}^{I}P_i = P_i$; ${}^{I}M_i = M_i$
I ↔ IV
${}^{IV}E_i = {}^{I}E_i - \mu_o e_{ijk}\dot{x}_j {}^{I}M_k$; ${}^{IV}P_i = {}^{I}P_i$
${}^{IV}B_i = \mu_o {}^{I}H_i + \mu_o {}^{I}M_i$; ${}^{IV}M_i = {}^{I}M_i - e_{ijk}\dot{x}_j {}^{I}P_k$
${}^{IV}t_{ij} = {}^{I}t_{ij} - \frac{1}{2}\delta_{ij}\mu_o M_k M_k$; ${}^{IV}q_i = {}^{I}q_i$
${}^{IV}(\rho U) = {}^{I}(\rho U) - B_i M_i + \frac{1}{2}\mu_o M_i M_i$
MODEL V. Lorentz
${}^{V}E_i = E_i - e_{ijk}\dot{x}_j B_k$;
${}^{V}B_i = B_i$;
${}^{V}P_i = P_i$; ${}^{V}M_i = M_i$
MODEL II. Two-dipole II
${}^{II}E_i = E_i - \mu_o e_{ijk}\dot{x}_j H_k$;
${}^{II}H_i = H_i + \epsilon_o e_{ijk}\dot{x}_j E_k$;
${}^{II}P_i = P_i$; ${}^{II}M_i = M_i$
I ↔ III
${}^{III}E_i = {}^{I}E_i - \mu_o e_{ijk}\dot{x}_j {}^{I}M_k$; ${}^{III}D_i = \epsilon_o {}^{I}E_i + {}^{I}P_i$
${}^{III}H_i = {}^{I}H_i + e_{ijk}\dot{x}_j {}^{I}P_k$; ${}^{III}B_i = \mu_o {}^{I}H_i + \mu_o {}^{I}M_i$
${}^{III}t_{ij} = {}^{I}t_{ij}$; ${}^{III}q_i = {}^{I}q_i$; ${}^{III}(\rho U) = {}^{I}(\rho U)$

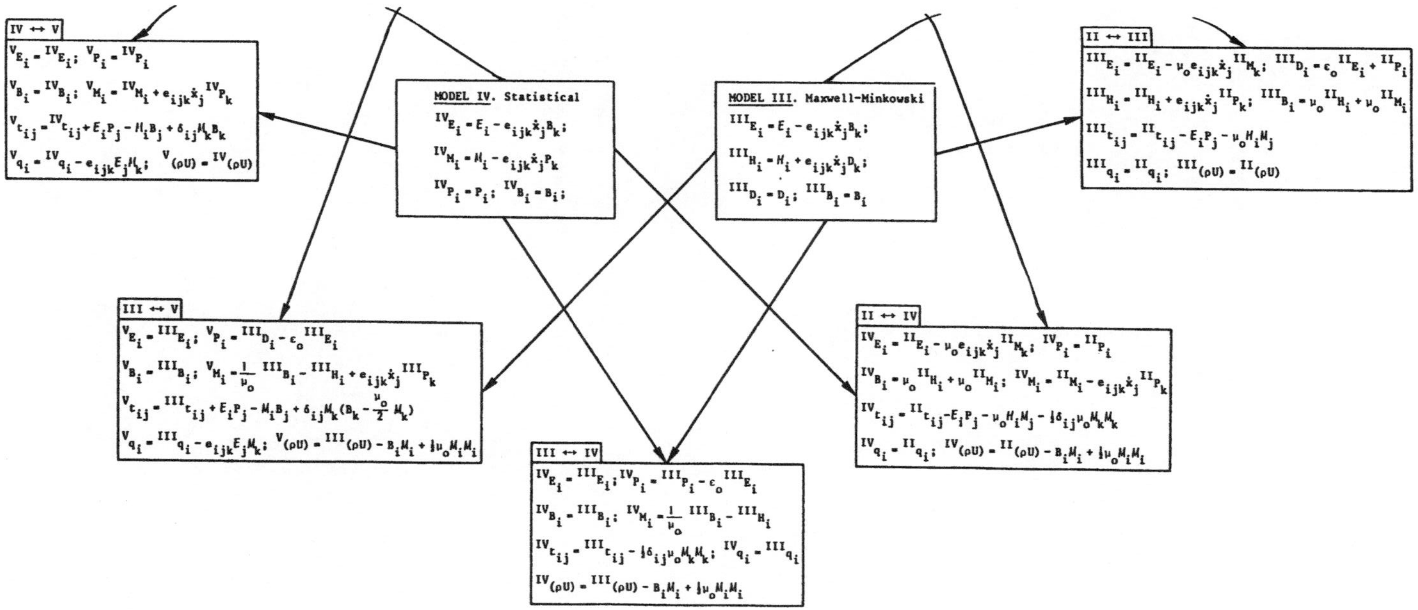
IV ↔ V
${}^{V}E_i = {}^{IV}E_i$; ${}^{V}P_i = {}^{IV}P_i$
${}^{V}B_i = {}^{IV}B_i$; ${}^{V}M_i = {}^{IV}M_i + e_{ijk}\dot{x}_j\,{}^{IV}P_k$
${}^{V}t_{ij} = {}^{IV}t_{ij} + E_iP_j - M_iB_j + \delta_{ij}M_kB_k$
${}^{V}q_i = {}^{IV}q_i - e_{ijk}E_jM_k$; ${}^{V}(\rho U) = {}^{IV}(\rho U)$
MODEL IV. Statistical
${}^{IV}E_i = E_i - e_{ijk}\dot{x}_jB_k$;
${}^{IV}M_i = M_i - e_{ijk}\dot{x}_jP_k$
${}^{IV}P_i = P_i$; ${}^{IV}B_i = B_i$;
MODEL III. Maxwell-Minkowski
${}^{III}E_i = E_i - e_{ijk}\dot{x}_jB_k$;
${}^{III}H_i = H_i + e_{ijk}\dot{x}_jD_k$;
${}^{III}D_i = D_i$; ${}^{III}B_i = B_i$
II ↔ III
${}^{III}E_i = {}^{II}E_i - \mu_o e_{ijk}\dot{x}_j\,{}^{II}M_k$; ${}^{III}D_i = \epsilon_o\,{}^{II}E_i + {}^{II}P_i$
${}^{III}H_i = {}^{II}H_i + e_{ijk}\dot{x}_j\,{}^{II}P_k$; ${}^{III}B_i = \mu_o\,{}^{II}H_i + \mu_o\,{}^{II}M_i$
${}^{III}t_{ij} = {}^{II}t_{ij} - E_iP_j - \mu_oH_iM_j$
${}^{III}q_i = {}^{II}q_i$; ${}^{III}(\rho U) = {}^{II}(\rho U)$
III ↔ V
${}^{V}E_i = {}^{III}E_i$; ${}^{V}P_i = {}^{III}D_i - \epsilon_o\,{}^{III}E_i$
${}^{V}B_i = {}^{III}B_i$; ${}^{V}M_i = \frac{1}{\mu_o}\,{}^{III}B_i - {}^{III}H_i + e_{ijk}\dot{x}_j\,{}^{III}P_k$
${}^{V}t_{ij} = {}^{III}t_{ij} + E_iP_j - M_iB_j + \delta_{ij}M_k(B_k - \frac{\mu_o}{2}M_k)$
${}^{V}q_i = {}^{III}q_i - e_{ijk}E_jM_k$; ${}^{V}(\rho U) = {}^{III}(\rho U) - B_iM_i + \frac{1}{2}\mu_oM_iM_i$
III ↔ IV
${}^{IV}E_i = {}^{III}E_i$; ${}^{IV}P_i = {}^{III}P_i - \epsilon_o\,{}^{III}E_i$
${}^{IV}B_i = {}^{III}B_i$; ${}^{IV}M_i = \frac{1}{\mu_o}\,{}^{III}B_i - {}^{III}H_i$
${}^{IV}t_{ij} = {}^{III}t_{ij} - \frac{1}{2}\delta_{ij}\mu_oM_kM_k$; ${}^{IV}q_i = {}^{III}q_i$
${}^{IV}(\rho U) = {}^{III}(\rho U) - B_iM_i + \frac{1}{2}\mu_oM_iM_i$
II ↔ IV
${}^{IV}E_i = {}^{II}E_i - \mu_o e_{ijk}\dot{x}_j\,{}^{II}M_k$; ${}^{IV}P_i = {}^{II}P_i$
${}^{IV}B_i = \mu_o\,{}^{II}H_i + \mu_o\,{}^{II}M_i$; ${}^{IV}M_i = {}^{II}M_i - e_{ijk}\dot{x}_j\,{}^{II}P_k$
${}^{IV}t_{ij} = {}^{II}t_{ij} - E_iP_j - \mu_oH_iM_j - \frac{1}{2}\delta_{ij}\mu_oM_kM_k$
${}^{IV}q_i = {}^{II}q_i$; ${}^{IV}(\rho U) = {}^{II}(\rho U) - B_iM_i + \frac{1}{2}\mu_oM_iM_i$